RELATIVISTIC ASTROPHYSICS
AND COSMOLOGY

THE SCIENCE AND CULTURE SERIES – ASTROPHYSICS

Series Editor: A. Zichichi

International School of Cosmic Ray Astrophysics
13th Course

RELATIVISTIC ASTROPHYSICS AND COSMOLOGY

Erice, Italy   2–14 June 2002

edited by

Maurice M. Shapiro
University of Maryland, USA

Todor Stanev
University of Delaware, USA

John P. Wefel
Louisiana State University, USA

 **World Scientific**

NEW JERSEY • LONDON • SINGAPORE • SHANGHAI • HONG KONG • TAIPEI • BANGALORE

Published by

World Scientific Publishing Co. Pte. Ltd.

5 Toh Tuck Link, Singapore 596224

USA office: Suite 202, 1060 Main Street, River Edge, NJ 07661

UK office: 57 Shelton Street, Covent Garden, London WC2H 9HE

British Library Cataloguing-in-Publication Data
A catalogue record for this book is available from the British Library.

International School of Cosmic Ray Astrophysics
13th Course: Relativistic Astrophysics and Cosmology

Copyright © 2004 by World Scientific Publishing Co. Pte. Ltd.

ISBN 981-238-727-7

Printed in Singapore by World Scientific Printers (S) Pte Ltd

Celebrating the 25th Anniversary of the School

and

Dedicated in Fond Memory

of

Rein Silberberg

PREFACE

"Relativistic Astrophysics and Cosmology" was the theme of the Thirteenth Course of the International School of Cosmic Ray Astrophysics held at the Ettore Majorana Centre in Erice, Sicily, Italy. The school, and the papers in this volume, focus on major areas of Astrophysics, their relation to Cosmic Ray Physics and our current understanding of the energetic processes in the Galaxy and the Universe that govern the acceleration and form the features of the cosmic rays that we detect at Earth.

One of the long standing exciting problems in astrophysics is the origin and acceleration of cosmic rays. Large and advanced new experiments have recently contributed to the solution of this problem. Other experiments, currently under design and construction, are aimed at detection of the highest energy particles in Nature. The significant experimental progress in the field was discussed together with the theoretical interpretation of the current results. At the same time the participants in the School learned about the most luminous astrophysical processes.

During the next several years we are expecting important results from the new astronomical observations in TeV gamma rays and neutrinos. Two third generation TeV Cherenkov telescopes will continue the search for sources and study the astrophysical processes involved in their production. At the same time the first neutrino telescopes are being built at the South Pole and in the Mediterranean. These devices, the expected results and the astrophysics involved are a vital part of the School and this volume.

This course was dedicated to the memory of Dr. Rein Silberberg. Rein was not only an excellent scientist but also an essential part of the International School of Cosmic Ray Astrophysics since its inception in 1978.

Seventy students, lecturers, and senior participants from 18 countries attended the thirteenth course, which celebrated the 25th anniversary of the school. V. Ptuskin from IZMIRAN, Moscow and T. Stanev from the Bartol Research Institute, University of Delaware, co-directed the course under the guidance of M.M. Shapiro, director of the school. Executive secretary, Arthur Smith worked tirelessly to ensure the efficient operation of the course.

The organizers are grateful to Prof. A. Zichichi, founder and director of the Majorana Centre in Erice for providing the infrastructure and support for this course. Ms. Maria Zaini and Ms. Fiorella Ruggiu have given essential administrative support and Mr. P. Acceto has been very helpful.

Maurice M. Shapiro, Todor Stanev, John P. Wefel

CONTENTS

Extensive air showers

Gamma ray and neutrino astronomy

UNDERSTANDING AND MODELING

THE UNIVERSE

AND ITS LUMINOUS SYSTEMS

AN ACCELERATING CLOSED UNIVERSE

J. OVERDUIN

Astrophysics and Cosmology Group, Department of Physics
Waseda University, Okubo 3-4-1, Shinjuku, Tokyo 169-8555, Japan
E-mail: overduin@gravity.phys.waseda.ac.jp

W. PRIESTER

Institut für Astrophysik und Extraterrestrische Forschung
Universität Bonn, Auf dem Hügel 71, D-53121 Bonn, Germany
E-mail: priester@astro.uni-bonn.de

The dark matter which makes up 99% or more of the Universe by mass is now believed to contain four separate ingredients: dark baryons, massive neutrinos, "exotic" cold dark-matter particles and vacuum energy, also known as the cosmological constant (Λ). Of these, only baryons fit within the current standard model of particle physics. The fourth (vacuum) component has come into new prominence in the past few years, largely at the expense of the third (exotic dark matter). We argue here that vacuum energy may be dominant enough to remove the need for exotic dark matter and close the accelerating Universe.

1. Dark Matter and the Evolution of the Universe

The shape and evolution of the Universe are governed by Einstein's gravitational field equations. Under the standard assumptions of large-scale homogeneity and isotropy, these reduce to a pair of differential equations in the cosmological scale factor R and its time derivatives, including the Hubble parameter $H \equiv \dot{R}/R$ (the expansion rate of the Universe). The equation for H is

$$\left[\frac{H(z)}{H_0}\right]^2 = \Omega_{\mathrm{R},0}(1+z)^4 + \Omega_{\mathrm{M},0}(1+z)^3 + \Omega_{\Lambda,0} - (\Omega_{\mathrm{TOT},0} - 1)(1+z)^2. \quad (1)$$

Here Ω_{R}, Ω_{M} and Ω_{Λ} refer to the densities[a] of radiation, pressureless matter and vacuum energy respectively; and $z \equiv (R/R_0)^{-1} - 1$ is the cosmological

---

[a]The symbol Ω denotes a physical density ρ expressed in units of the *critical density* $\rho_{\mathrm{crit}}(t) \equiv 3H^2(t)/8\pi G$. This latter quantity depends on the Hubble parameter $H(t)$ and takes the value $\rho_{\mathrm{crit},0} = 3H_0^2/8\pi G$ at the present time. If H_0 lies in the range $70 - 90$ km s^{-1} Mpc^{-1}, then $\rho_{\mathrm{crit},0}$ is equivalent to between 5.5 and 9.1 protons m^{-3}.

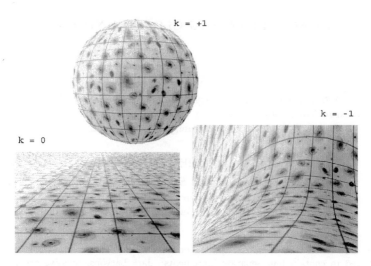

k = +1

k = -1

k = 0

Figure 1.   Three possibilities for the spatial geometry of the Universe, with corresponding values of the curvature constant k (graphic courtesy E. C. Eekels).

redshift. The subscript "0" denotes quantities measured at the *present
time*; i.e. at redshift $z = 0$. These are constants, and must be carefully
distinguished from functions of time (or redshift) such as $\Omega_{\rm R}, \Omega_{\rm M}$ and Ω_{Λ}.
The constant $\Omega_{\rm TOT,0} \equiv \Omega_{\rm R,0} + \Omega_{\rm M,0} + \Omega_{\Lambda,0}$ is of particular interest, because
it separates spatially spherical models from hyperbolic ones (Fig. 1). If
$\Omega_{\rm TOT,0} > 1$, then the Universe is closed and finite in extent, with a curvature
constant $k \equiv (H_0 R_0/c)^2 (\Omega_{\rm TOT,0} - 1)$ whose value can be set to $k = +1$ by
choice of units for R_0. Conversely, if $\Omega_{\rm TOT,0} < 1$, then it is open and infinite
in extent ($k = -1$). Only if $\Omega_{\rm TOT,0} = 1$ *exactly* do we live in an infinite
Universe which is spatially flat or "Euclidean" with $k = 0$.

The shape of the Universe thus depends on the values of the density
parameters $\Omega_{\rm R,0}, \Omega_{\rm M,0}$ and $\Omega_{\Lambda,0}$. These determine also its evolution in time,
including for instance the question of whether it is static ($H = $ constant,
as originally supposed by Einstein), decelerating ($\dot{H} < 0$, as was widely
believed until recently), or accelerating ($\dot{H} > 0$).

Much can already be learned by inspection of Eq. (1). The first term
on the right-hand side, $\Omega_{\rm R,0}(1 + z)^4$, shows that radiation increases the
expansion rate $H(z)$ with z (i.e. slowing it down with time). This term
dominates at the highest redshifts (i.e. earliest times), so that the Universe
during its childhood must have been governed by the density of photons and
relativistic particles. Nowadays, however, the value of $\Omega_{\rm R,0}$ is low enough
compared to $\Omega_{\rm M,0}$ and $\Omega_{\Lambda,0}$ that we may ignore this term.

The second term, $\Omega_{M,0}(1 + z)^3$, shows that matter also brakes the expansion, but does so at a rate which drops less steeply with time than that of radiation. (This is related to the fact that matter is pressureless, and pressure in general relativity exerts gravitational attraction just as density does.) Matter thus governs the *adolescence* of the Universe. The duration of this phase is determined by the value of $\Omega_{M,0}$, which is uncertain because most of it comes from dark matter which has not been seen.

The third term, $\Omega_{\Lambda,0}$, is independent of redshift, which means that its influence is not diluted with time. Vacuum or "dark energy" must therefore eventually dominate the *old age* of any Universe in which $\Lambda > 0$. In the limit $t \to \infty$, the other terms drop out of Eq. (1) altogether and the vacuum density parameter may be expressed as $\Omega_{\Lambda,0} = (H_\infty/H_0)^2$ or

$$\Lambda c^2 = 3H_\infty^2. \tag{2}$$

Here H_∞ is the limiting value of the Hubble parameter as $t \to \infty$ (assuming that this latter quantity exists; i.e. that the Universe does not recollapse in the future). Thus Λ (a constant of nature in Einstein's theory) is connected to the asymptotic expansion rate (a dynamical parameter). If $\Lambda > 0$, and if we are living at sufficiently late times, then Eq. (2) immediately predicts that we will measure $\Omega_{\Lambda,0} \sim 1$.

The fourth term in Eq. (1), finally, shows that an excess of $\Omega_{TOT,0}$ over one (i.e. a positive curvature) acts to offset the contribution of the first three terms to the expansion rate, while a deficit (i.e. a negative curvature) enhances them. Open models, in other words, expand more quickly at any given redshift z (and therefore last longer) than closed ones. This curvature term, however, goes only as $(1+z)^2$, which means that its importance drops off relative to the matter and radiation terms at early times, and becomes negligible compared with that of the vacuum term at late ones.

Until recently, many cosmologists routinely dropped not only the first (radiation) term in Eq. (1), but the third (vacuum) and fourth (curvature) terms as well. There are four principal reasons for this. First, these terms differ sharply from each other (and from the matter term) in their dependence on redshift z, and the probability that we should happen to find ourselves in an era when all four are of similar size would seem *a priori* very remote. By this argument, originally due to Dicke, it was felt that only one term ought to dominate at any given time. Second, the vacuum term in particular was regarded as suspicious because its value could not be related to existing quantum field theory in a sensible way (this is still true today, but no longer seen as an excuse to ignore the term). Third, a period of cosmic *inflation* was asserted to have driven $\Omega_{TOT}(t)$ to unity in the

6

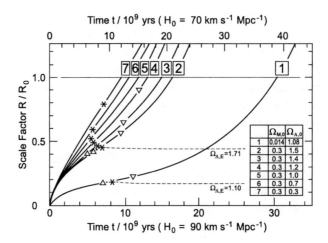

Figure 2. Evolution of the cosmological scale factor $R(t)/R_0$ as a function of time in several models with $\Lambda > 0$.

early Universe. (This is not necessarily true of all inflationary models.[1,2]) And finally, this "standard Einstein-de Sitter" (EdS) model was favoured on grounds of simplicity. These arguments are no longer valid today. We are justified in neglecting the radiation term, and *only* the radiation term in Eq. (1), leaving

$$H^2(z) = H_0^2 \left[\Omega_{M,0}(1+z)^3 + \Omega_{\Lambda,0} - (\Omega_{TOT,0} - 1)(1+z)^2 \right]. \tag{3}$$

This is the modern version of what is usually called *Friedmann's equation* in cosmology. It may be integrated numerically for the cosmological scale factor $R(t)$ as a function of time. We show several examples in Fig. 2, including closed models (1 through 5) and one flat (6) and open model (7). Model 1, with $\Omega_{M,0} = 0.014$ and $\Omega_{\Lambda,0} = 1.08$, has been proposed in[3,4] and will be discussed below. The others all have $\Omega_{M,0} = 0.3$, a figure widely quoted today for the total density of gravitating matter. Model 6, with $\Omega_{\Lambda,0} = 0.7$ (known as the ΛCDM model), has been singled out as the latest "standard model" of cosmology. Two timescales are plotted (top and bottom), depending on the present value H_0 of Hubble's parameter.

Along each of the curves, we have marked the points where $\Omega_M(z)$ takes on maximum values ($\triangle$), the points of inflection ($*$), and the points where $\Omega_\Lambda(z)$ takes on maximum values and $H(z)$ reaches a minimum ($\triangledown$). R takes the special value $R_\Lambda \equiv 1/\sqrt{\Lambda}$ at the points marked ($\triangle$). The cosmological constant thus corresponds physically to the curvature of space (in closed models) at the time when the matter density parameter goes through its maximum.[5] Joining the points of inflection in Fig. 2 are two dashed lines

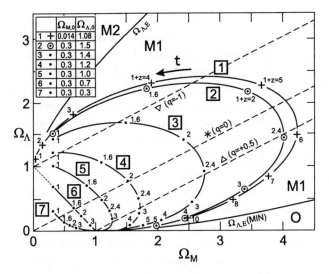

Figure 3. Evolution in phase space defined by $\Omega_M(z)$ and $\Omega_\Lambda(z)$ of the models whose scale factors are plotted in Fig. 2.

labelled $\Omega_{\Lambda,E}$ (for "Einstein limit"). One must have $\Omega_{\Lambda,0} < \Omega_{\Lambda,E}$ (a function of the matter density $\Omega_{M,0}$) if the current expansionary phase originated in a singularity with $R = 0$. Solutions with $\Omega_{\Lambda,0} = \Omega_{\Lambda,E}$ go over to Einstein's static model as $t \to \infty$. When $\Omega_{\Lambda,0} > \Omega_{\Lambda,E}$, $R(t)$ drops to a nonzero minimum and starts to climb again in the past direction; these are Eddington-Lemaître (or "bounce") models. The importance of the Einstein limit in cosmology has been discussed in.[6,7,8]

The differences between the models shown in Fig. 2 become apparent when their evolution is plotted on a phase diagram such as Fig. 3, with matter density parameter along one axis and vacuum density parameter along the other. The key equations[8] are

$$\Omega_\Lambda(z) = \Omega_{\Lambda,0} / \left[\Omega_{M,0}(1 + z)^3 + \Omega_{\Lambda,0} - (\Omega_{M,0} + \Omega_{\Lambda,0} - 1)(1 + z)^2 \right]$$
$$\Omega_M(z) = (\Omega_{M,0}/\Omega_{\Lambda,0}) \, \Omega_\Lambda(z)(1 + z)^3. \tag{4}$$

Fig. 3 depicts the same family of models as Fig. 2, with redshift factors $[1 + z = R_0/R(t)]$ labelled at intervals along the curves. Also marked are contours of constant *deceleration*, defined by $q \equiv -\ddot{R}R/\dot{R}^2 = \Omega_M/2 - \Omega_\Lambda$. This parameter takes values of 0.5 at each point of maximum matter density parameter ($\triangle$), zero at the inflection points ($*$), and -1 at the points of minimum expansion rate (∇). All positive-Λ models begin in Fig. 3 at the point $(1,0)$ and evolve asymptotically toward $(0,1)$ as $t \to \infty$. Flat (Euclidean) models follow a straight line; any deviation from critical density produces a curved path. Those to the right of Model 6 are all closed. Models 5 through 2 are increasingly unlikely insofar as they violate observational upper bounds on the total density parameter, $\Omega_{\mathrm{TOT},0} \lesssim 1.2$.[9] Model 2 in particular cannot describe the real Universe. However, the model immediately adjacent to it in phase space (Model 1) is perfectly acceptable in this regard, since it has $\Omega_{\mathrm{TOT},0} = 1.094$. Very different combinations of $\Omega_{\mathrm{M},0}$ and $\Omega_{\Lambda,0}$, in other words, can produce almost identical trajectories in phase space. Indeed, from the perspective of Fig. 3, the popular Euclidean models appear implausibly fine-tuned.

The slow expansion rate and high matter density parameter between the points marked ($\triangle$) and ($\triangledown$) single out this stage of evolution for *structure formation*. If $\Omega_{\Lambda,0}$ is of the same order as (or less than) $\Omega_{\mathrm{M},0}$, however, the process must occur very quickly. Consider ΛCDM, represented by Model 6 in Figs. 2 and 3. Observations suggest that the number density of galaxies at redshifts $z \approx 4 - 6$ (i.e. at scale factors $R/R_0 \approx 0.1 - 0.2$) is equal to that at $z = 0$.[10] If so, then these objects were in place by $z = 4$, or (consulting Fig. 2 for Model 6, and using either the top or bottom scale for H_0) within 1.2-1.5 Gyr after the big bang. This poses a serious challenge for the model, since overdense regions must not only decouple from the Hubble expansion very quickly, but do so at a time when the expansion rate (the slope of the curve in Fig. 2) is some six times its present value. The problem is even worse in models with lower values of $\Omega_{\Lambda,0}$ (e.g. Model 7).

The standard way to address this has been to suppose that most of the matter density is in an exotic new form (CDM) which is able to decouple from the primordial fireball before the baryons, preparing potential wells for them to fall into. This approach successfully accelerates structure formation on large scales, but is perhaps *too* successful on smaller ones. Galaxy-sized regions are formed with excessively peaked central masses (the "density cusp problem") and too many low-mass fragments (the "substructure problem"). These difficulties may be resolved within the CDM picture by refining the properties of the exotic matter.

Alternatively, problems with the growth of large-scale structure are substantially eased in models with larger ratios of $\Omega_{\Lambda,0}$ to $\Omega_{\mathrm{M},0}$. In Model 1, for instance, Fig. 2 shows that redshift $z = 4$ corresponds to between 9.3 and 11.9 Gyr after the big bang (depending on the value of H_0), giving the galaxies seven times longer to form. The expansion rate at this redshift is only 0.7 times its present value. Nor is the low (present) density of gravitating matter a hindrance in this model, because $\Omega_{\mathrm{M}}(z)$ reaches levels as high as four times the critical density at redshifts near $z \approx 5$ (Fig. 3). It is natural to associate this redshift with the onset of galaxy formation, and it would be of great interest to test "slightly-closed" cosmologies of this kind via numerical simulations. Studies carried out to date (e.g. by the VIRGO collaboration[11]) have been restricted to flat and open models, with Λ-dominated models giving a distribution of mass which comes closest to the observed distribution of light in the Universe.

2. The Matter Density Parameter $\Omega_{M,0}$

We wish to review the current status of the parameters $\Omega_{M,0}$ and $\Omega_{\Lambda,0}$. Let us begin our inventory with the contribution to $\Omega_{M,0}$ from *luminous* baryons (i.e. those in stars), whose density can be inferred from the observed luminosity density of the Universe. A recent estimate[12] is:

$$\Omega_{LUM} = (0.0027 \pm 0.0014)h_0^{-1}, \tag{5}$$

where h_0 is the present value of H_0 expressed in units of 100 km s^{-1} Mpc^{-1}. This latter parameter is unfortunately not yet fixed by observation, and we pause to discuss its value before proceeding. Using various relative-distance methods, all calibrated against Cepheid variable stars in the Large Magellanic Cloud (LMC), the Hubble Key Project (HKP) team has determined that $h_0 = 0.71 \pm 0.06$.[13] "Absolute" methods (e.g. gravitational lensing time delays [GLTDs] and the Sunyaev-Zeldovich or SZ effect) have higher uncertainties but are roughly consistent with this, $h_0 \approx 0.65 \pm 0.10$. The near convergence of these approaches has been widely hailed, with many asserting that precision values of h_0 are nearly upon us.

On the other hand, a recalibrated LMC Cepheid period-luminosity relation based on a much larger Cepheid sample (from the OGLE collaboration) leads to a considerably higher value of $h_0 = 0.85 \pm 0.05$.[14] And data on water masers in the galaxy NGC4258 leads to a purely geometric distance[15] which also implies that the traditional calibration is off, boosting Cepheid-based estimates by $12 \pm 9\%$.[16] This would raise the HKP value to $h_0 = 0.80 \pm 0.09$. Independent support for such a recalibration comes from observations of eclipsing binaries and "red clump stars" in the LMC. The absolute methods are also vulnerable to systematic effects: GLTD-based values of h_0, which are routinely computed assuming EdS, rise by about 7% (on average) in ΛCDM, and 9% in open models.

"Hubble fatigue" may therefore have prompted cosmologists to embrace prematurely small levels of uncertainty in h_0. We consider two two possibilities in what follows: a "low value" of $h_0 = 0.7$ and a "high value" of $h_0 = 0.9$. This modest range turns out to discriminate powerfully between the cosmological models considered here. To a large extent this is a function of their ages. Fig. 2 reveals, for example, that ΛCDM (represented by

Model 6) is 13.5 Gyr old if $h_0 = 0.7$, or 10.5 Gyr if $h_0 = 0.9$. The oldest stars seen in the Milky Way have an age of 15.6 ± 4.6 Gyr,[17] setting a firm lower limit of 11.0 Gyr on the age of the Universe. This is enough to rule out ΛCDM with the high value of h_0, but not the low one. Model 1 faces the opposite problem: Fig. 2 shows that it has a total age of 30.2 Gyr if $h_0 = 0.9$, or 38.8 Gyr if $h_0 = 0.7$. Both numbers are larger than most cosmologists are prepared to accept. However, upper limits on the age of the Universe are not as secure as lower ones. One must take into account, for instance, that the galaxy formation associated above with redshifts $z \approx 4$ occured between 9 and 12 Gyr after the big bang in this model, so that galaxies would not be older than 24 ± 3 Gyr in any case.

There are various ways to test such a hypothesis. One might expect a greater *spread* in galaxy ages (and hence colors) at $z \approx 4$, given their longer formation time. Galaxies in Model 1 had ~ 6 Gyr to form, or about $\sim 25\%$ of their nominal lifetime, according to Fig. 2 (with $h_0 = 0.9$). The corresponding fraction in Model 6 is $\lesssim 10\%$. This may not necessarily translate into a difference between observed color spreads, however, since high-redshift galaxies are seen principally during (relatively brief) episodes of star formation. Very old galaxies, if they exist, should also be present at lower redshifts. They would be inherently faint and reddened, making them difficult to find and distinguish from younger objects which are simply obscured by dust. If our own Milky Way is not unusually young, we should also expect to find large numbers of dead stars in the galactic halo. These would act as *microlenses*, inducing variability in background stars and quasars, even if they were too dim to be seen directly. Such objects have now been detected in the direction of the LMC (see below).

Returning to the density of luminous matter, we find with our values for h_0 that Eq. (5) gives

$$\Omega_{\text{LUM}} = 0.0034 \pm 0.0018. \tag{6}$$

The visible Universe, in other words, constitutes an insignificant 0.5% or less of the critical density.

Some baryons, of course, may not be visible. The theory of cosmic nucleosynthesis provides us with an independent method for determining the density of *total* baryonic matter in the Universe, based on the assumption that the light elements we see today were forged in the furnace of the hot big bang. Results using different light elements agree tolerably well, which is impressive in itself. The primordial abundances of ^{4}He (by mass) and ^{7}Li (relative to H) imply a baryon density of $\Omega_{\text{BAR}} = (0.011 \pm 0.005)h_0^{-2}$,[18] whereas measurements based exclusively on the primordial D/H abundance give a higher value: $\Omega_{BAR} = (0.019 \pm 0.002)h_0^{-2}$.[19] These do not overlap. In our view it is premature at present to exclude either one. We therefore adopt the same strategy here as with Hubble's parameter, retaining a "low" value of $0.01h_0^{-2}$ and a "high" one of $0.02h_0^{-2}$. Combining this with our range of values for h_0, we find

$$\Omega_{BAR} = 0.012 - 0.041. \tag{7}$$

One can obtain the same numbers by the entirely independent method of adding up individual contributions from all known repositories of baryonic matter via their estimated mass-to-light (M/L) ratios.[12] If $\Omega_{TOT,0}$ is close to unity (as is now believed), then it follows from Eq. (7) that baryons — and everything that would have been recognized as "matter" before 1930 — make up less than 5% of the Universe by mass.

Most of these baryons, moreover, are invisible. Using Eq. (5) together with our ranges of values for h_0 and Ω_{BAR}, we find a baryonic dark-matter fraction $f_{BDM} = 1 - \Omega_{LUM}/\Omega_{BAR} = 77\% - 95\%$. Where could the unseen baryons be? One possibility is that they are smoothly distributed in the form of a gaseous intergalactic medium, which would have to be strongly ionized in order to explain why it has not left a more obvious signature in quasar absorption spectra. Alternatively, they could be bound up in dark-matter clumps such as substellar objects (jupiters, brown dwarfs) or stellar remnants (white, red and black dwarfs, neutron stars, black holes). Objects at the low-mass end of the spectrum (e.g. brown dwarfs) are too light to make a significant contribution, while those at the high-mass end (e.g. supermassive black holes) cannot be very important or they would have made their presence obvious through tidal disruptions and lensing effects. Attention has therefore focused on dark clumps of approximately solar mass. Microlensing of stars in the LMC has shown that such objects do exist in the halo of the Milky Way (where they are known as massive compact halo objects, or MACHOs). They appear to make up no more than 50% of the mass of the halo under natural assumptions.[20] The identity of these objects has not yet been established.

The suggestion that $\Omega_{M,0}$ receives contributions from a second species of invisible matter, the exotic *cold dark matter* (CDM), has been primarily motivated in three ways: (1) a range of observational arguments imply that there is more gravitating matter on galactic and larger scales than can be accounted for baryons and neutrinos alone; (2) our current understanding of large-scale structure formation requires the process to be helped along by some form of cold (i.e. nonrelativistic), weakly-interacting matter in the early universe with properties different from those of baryons and neutrinos; and (3) theoretical physics supplies several plausible (albeit still hypothetical) candidate CDM particles. Since our ideas on structure formation may yet change, and the candidate particles may not materialize, the case for exotic CDM turns on the observational arguments. The lower limit is crucial: only if $\Omega_{M,0} > \Omega_{BAR} + \Omega_\nu$ do we require $\Omega_{CDM} > 0$.

The arguments can be broken into two classes: those which are purely empirical, and those which assume in addition the validity of the gravitational instability (GI) picture of structure formation. Empirical arguments employ such things as galactic rotation curves, mass-to-light ratios, the baryon fraction in galaxy clusters, and radio galaxy lobes as standard rulers. These are compatible in some cases with $\Omega_{M,0} = \Omega_{BAR}$ (i.e. $\Omega_{CDM} = 0$), and generally imply that $\Omega_{M,0} \lesssim 0.4$ or *lower*. Arguments based on GI theory involve phenomena such as the evolution of cluster number density with redshift, changes in the power spectra of large-scale structures, and the distribution of galaxy peculiar velocities. They are "circular" in the sense that structure could not have evolved in the way envisioned *unless* $\Omega_{M,0}$ is considerably larger than Ω_{BAR}. But inasmuch as GI theory is currently the only structure-formation theory we have which is both fully worked out and in reasonable agreement with observations (at least on large scales), they deserve to be taken seriously. GI-based arguments suggest that $\Omega_{M,0} \gtrsim 0.2$ or *higher*. In the literature one often finds the results of both empirical and GI-based arguments combined into a single bound of the form $\Omega_{M,0} \approx 0.3 \pm 0.1$. This constitutes a "proof" of the existence of exotic CDM, since $\Omega_{BAR} < 0.05$ from Eq. (7). (Neutrinos do not affect this as we note below.) However, we believe such a limit to be overly optimistic. Taking the uncertainties in the individual arguments into account, our reading of the evidence suggests to us[21] that

$$\Omega_{CDM} = \begin{cases} 0.1 - 0.5 & \text{(GI theory)} \\ 0 - 0.4 & \text{(otherwise).} \end{cases} \tag{8}$$

If our understanding of structure formation via gravitational instability is correct, then exotic CDM must exist. Conversely, if exotic CDM does not exist, then our understanding of structure formation is incomplete.

The question, of course, becomes moot if exotic CDM (with $\Omega_{CDM} \sim 0.3$) is discovered in the laboratory. Theorists have proposed a long list of particle candidates with varying degrees of testability (see[22] for a review). Two, the axion and the weakly interacting massive particle (WIMP), have emerged as most plausible because they are: (1) *cold* (i.e. nonrelativistic in the early Universe), (2) *weakly-interacting* enough to drop out of equilibrium with the primordial fireball well before baryons; and (3) *expected* to have a collective density close to the critical one. Experimental detection efforts around the world are now directed at both particles. While they have not turned up anything so far, most of the theoretical parameter space remains unexplored.

The third contribution to $\Omega_{M,0}$ is now believed to come from *neutrinos*. With their existence beyond dispute, and a number density comparable to that of photons, these dark-matter candidates were indeed once favoured to provide the entire critical density. Any such hopes have been dashed by analyses of structure formation in relativistic or hot dark-matter (HDM) dominated models. Neutrinos are able to stream freely out of density perturbations in the early Universe, erasing them before they have a chance to grow. The observed abundance of large-scale structures imposes an upper bound on the neutrino rest energy of $m_\nu c^2 \lesssim (9.2 \text{ eV})\Omega_{\text{CDM}}$ for flat models with $0 \leqslant \Omega_{\text{CDM}} \leqslant 0.6$.[23] Dividing $m_\nu n_\nu$ by the critical density of the Universe gives a neutrino density $\Omega_\nu = (\sum m_\nu n_\nu / 94 \text{ eV}) h_0^{-2}$, so that $\Omega_\nu < 0.12\Omega_{\text{CDM}}$ (if $h_0 = 0.9$) or $0.20\Omega_{\text{CDM}}$ (if $h_0 = 0.7$).

Lower limits on Ω_ν have come from experiments indicating that neutrino species "oscillate," a process which can only take place if $m_\nu > 0$. The strongest evidence comes from Super-Kamiokande, which has reported that the square of the difference of of τ- and μ-neutrino rest energies lies between $5 \times 10^{-4} \text{ eV}^2$ and $6 \times 10^{-3} \text{ eV}^2$.[24] If neutrino masses are hierarchical (like those of other fermions) then $m_{\nu_\tau} \gg m_{\nu_\mu}$ and $m_{\nu_\tau} c^2 > 0.02 \text{ eV}$. This implies that $\Omega_\nu > 0.0003$ (if $h_0 = 0.9$) or 0.0005 (if $h_0 = 0.7$). If, instead, neutrino masses are nearly degenerate, then their combined contributions to Ω_ν would be larger than this, but could not exceed the above structure-formation bound in any case. We conclude that

$$\Omega_\nu = 0.0003 - 0.2\Omega_{\text{CDM}}. \tag{9}$$

The neutrino contribution to $\Omega_{M,0}$ is anywhere from an order of magnitude below that of the visible stars and galaxies to as much as one-fifth of that attributed to exotic CDM.

3. The Vacuum Energy Density Parameter $\Omega_{\Lambda,0}$

Let us move on to the vacuum density parameter $\Omega_{\Lambda,0}$, which is just Einstein's cosmological constant in new clothes ($\Omega_{\Lambda,0} = \Lambda c^2/3H_0^2$). When Einstein realized in 1931 that the value of Λ could not be determined from observations available at that time, he recommended dropping the term "aus Gründen der logischen Ökonomie" (for reasons of logical economy). The saying famously attributed to him by Gamow, "Die Einführung von Lambda war vielleicht die größte Eselei in meinem Leben" ("perhaps the biggest blunder in my life") prompted most observational astronomers to drop the Λ-term entirely *a priori*. This attitude prevailed until very recently, leading cosmology almost into a dead end. In this context Steven Weinberg wrote in 1993, "The experience of the past three quarters of our century has taught us distrust such assumptions. We generally find that any complication in our theories that is not forbidden by some symmetry or other fundamental principle actually occurs."[25]

But what is the physical meaning of Λ? We have seen above that this term is connected with three important macroscopic quantities: the asymptotic expansion rate of the Universe ($\Lambda c^2 = 3H_\infty^2$), the curvature of the closed Universe at the epoch of maximum matter density ($R_\Lambda \equiv 1/\sqrt{\Lambda}$) and the energy density of the cosmological vacuum ($\rho_\Lambda c^2 = \Lambda c^4/8\pi G$). At a microscopic level, it remains an open problem how to account for such a vacuum energy density in terms of the ground states of quantized fields (see Hans Blome's contribution to these proceedings).

Let us pass now to the observational data on $\Omega_{\Lambda,0}$. As remarked above, numerical simulations of large-scale structure formation appear to favour Λ-dominated models. Suggestions of a lower limit on $\Omega_{\Lambda,0}$ have also come from galaxy counts at faint magnitudes. *Upper* limits have been reported based on the statistics of gravitational lenses, and (for closed models) the redshift of the antipodes, which must be greater than the redshifts of normally-lensed objects and less than the redshift of the last scattering surface. (Model 1 in Figs. 2 and 3 has its antipodes at $z \approx 12$ and meets both requirements; Fig. 4.) We have discussed these tests in more detail elsewhere.[21] Observations of the extragalactic background light (EBL) are also improving to the point where it may become feasible to constrain the value of $\Omega_{\Lambda,0}$ by this method in the near future.[22]

In 1998-99, two independent teams reported measurements of $\Omega_{\Lambda,0}$ based on the use of Type Ia supernovae (SNIa) in the classical magnitude-redshift relation.[26,27] Their results came as a surprise to many, implying that

$$\Omega_{\Lambda,0} \approx \frac{4}{3}\,\Omega_{\mathrm{M},0} + \frac{1}{3} \pm \frac{1}{2}. \tag{10}$$

This finding, more than any other, appears to have convinced the majority of cosmologists to reconsider the possibility of a vacuum-dominated Universe. We caution, however, that there are a number of possible systematic effects whose ramifications have yet to be fully worked out.

Beginning in 2000, more precise constraints on $\Omega_{\Lambda,0}$ began to come in from measurements of the angular power spectrum of fluctuations in the CMB.[28,29,30] The location of the first peak in this spectrum is a direct measure of the largest size of fluctuations in the primordial plasma at the

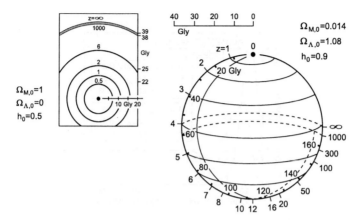

Figure 4. Two-dimensional slices in time of the flat EdS model ($\Omega_{M,0} = 1, \Omega_{\Lambda,0} = 0$) with $h_0 = 0.5$ (left), and the closed Model 1 of Figs. 2 and 3 with $h_0 = 0.9$ (right). There is nearly ten times as much linear distance between redshifts $z = 3$ and 4 in the closed model as there is in the flat one (both figures to same scale).

moment of last scattering, as seen through the "lens" of a curved Universe, and implies that[9]

$$\Omega_{\Lambda,0} = 1.11^{+0.13}_{-0.12} - \Omega_{M,0}. \tag{11}$$

This result is more reliable than the others discussed so far because it bypasses "local" phenomena such as supernovae, galaxies, and even lensed quasars; taking us directly back to the radiation-dominated era when physics was simpler. Let us therefore use Eq. (11) to calculate the energy density of the vacuum. Summing the baryon, exotic CDM, and neutrino densities — Eqs. (7), (8) and (9) respectively — gives the total matter density $\Omega_{M,0}$. Substituting this into Eq. (11), we find

$$\Omega_{\Lambda,0} = \begin{cases} 0.3 - 1.1 & \text{(GI theory)} \\ 0.4 - 1.2 & \text{(otherwise)}. \end{cases} \tag{12}$$

Vacuum energy thus makes up most of the Universe, while baryons — the constituents of our familiar world — appear almost incidental.

The basis for ΛCDM as the new favourite among cosmological models lies in the approximate orthogonality of the CMB and supernova bounds, Eqs. (10) and (11). Indeed, if we take *both* of these results at face value, we can substitute one into the other and solve to find $\Omega_{\Lambda,0} = 0.78 \pm 0.23$ and $\Omega_{M,0} = 0.33 \pm 0.22$. The fact that this latter number is very near the center of the range of allowed values for $\Omega_{M,0}$ in Eq. (8) has been taken as a further sign of the basic correctness of both the ΛCDM model in particular and the GI theory of structure formation in general.

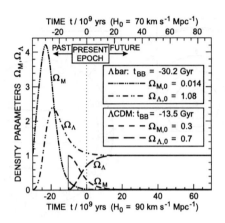

Figure 5. Evolution of Ω_M and Ω_Λ in the Λbar and ΛCDM models (Models 1 and 6 in Figs. 2 and 3). Time is set to zero at the present epoch; t_{BB}, the time of the big bang, is calculated using $h_0 = 0.9$ for Λbar (bottom scale) and $h_0 = 0.7$ for ΛCDM (top scale).

While this is a self-consistent account, and one that agrees with most observations, it suffers from one flaw: it is inherently improbable. The densities of baryonic matter (and exotic CDM, should it exist) evolve at a very different rate from neutrinos; and both of these components evolve at very different rates from vacuum energy. So one has three kinds of matter which should not have anything like the same density parameters at any given time — and yet two of them (at least) do. In the ΛCDM picture, in particular, it seems that we happen to live at a time when $\Omega_{\Lambda,0}$ and $\Omega_{M,0}$ are separated by a mere factor of two. Over the first 80 Gyr or so of the age of the Universe (i.e. the lifetime of the galaxies), Fig. 5 shows that this does indeed seem to put us "preposterously" close to the brief moment (cosmologically speaking) when $\Omega_{\Lambda,0} \approx \Omega_{M,0}$.

In Fig. 5, we have also plotted the evolution of $\Omega_{\Lambda,0}$ and $\Omega_{M,0}$ for the closed Model 1 of Figs. 2 and 3, which we term here the "Λbar" model (for Λ plus baryons only). This model, which was first proposed by Liebscher *et al* in 1992,[3,4] is less preposterous than ΛCDM in the sense that a factor of ~ 80 (rather than two) separates the presently observed values of vacuum energy and matter density. Indeed these parameters are much closer to their "cosmological average" values of one and zero respectively. While this in itself does not constitute a case for the model, it prompts us to wonder whether Λ might not be more important than most cosmologists have been willing to consider. Could vacuum energy be not just the dominant, but the *only significant component of the dark matter?*

Such an idea would have been unthinkable only a few years ago, when it

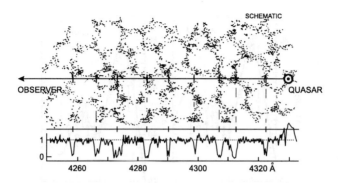

Figure 6.   Schematic illustration of a network of cell-like structures, with Lyα absorption lines seen by an observer in the spectrum of a distant quasar due to the fact that the line of sight passes through the cell walls.

was still routine to set $\Omega_{\Lambda,0} = 0$ and cosmologists had two main choices: the "true faith" ($\Omega_{M,0} \equiv 1$), or the "reformed" (with each believer being free to choose his or her own value near $\Omega_{M,0} \approx 0.3$). All this has been irrevocably altered by the CMB experiments. If there is a single guiding principle now, it is no longer $\Omega_{M,0} \approx 0.3$, and certainly not $\Omega_{\Lambda,0} = 0$; it is $\Omega_{M,0} + \Omega_{\Lambda,0} \approx 1$ from the power spectrum of the CMB. With this in mind, we will conclude with a few words about the Λbar model, which has $\Omega_{M,0} + \Omega_{\Lambda,0} = 1.094$ in excellent agreement with Eq. (11).

The first evidence in favour of such a model emerged from high-resolution quasar spectra in the Lyman α forest.[3,4] The method was described by one of us at the 1994 Erice School[31] so we give only an outline here. One supposes that Lyα absorbers, like galaxies, are distributed with a cell-like structure, and that absorption lines are produced when the line of sight to a distant quasar cuts through the cell walls (Fig. 6). The assumption is then made that the cells expand with the Hubble flow, and that evolution *within* them can be neglected. This is logical, given that the expansion velocity of a typical cell would be of order ~ 3000 km/s, whereas peculiar motions inside the cell walls would not exceed ~ 300 km/s. One then counts the lines and measures the mean spacing $\Delta\lambda$ between them. This gives the redshift spacing $\Delta z(z)$ of the cells as a function of redshift, which may in turn be related to their comoving coordinate size $\Delta\chi$ by $[\Delta z(z)]^2 = (R_0 \Delta\chi/c)^2 H^2(z)$, where χ is the radial coordinate distance. Here Δz is an observable and $H^2(z)$ is given as a function of z by Eq. (3), so one has a simple regression problem; and one, moreover, with *no linear term*. A least-squares fit leads directly to the values[4]

$$\Omega_{M,0} = 0.014 \pm 0.006, \quad \Omega_{\Lambda,0} = 1.08 \pm 0.03. \tag{13}$$

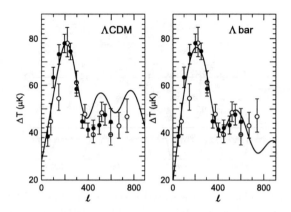

Figure 7. Observational data on the CMB power spectrum from the BOOMERANG (filled circles) and MAXIMA experiments (open circles), together with theoretical expectations based on a ΛCDM model (left) with $\Omega_{BAR} = 0.039, \Omega_{CDM} = 0.317, \Omega_{\Lambda,0} = 0.644$ and $h_0 = 0.7$; and a closed Λbar model (right) with $\Omega_{BAR} = 0.034, \Omega_{CDM} = 0, \Omega_{\Lambda,0} = 1.006$ and $h_0 = 0.75$. Figure courtesy S. McGaugh.

This result passes several basic consistency tests. First, the sum of $\Omega_{M,0} + \Omega_{\Lambda,0}$ matches that seen in the CMB experiments, Eq. (11). Second, the value of $\Omega_{M,0}$ is within the bounds imposed by cosmic nucleosynthesis, favouring low values of Ω_{BAR} and high values of h_0 ($\Omega_{BAR} \leqslant 0.016h_0^{-2}$ for $h_0 = 0.9$, or $h_0 \geqslant 0.71$ for $\Omega_{BAR} = 0.01h_0^{-2}$). And thirdly, the regression fit is found to pass through $z = 0$ at $\Delta z \approx 0.009$, in good agreement with the distribution of galaxy structure seen in our own cosmic neighborhood.[32] These phenomena involve independent physics on widely different scales, and we would regard it as remarkable for a simple procedure like the one described above to agree with all three by chance alone. Other aspects of this approach have been reviewed elsewhere.[21,31,33]

New evidence for the Λbar-type universe has come from the angular power spectrum of the CMB (Fig. 7). The odd-numbered peaks in this spectrum are produced by regions of the primordial plasma which are maximally compressed by infalling material, while the even ones correspond to maximally rarefied regions which have rebounded due to photon pressure. A high baryon-to-photon ratio enhances the compressions and retards the rarefractions, thus suppressing the height of, e.g. the second peak relative to the first. The strength of this effect depends on the fraction of baryons (relative to the more weakly-bound neutrinos and CDM particles) in the overdense regions. Data taken by the BOOMERANG and MAXIMA experiments show a strong suppression of the second peak relative to the first, inconsistent with expectations based on the ΛCDM model (Fig. 7, left-hand side). The ratio of baryons to CDM in the plasma thus appears to be higher than predicted. If the assumed CDM density is correct, then the implied baryon density exceeds nucleosynthesis limits. It may be possible to avoid this conclusion by tilting the spectrum of initial perturbations to disfavour smaller-scale (higher-order) peaks, or suppressing these peaks with processes such as delayed recombination.

The alternative is to take the apparent lack of CDM at face value. This can either be done in a half-hearted or whole-hearted way. The half-hearted way is to retain a minimum density of CDM with a statistical "prior." Thus, requiring that $\Omega_{\rm CDM} > 0.1$, but otherwise fitting the combined BOOMERANG, MAXIMA and COBE data, one obtains a model[9] with best-fit parameters $\Omega_{\rm BAR} = 0.032 h_0^{-2}$ and $\Omega_{\rm CDM} = 0.14 h_0^{-2}$. The whole-hearted approach, which may however require extending the standard picture of structure formation, is to drop the requirement of CDM altogether. Results are shown in Fig. 7 (right-hand side), which is a statistical fit to $\Omega_{\rm BAR}$ with $\Omega_{\rm CDM} = 0$ and $h_0 = 0.75$.[34] The best-fit model passes neatly through both peaks and has $\Omega_{\rm M,0} = \Omega_{\rm BAR} = 0.034, \Omega_{\Lambda,0} = 1.006$, in agreement with nucleosynthesis as well as Eq. (13). This model has an age of 22.2 Gyr (with $h_0 = 0.75$), and a total density parameter slightly above one. Similar models have been considered in other analyses of the BOOMERANG and MAXIMA data.[35,36] Thus the shape of the CMB power spectrum, along with the spectra of Lyα absorbers, suggest to us that we may live in a closed accelerating Universe, dominated by vacuum energy and with no significant contributions from CDM or neutrinos.

4. The (Four?) Elements of Modern Cosmology

Collecting Eqs. (7), (8), (9), (11) and (12), we may summarize the present contributions of baryons, exotic CDM, neutrinos and vacuum energy to the total density of the Universe as follows:

$$\Omega_{\rm BAR} = 0.012 - 0.041$$

$$\Omega_{\rm CDM} = \begin{cases} 0.1 - 0.5 & \text{(GI theory)} \\ 0 - 0.4 & \text{(otherwise)} \end{cases}$$

$$\Omega_\nu = 0.0003 - 0.2 \Omega_{\rm CDM}$$

$$\Omega_{\Lambda,0} = \begin{cases} 0.3 - 1.1 & \text{(GI theory)} \\ 0.4 - 1.2 & \text{(otherwise)} \end{cases}$$

$$\Omega_{\rm TOT,0} = 0.99 - 1.24, \,^9 \tag{14}$$

where "GI" refers to gravitational instability theory. Baryons, the stuff of which we are made, are apparently little more than a cosmic afterthought.

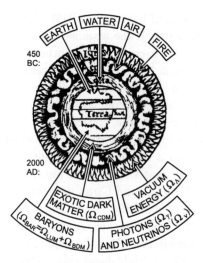

Figure 8. The "four elements" of cosmology, past and present. (Adapted from a 1519 edition of Aristotle's *De caelo* by Johann Eck; see[21] for discussion.)

This has rightly been described as a "second Copernican revolution" and lends a double meaning to the identification of baryons with "earth," the first element of modern cosmology (Fig. 8). Neutrinos ("air") and exotic CDM ("water") may both play more significant roles in determining the past and future evolution of the Universe, though this is not certain. What is clear is that all three forms of matter are dwarfed in importance by a newcomer whose physical origin remains shrouded in obscurity: the energy of the vacuum ("fire"). We have argued that this may in fact dominate so completely that there is no room for significant amounts of exotic dark matter at all. This would simplify our picture of the Universe, ease problems with the "preposterously" fine-tuned values of the observed cosmological parameters, and allow more time for galaxies and other structures to form. It would also, however, require that we modify the GI paradigm. We have reviewed the various lines of observational argument, both for and against such an idea. It appears to us quite possible that the vacuum density $\Omega_{\Lambda,0}$ is close to one, that the sole contributions to the matter density $\Omega_{M,0}$ come from a small amount of baryons and neutrinos, and that $\Omega_{\Lambda,0}$ and $\Omega_{M,0}$ together are enough to "just close" the Universe.

Acknowledgments

This work is based partly on Ref. 21. J.O. thanks the Alexander von Humboldt Stiftung and Japan Society for Promotion of Science for support.

References

1. G. F. R. Ellis, *Class. Quant. Grav.* **5**, 891 (1988).
2. P. Hübner and J. Ehlers, *Class. Quant. Grav.* **8**, 333 (1991).
3. D.-E. Liebscher, W. Priester and J. Hoell, *Astron. Astrophys.* **261**, 377 (1992).
4. D.-E. Liebscher, W. Priester and J. Hoell, *Astron. Nachr.* **313**, 265 (1992).
5. W. Priester, J. Hoell and H.-J. Blome, *Comments Astrophys.* **17**, 327 (1995).
6. H.-J. Blome and W. Priester, *Astrophys. Sp. Sci.* **117**, 327 (1985).
7. W. Priester and C. van de Bruck, *Naturwissenschaften* **85**, 524 (1998).
8. H.-J. Blome, J. Hoell and W. Priester, *Bergmann-Schaefer: Lehrbuch der Experimentalphysik* v. 8, 2nd ed., 439-582 (Berlin: W. de Gruyter, 2002).
9. A. H. Jaffe *et al*, *Phys. Rev. Lett.* **86**, 3475 (2000).
10. T. Shanks *et al*, in *The Extragalactic Infrared Background and its Cosmological Implications*, ed. M. Harwit and M. G. Hauser (San Francisco: Astron. Soc. Pac., 2001).
11. A. Jenkins *et al*, *Astrophys. J.* **499**, 20 (1998).
12. M. Fukugita, C. J. Hogan and P. J. E. Peebles, *Astrophys. J.* **503**, 518 (1998).
13. J. Mould *et al*, *Astrophys. J.* **529**, 786 (2000).
14. J. A. Willick and P. Batra, *Astrophys. J.* **548**, 564 (2000).
15. J. R. Herrnstein *et al*, *Nature* **400**, 539 (1999).
16. E. Maoz *et al*, *Nature* **401**, 351 (1999).
17. J. J. Cowan *et al*, *Astrophys. J.* **521**, 194 (1999).
18. K. A. Olive, *Nucl. Phys. Proc. Suppl.* **80**, 79 (2000).
19. D. Tytler *et al*, *Physica Scripta* **85**, 12 (2000).
20. C. Alcock *et al*, *Astrophys. J.* **542**, 281 (2000).
21. J. Overduin and W. Priester, *Naturwissenschaften* **88**, 229 (2001). Preprint version available at astro-ph/0101484.
22. J. M. Overduin and P. S. Wesson, *Dark Sky, Dark Matter* (Oxford: Institute of Physics Press, 2002).
23. R. A. C. Croft, W. Hu and R. Davé, *Phys. Rev. Lett.* **83**, 1092 (1999).
24. Y. Fukuda *et al*, *Phys. Rev. Lett.* **81**, 1562 (1998).
25. S. Weinberg, *Dreams of a Final Theory* (New York: Pantheon Books, 1993).
26. A. G. Riess *et al*, *Astron. J.* **116**, 1009 (1998).
27. S. Perlmutter *et al*, *Astrophys. J.* **517**, 565 (1999).
28. P. de Bernardis *et al*, *Nature* **404**, 955 (2000).
29. S. Hanany *et al*, *Astrophys. J.* **545**, L5 (2000).
30. C. Pryke *et al*, *Astrophys. J.* **568** 46 (2001).
31. W. Priester, in *Currents in High-Energy Astrophysics*, ed. M. M. Shapiro, R. Silberberg and J. P. Wefel (Dordrecht: Kluwer Academic, 1995) 291-312.
32. M. Geller and J. Huchra, *Science* **246**, 897 (1989).
33. C. van de Bruck and W. Priester, in *Dark Matter in Astrophysics and Particle Physics 1998*, ed. L. Baudis and H. V. Klapdor-Kleingrothaus (Oxford: Institute of Physics Press, 1999).
34. S. S. McGaugh, *Int. J. Mod. Phys.* **A16**, 1031 (2001).
35. M. White, D. Scott and E. Pierpaoli, *Astrophys. J.* **459**, 415 (2000).
36. L. M. Griffiths, A. Melchiorri and J. Silk, *Astrophys. J.* **553**, L5 (2001).

THE ENTANGLED UNIVERSE

THOMAS L. WILSON

National Aeronautics and Space Administration
Houston, Texas 77058, USA
E-mail: twilson@ems.jsc.nasa.gov

HANS-JOACHIM BLOME

University of Applied Sciences, AeroSpace Department
D-52064 Aachen, Germany
E-mail: blome@fh-aachen.de

We review the importance of coherence and nonlocal entanglement in quantum cosmology. In particular, the origin of quantum temperature for the Universe itself is explored using the thermalization theorems in quantum gravity. The resulting temperatures for accelerating Friedmann-Lemaitre-Robertson-Walker models are compared with the classical Gamow temperature which follows from general relativity for the matter- and radiation-dominated eras of the Big Bang model.

1. Introduction

The temperature of the Universe is an important concept in cosmology because of the significance it has for our understanding of the origin, history, and evolution of what astrophysicists observe today. Measurements of the remnant photon gas[1,2] known as the cosmic microwave background radiation (CMBR) have found a photon temperature of 2.73 K. Maps of anisotropies in that background temperature are disclosing much about early fluctuations that may have contributed to the formation of large-scale structure. In a related development, recent observations[3,4] also seem to indicate that the Universe is not only expanding but accelerating as well.

The interesting physics question of deriving these temperatures for an accelerating model from first principles is not usually addressed because cosmology involves a complex mixture of classical relativity and quantum physics. The subject of quantum cosmology *per se* began in 1931 when Lemaitre[5] introduced the startling idea that perhaps the entire Universe originated from the explosion of some sort of quantum object or primeval atom. In spite of the marvelous successes in both theory and observation of what is now called the standard Big Bang model, many questions remain unanswered. One basic aspect of quantum physics is passionately avoided or disregarded completely in cosmology. This is the subject of quantum entanglement[6], in spite of the fact that it has been well established in experimental laboratories. We will attempt to show here the importance of entanglement as a fundamental feature of cosmology and relate it to the derivation of the quantum temperature of accelerating Universe models.

Nonlocality and entanlement in quantum field theory will be used interchangeably throughout. The focus will involve both the pre-Planck era of the very early Universe and the future Universe as well. Similar to suggestions made in string theory[7], the point of view will be that there is physics in the pre-Planck era and beyond the Planck scale. However, the further we look back in cosmic time, the more nonlocal that physics becomes.

2. Horizons and Nonlocal Quantum Cosmology

Horizons first came into prominence in the study of gravitational collapse and cosmology. However, they are ubiquitous and encompass every realm of physics from the Planck scale to the macroscopic Universe. They are classical in origin and basically split the structure of spacetime up into distinct and disjoint parts. Quantum physics, on the other hand, is a formalism concerned with the physical behavior of elementary particles, atoms, and molecules characterized by a wave function $\psi(x)$ which is strongly dependent upon the property of coherence. A quantum object $\psi(x)$ is subject to axiomatic constraints that derive from this coherence in Hilbert space. Unlike spacetime, there are no horizons in Hilbert space. So how do the classical and quantum worlds relate in cosmology?

The first problem presented by horizons is that they result in classical spacetime being broken up into causally disconnected regions with different vacua that cannot communicate with one another. They basically arise because of relativity, and a short phase of exponential expansion in the very early Universe has been suggested to cope with this problem[8] although some[9] contend that such inflation is not a general solution.

The horizon problem is illustrated in Fig. 1 for arbitrary spacetimes, adapted from the treatment due to Penrose[10]. The light cones in Minkowski spacetime become the past null cones defining the particle and event horizons introduced by Rindler[11]. These will be depicted as Penrose diagrams for curved spacetime backgrounds in what follows.

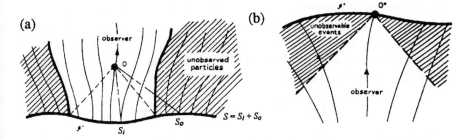

Figure 1. Spacetime horizons as introduced by Rindler and Penrose. Dashed lines represent future and past lightcones for a spacelike hypersurface S. That part of S outside the lightcone is S_o. *(a)* When I^- is spacelike, a particle horizon occurs. *(b)* When I^+ is spacelike an event horizon occurs.

Quantum entanglement can also be characterized using Fig. 1. When there exists global coherence in Hilbert space, nonlocal interactions necessarily couple a $\psi(x)$ inside its lightcone at S_i with other $\psi(x)$'s outside its lightcone at S_o by means of a quantum interaction potential. The neglect of this coherence might be described as the quantum entanglement problem.

3. Examples: Nonlocality in Nuclear Physics and Quantum Mechanics

What are some specific examples of nonlocality and quantum entanglement in physics? The answer will provide several of the principal tools for physics at the Planck scale.

One of the best examples appears in nuclear physics. It follows directly from Heisenberg's uncertainty principle

$$\Delta x \Delta p_x \geq \hbar/2 \tag{1}$$

where $\hbar$ is Planck's constant divided by 2π. When classical physicists speak of a point $x = \{\mathbf{x}, ct\}$ in spacetime (with boldface Latin characters and indices j representing 3-space vectors and Greek indices μ running as $\mu = 0, 1, 2, 3$) they mean a precise mathematical point. However, an exact specification of point x in (1) creates an indeterminacy in the quantum momentum $p_x = -i\hbar\partial_x$ which becomes an infinite nonlocal operator (one which cannot be expressed as a simple finite polynomial in the space derivatives ∂_j). Such an operator necessarily requires spacelike support outside the lightcone of classical relativity as shown in Fig. 1 for *all* of S, not simply $S = S_i$. This is the first new piece of physics that circumvents the restrictions of Einstein's special theory of relativity, and it arises from the assumption of coherence of the wave function ψ in quantum mechanics. How does this come about? The general eigenvalue problem for energy E in terms of a nonlocal potential $V(\mathbf{r}, \mathbf{r}')$ is described by an integro-differential Schrödinger equation

$$-\frac{\hbar^2}{2m}\nabla^2\Psi(\mathbf{r}) + \int d^3r' V(\mathbf{r}, \mathbf{r}')\Psi(\mathbf{r}') = E\Psi(\mathbf{r}) \tag{2}$$

where knowledge of Ψ for all $\mathbf{r}'$ on the spacelike hypersurface $S = S_i + S_o$ in Fig. 1a is required to determine the interaction at $\mathbf{r}$. A local potential is simply a special case of the nonlocal one, with

$$V(\mathbf{r}, \mathbf{r}')|_{local} = \delta(\mathbf{r} - \mathbf{r}')V(\mathbf{r}) \tag{3}$$

whereby (2) reduces to the conventional (local) Schrödinger equation using (3). Historically, Schrödinger implicitly assumed (3) which made a wavefunction $\psi(r)$ at the point $\mathbf{r}$ be determined by the potential $V(\mathbf{r})$ at the same point and nowhere else, such as $\mathbf{r}' \neq \mathbf{r}$.

It can be shown[12] that the appearance of $V(\mathbf{r}, \mathbf{r}')$ in (2) is equivalent to introducing a momentum-dependent potential, as $V(\mathbf{r}, \mathbf{r}') \to V(\mathbf{r}, \mathbf{p})$ or $V(x, x') \to V(x, p)$, resulting in the presence of off-diagonal matrix elements in the energy eigenvalue problem. It is the occurrence of powers of p in $V(\mathbf{r}, \mathbf{p})$ from which it follows that all derivatives of V (e.g., V as a continuous analytic function) are tacitly assumed (in classical relativity) to be known at a spacetime point $\mathbf{r}'$ whereby the value of V anywhere depends upon the value of V everywhere else. This mathematical truth about Schrödinger's Eq. (2) has nothing to do with acausal propagation, because the potential V is not an observable in quantum theory. Rather, it is the quantum coherence introduced by the requirement that ψ be complete in the Hilbert space and be defined over all spacetime which is the new experimental result. Momentum-dependent potentials have been used successfully in the optical model of nuclear physics[13-16].

4. BCS Theory and Quantum Cosmology

Another experimental example is the BCS (Bardeen, Cooper, Schrieffer) theory of superconductivity[17,18]. To understand this connection, accelerated frames must now be introduced in the context of spacetime horizons. The fundamental concept is that of a Rindler frame[19], shown in Fig. 2.

When an observer in Minkowski space is subjected to a uniform proper acceleration g, spacetime is split up into four quadrants I, II, F, and P due to

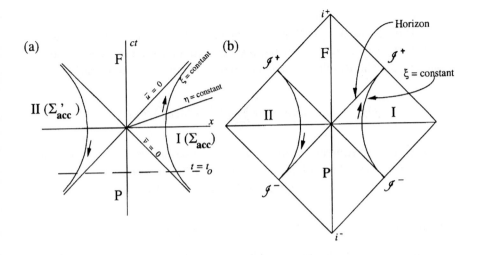

Figure 2. Nonlocal quantum field theory in accelerated Rindler space. *(a)* Two-dimensional Minkowski spacetime (x, ct) consists of four quadrants, covered by separate coordinate patches: Two wedges I (when $x > |ct|$) and II (when $x < |ct|$) created by Rindler coordinates $(\xi, c\eta)$, plus a future (F) and past (P) region. The horizontal dashed line is a spacelike hypersurface S at $t = t_o$. *(b)* Conformal Penrose diagram of (a).

the appearance of horizons $\bar{u} = 0$ and $\bar{\nu} = 0$ illustrated in Fig. 2a. This can be shown by considering a two-dimensional Minkowski space with metric $ds^2 = d\bar{u}d\bar{\nu} = c^2dt^2 - dx^2$. The Minkowski frame $\Sigma(x, ct)$ coordinates transform to the accelerated Rindler frame $\Sigma_{acc}(\xi, c\eta)$ as follows[20]

$$x = c^2 g^{-1} e^{g\xi/c^2} \cosh\, g\eta/c \qquad (4)$$

$$ct = c^2 g^{-1} e^{g\xi/c^2} \sinh\, g\eta/c \qquad (5)$$

where $g > 0$ is the constant uniform acceleration, while the Rindler "time" is coordinate $\eta > -\infty$ and Rindler "space" is coordinate $\xi < \infty$. Equivalently,

$$\bar{u} = -cg^{-1}e^{-gu/c} \qquad (6)$$

$$\bar{\nu} = +cg^{-1}e^{+g\nu/c} \qquad (7)$$

where $u = c\eta - \xi$, $\nu = c\eta + \xi$, transform the Rindler metric $ds^2 = d\bar{u}d\bar{\nu}$ into

$$ds^2 = e^{2g\xi/c}dud\nu = e^{2g\xi/c}(c^2d\eta^2 - d\xi^2). \qquad (8)$$

The conformal Penrose diagram of Rindler space in Fig. 2b illustrates that the infinite Minkowski frame for any observer on ξ=constant maps into two causally disjoint wedges (I and II).

Lee[21-23] and Coleman cited therein have shown that in such an accelerated Rindler frame the Minkowski vacuum $|O_M>$ is a coherent state with a BCS-type pair correlation between two Rindler quanta, one inside and one outside the Rindler horizon created by the uniform acceleration g

$$|O_M> = constant[exp(\sum_k e^{-\pi\omega_k/g} a_k^\dagger a_k'^\dagger)]|O_R> .\qquad(9)$$

$|O_R>$ is the Rindler vacuum, $a_k^\dagger$ is a creation operator, and primes designate from which of two accelerated Rindler wedges $\Sigma_{acc}(\xi, c\eta)$ or $\Sigma'_{acc}(\xi', c\eta')$ the quanta originate in Fig. 2 for any hypersurface $t = t_o$.

5. Rindler Wedges and Time-Like Killing Vectors

An important pedagogical concept follows from the previous section concerning the Rindler vacuum, with reference now to Fig. 3. To say that $|O_M>$ is a coherent state of Rindler pairs is to picture it as a vacuum condensate. Every particle traveling in one direction on the hyperbola $\xi =$ constant in wedge I of Fig. 2a has a 'twin' traveling on the equivalent hyperbola in wedge II but in the opposite direction. This is a consequence of the global coherence condition for the vacuum and is the origin of matter-antimatter pair production in quantum physics. Although these pairs exist in causally disconnected spacetimes (lying outside each other's lightcones), they are intimately connected in Hilbert space.

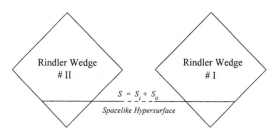

Figure 3. Rindler wedges I and II in Fig. 2b, separated for emphasis. The portion of dashed line contains no mathematical points mapped from Minkowski space.

The BCS-type twins have related time-like Killing vectors defined by $+\partial_\eta$ in wedge I and $-\partial_\eta$ in wedge II. Positive frequency modes with respect to $+\partial_\eta$ and negative frequency modes with respect to $-\partial_\eta$ appear, a mixture is formed with nonzero Bogoliubov coefficients[24], and pairs of particles are produced as thermal radiation. In F and P the time-like Killing vectors become spacelike and no radiation results there. Time-like Killing vector fields are thoroughly discussed[20,25] and given for the de Sitter space[26] which concerns us later.

When one allows the uniform Rindler acceleration g to become zero ($g \to 0$), the Rindler horizons and wedges disappear, a conformal Minkowski spacetime is recovered, and a marvelous piece of physics occurs that saves 'causality' by what has been called a flat-space miracle[27]. The quantum field theory contributions for the two twin particles exactly cancel out.

If instead we excite or accelerate the vacuum into a Rindler state, we recognize that it is actually a BCS-type condensate from which emerges pair production and the quantum field theory contributions do not cancel out. This is the quantum origin of thermal radiation that is of interest in cosmology and astrophysics. Furthermore, because of global coherence in Rindler space an excitation of one wedge must necessarily produce interference with the quantum state in the other wedge via Hilbert space.

6. Lee's Theorem and Quantum Gravity

At this point, it is necessary to introduce the so-called thermalization theorems. The most fundamental is due to Bisognano and Wichmann[28,29], and the subject has been reviewed by Takagi[30]. Lee's Theorem[21−23] will be used in the derivations here because of the simplicity with which it relates directly to quantum gravity[31].

Lee begins with the axiomatic requirement in quantum theory that a globally coherent wave function must have support from the entire spacelike hypersurface S in Fig. 1 and Fig. 2 and therefore from inside (S_i) and outside (S_o) the observer's horizon: $S = S_i + S_o$. This is the *global coherence condition* which we have already discussed. Using Euclidean quantum field theory (Wick-rotating the Minkowski space Green's functions e^{-itH} with $\hat{t} = it$ into $e^{-\hat{t}H}$, taking state vectors over the entire spacelike hypersurface S), the horizons vanish and a complete, coherent quantum eigenstate is constructable in Hilbert space.

Lee's theorem then follows. Defining $|O_M >$ as the ground state of the total Hamiltonian H_M in the Minkowski frame $\Sigma(x, ct)$ shown in Fig. 2a, the coordinate q-representation $< q, q'|O_M >$ of that ground state is related to a matrix element of the Rindler Hamiltonian H in the frame $\Sigma_{acc}(\xi, c\eta)$, undergoing uniform acceleration g, by

$$< O_M|q, q' >= \frac{< q'|e^{-\pi H/g}|q >}{(Trace \quad e^{-2\pi H/g})^{1/2}}. \tag{10}$$

For any observable $\oslash = \oslash(q, \delta/\delta q)$ in Σ_{acc} one obtains

$$< O_M| \oslash |O_M >= \frac{Trace \oslash \rho_v}{(Trace \quad \rho_v)}, \tag{11}$$

where ρ_v is the corresponding vacuum-state density matrix $\rho_v = e^{-2\pi H/g}$.

7. Examples of Lee's Theorem

The vacuum expectation value of the occupation number operator $n_k = a_k^\dagger a_k$ follows from (10) and (11) [Ref. 21, App. A and Eq. 2.24]

$$< O_M|a_k^\dagger a_k|O_M >= (e^{2\pi\omega/g} - 1)^{-1} \tag{12}$$

which is the Rindler radiation formula derived using Bogoliubov transformations [24,22,20] in flat spacetime BCS theory. It corresponds to the Rindler-Unruh temperature

$$T_U = \hbar(2\pi k_B c)^{-1} g \tag{13}$$

of isotropic scalar radiation in the Davies-Unruh effect[32,33], observable by an accelerated detector in Σ_{acc} where k_B is the Boltzmann constant. Lee's proof of (12) and (13) is for scalar Spin-0 radiation but is readily extended to radiation fields of arbitrary spin where the Hamiltonian H is a quadratic function of the field, in any space-dimension with arbitrary interactions. Friedberg, Lee, and Pang[34] have done this for Spin-1/2 radiation fields.

Similarly, the theorem is extendable to metrics other than Rindler space which are static or quasi-stationary curved spacetimes. The first example is the Schwarzschild metric

$$ds^2 = c^2(1 - \frac{2Gm}{rc^2})dt^2 - (1 - \frac{2Gm}{rc^2})^{-1}dr^2 - r^2(d\theta^2 + sin^2\theta d\phi^2) \tag{14}$$

which represents a spherically symmetric curved background and accelerated frame about a mass m with Newtonian gravitational constant G and c the speed of light. In the classical, weak-field Newtonian approximation for spherically symmetric metrics in general relativity, the coefficient of $c^2 dt^2$ in (14) is equivalent to $(1 + 2\Phi_N/c^2)$ where Φ_N is the effective Newtonian potential $\Phi_N = -Gm/r$. From this one derives the gravitational acceleration as $g = -\nabla_r\Phi_N = Gm/r^2$. The gravitational horizon appears when that coefficient disappears, or $\Phi_N = -\frac{1}{2}c^2$, which occurs at the Schwarzschild radius $r_s = 2Gm/c^2$ and $g = c^4(4Gm)^{-1}$ is called the surface gravity. By replacing the Rindler acceleration g in (10) and (11) with the gravitational acceleration at the Schwarzschild radius, the theorem gives

$$< O_S| \oslash |O_S >= \frac{Trace[e^{-8\pi GmH} \oslash]}{Trace[e^{-8\pi GmH}]} \qquad (15)$$

where now $|O_S >$ is the ground state of the scalar Hamiltonian H outside the Schwarzschild radius $(r > 2Gm/c^2)$, and

$$T_H = \hbar c^3 (8\pi Gmk_B)^{-1} \qquad (16)$$

is the Hawking[35] black-body temperature. That is, both Rindler-Unruh-Davies and Hawking radiation follow from Lee's theorem (10) and (11). Gibbons and Hawking[36] extended (16) by showing that each geodesic observer in a de Sitter spacetime with a cosmological constant $\Lambda \neq 0$, detects a temperature

$$T_{GH} = \hbar(\pi\sqrt{12})^{-1}\sqrt{\Lambda} \, . \qquad (17)$$

(17) appears to be the first attempt to establish that cosmological horizons radiate at some Rindler-type temperature.

8. Derivation of the Quantum Temperature of the Accelerating Universe

We now discuss an extension of the above results to be published elsewhere[37], in which Lee's theorem (10) and (11) is used to determine the temperature of scalar radiation in the Friedmann-Lemaitre (FL) models of "big bang" cosmology, with $k = 0$, $\Lambda \neq 0$, and metric in isotropic Robertson-Walker (RW) form

$$ds^2 = c^2 dt^2 - a(t)^2 [\frac{dr^2}{1 - kr^2}] - r^2(d\theta^2 + sin^2\theta d\phi^2) \, . \qquad (18)$$

k is the curvature parameter and $a = a(t)$ is the FL expansion factor determinable from Einstein's field equations the once a matter distribution is specified. In order to draw a similar conclusion regarding Rindler acceleration g in Lee's theorem (10) and (11) as in the Schwarzschild case (14), the RW metric (18) must be transformed into a spherically symmetric form. This is a straightforward procedure[38–40] using the transformation

$$R = ra(t) \qquad (19)$$

$$T = t - t_o + \frac{1}{2}\frac{r^2}{c^2}a\dot{a} + O(\frac{r^4}{c^4}) \qquad (20)$$

and retaining only quadratic terms in r^2 and R^2 so that third-order mixing terms vanish. The result is

$$ds^2 \sim c^2[1 - \frac{\ddot{a}}{a}\frac{R^2}{c^2}]dT^2 - [1 + (\frac{kc^2}{a^2} + \frac{\dot{a}^2}{a^2})\frac{R^2}{c^2}]dR^2 - R^2(d\theta^2 + sin^2\theta d\phi^2) . \quad (21)$$

Note in (21) that the left-hand sides of the Einstein-Friedmann equations

$$\frac{\dot{a}^2 + kc^2}{a^2} = \frac{8\pi G}{3}\rho + \frac{1}{3}\Lambda c^2 \quad (22)$$

$$\frac{\ddot{a}}{a} = \frac{4\pi G}{3}(\rho + \frac{3p}{c^2}) + \frac{1}{3}\Lambda c^2 \quad (23)$$

have appeared in this approximation. Notation is matter density ρ and pressure p with cosmological constant Λ.

Following the Hubble parameterization, we define $\mathcal{H} = \dot{a}/a$ along with the deceleration parameter $q = -\ddot{a}a/\dot{a}^2$, and the isotropic metric (21) can now be stated as

$$ds^2 \sim c^2[1 + q\mathcal{H}^2\frac{R^2}{c^2}]dT^2 - [1 + (\mathcal{H}^2 + \frac{kc^2}{a^2})\frac{R^2}{c^2}]dR^2 - R^2 d\Omega^2 , \quad (24)$$

having defined $d\Omega^2 = (d\theta^2 + sin^2\theta d\phi^2)$.

We are concerned here with those FL models for which metric (24) approximates the familiar form similar to (14)

$$ds^2 \sim c^2[1 - \gamma\frac{R^2}{c^2}]dT^2 - [1 - \gamma\frac{R^2}{c^2}]^{-1}dR^2 - R^2 d\Omega^2 , \quad (25)$$

with $\gamma R^2/c^2 \ll 1$ and $[1 - \gamma R^2/c^2]^{-1} \approx [1 + \gamma R^2/c^2]$. (25) is the de Sitter metric[41] or an empty ($m = 0$) Kottler-Schwarzschild solution[42]. It is also referred to as the metric for a "harmonic oscillator potential"[43].

For $q \equiv -1$ with $k = 0$ (an accelerating, open FL model where $\ddot{a} > 0$) in (24) and (25), it necessarily follows that $\gamma = \mathcal{H}^2$, making (24) and (25) equivalent to the order of magnitude assumed in (19) and (20). This is the metric of an effective cosmological term $\gamma = \mathcal{H}^2$ in (22) and (23) when $\rho = p = 0$, and $\mathcal{H}^2 = \Lambda c^2/3$. Taking the negative gradient of the "Newtonian" approximation potential $\Phi_N = +\frac{1}{2}q\mathcal{H}^2 R^2$ in (24), one obtains acceleration $g = -\nabla_R\Phi_N = -q\mathcal{H}^2 R$. The horizon appears when $q\mathcal{H}^2 R^2 = -c^2$, or $R = c/\sqrt{|-q|}\mathcal{H}$ and $g = \sqrt{|-q|}\mathcal{H}c$ for defining the acceleration in (10) and (11) or (9). For $q \equiv -1$ then

$$< O_{FL}| \oslash |O_{FL} >= \frac{Trace \oslash [e^{-2\pi H/\mathcal{H}c}]}{Trace[e^{-2\pi H/\mathcal{H}c}]} . \quad (26)$$

With $c \neq 1$ cancelling, the temperature for $q \equiv -1$ follows as

$$T_{FL} = \hbar(2\pi k_B)^{-1}\mathcal{H} . \quad (27)$$

Since $\mathcal{H} = \sqrt{\Lambda c^2/3}$, (27) gives

$$T_{FL} = \hbar(2\pi k_B \sqrt{3})^{-1}\sqrt{\Lambda c^2} \qquad (28)$$

which is equivalent to the Gibbons-Hawking result (17) in units $c = k_B = 1$.

The result in (27) and (28) can be arrived at using an entirely different method, known as the KMS condition[44,45,29] (based on thermal equilibrium). When the thermal Green's functions are periodic in imaginary (Euclidean) time, thermal fluctuations arise in quantum field theory. If the metric can be put in periodic Euclidean form, the temperature follows directly. The de Sitter space metric (25) can be written[46]

$$ds^2 = c^2 dt^2 - \mathcal{H}^{-2}\cosh^2(\mathcal{H}t)[d\chi^2 - \sin^2\chi d\Omega^2]. \qquad (29)$$

Under analytic continuation $t \to i\hat{t}$, then $\cosh(\mathcal{H}t) \to \cos(\mathcal{H}\hat{t})$ with period $2\pi/\mathcal{H}$, and (27) follows directly.

However, de Sitter space has no deceleration parameter q. (24) was derived here from the RW metric where q is not a constant. The flat FL models are strongly regulated by $a = a(t)$ in metric (18) or (21). Since $a = a(t)$ is asymptotically increasing, $q = q(t)$ modulates the Newtonian potential by a small scale factor of $\frac{1}{2}$. For early cosmic time ($t \ll 1$), $|q| = 0.5$ and for late cosmic time ($t \gg 1$), $|q| = 1.0$ which will be shown analytically below.

Including all q in the Newtonian harmonic oscillator argument that led to (26) does not change the existence of time-like Killing vectors for this spacetime except when $q = 0$ and there is no Newtonian potential in (24). We conclude that if (27) is a temperature for the $q < 0$ limit (accelerating FL models, or $\ddot{a} > 0$), dimensionally the function is still a temperature for all q (including decelerating FL models when cosmic time t is small).

Repeating the derivation of (26) with q produces the direct result

$$< O_{FL}|\oslash|O_{FL} > = \frac{Trace \oslash e^{-2\pi H/\sqrt{|1-q|}\mathcal{H}c}}{Trace \, e^{-2\pi H/\sqrt{|1-q|}\mathcal{H}c}} \qquad (30)$$

where $|O_{FL} >$ is the vacuum ground state of the scalar Hamiltonian H outside the effective radius $R = c/\sqrt{|1-q|}\mathcal{H}$ for FL models. Relation (30) corresponds to the temperature

$$T_{FL} = \hbar(2\pi k_B)^{-1}\sqrt{|1-q|}\mathcal{H} \qquad (31)$$

or directly from (21) and (24) using $k = 0$

$$T_{FL} = \hbar(2\pi k_B)^{-1}\sqrt{\left|\frac{\ddot{a}}{a}\right|}. \qquad (32)$$

9. Quantum Temperature of Flat, Accelerating FL Models

The analytic solutions for the Einstein-Friedmann equations (22) and (23) with $k = 0$ and $\Lambda \neq 0$ are well known[47−49]. These are historically called the FL models in which $a(t)$ is

$$a(t) = \sqrt[3]{\Gamma(coshy - 1)} \tag{33}$$

where $y = t/\tau$ with $\tau = 1/\sqrt{3c^2\Lambda}$, $\Gamma = \Omega_m/\Omega_\Lambda$, and Ω_m and Ω_Λ are the contributions of baryonic matter m and Λ to the closure parameter Ω respectively. Straightforward derivations yield the following relationships

$$q = \frac{2 - coshy}{1 + coshy} \tag{34}$$

$$\mathcal{H} = \frac{1}{3\tau}\frac{sinhy}{coshy - 1} \tag{35}$$

and one obtains for temperature (27) using (35)

$$T_{FL} = \hbar(2\pi k_B)^{-1}\frac{1}{3\tau}\frac{sinhy}{coshy - 1} \tag{36}$$

while (31) and (32) become

$$T_{FL} = \hbar(2\pi k_B)^{-1}\frac{1}{3\tau}\sqrt{\frac{|2 - coshy|}{coshy - 1}} . \tag{37}$$

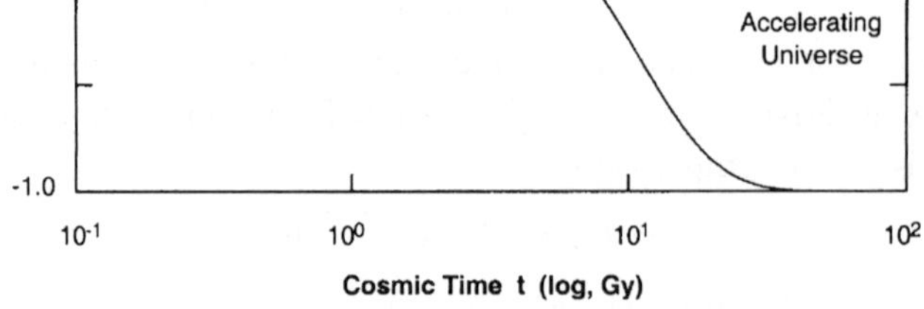

Deceleration Parameter q (Friedmann-Lemaitre Models)

Figure 4. On the cosmic scale of time the deceleration parameter quickly reverts itself from a deceleration to an acceleration phase.

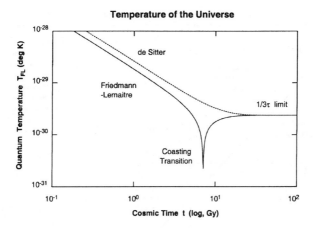

Temperature of the Universe

Figure 5. Comparison of FL temperature (37) with the asymptotic (dotted curve) de Sitter temperature (36) when q passes through zero.

Fig. 4 illustrates the deceleration parameter q in (34). Similarly Fig. 5 is a comparison of (36) and (37) when q changes sign. Relation (37) is also plotted in Fig. 6 and compared with Gamow's radiation-dominated era[51,52] where $T = 1.5 \times 10^{-10}t^{-1/2}$ and the matter-dominated era[52] where $T \sim t^{-2/3}$. Λ can be adjusted to fit the data[3,4] but we choose $\Lambda = 1.20 \times 10^{-56}cm^{-2}$ like Model 6 of Priester[48], giving $\tau = 5.57Gy = 1.76 \times 10^{17}s$.

Both (36) and (37) have the same asymptotic limit $1/3\tau$ for large $y \gg 1$. However, (36) is the temperature of a de Sitter space and in the small $y \ll 1$ limit (early Universe) there is a distinct difference from (37) visible in Fig. 5. Both vary as $T \sim t^{-1}$. But $T_{FL} \sim \frac{\sqrt{2}}{3}(1.2 \times 10^{-12})t^{-1}$, while $T_{dS} \sim \frac{2}{3}(1.2 \times 10^{-12})t^{-1}$. For intermediate cosmic time $y \sim 1$, the FL temperature (37) goes to zero when $q \to 0$ (neither accelerating nor decelerating FL models, with $\ddot{a} \sim 0$ at $coshy = 2$ or $y = 1.317$) and the Universe briefly "coasts." Since $y = t/\tau$ then $q = 0$ happens at $t = 1.317\tau = 7.33Gy$.

10. Conclusion

We have shown how the concept of global coherence in quantum gravity naturally leads to a temperature of the Universe and theoretical cosmology at any fundamental length scale. The result is the temperature T_{FL} in (27), (31), (32), (36) and (37). This followed by treating a cosmological scalar radiation field as a coherent state with nonlocal BCS-type pair correlation between two quanta, one inside (S_i) and one outside (S_o) the FL horizon

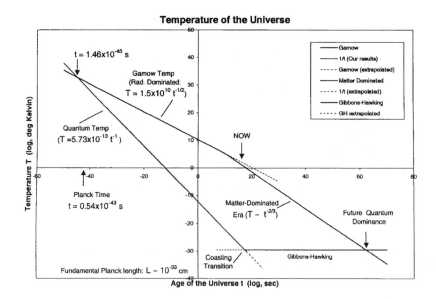

Figure 6. Quantum temperature of FL models compared against classical matter- and radiation-dominated Gamow temperatures for the standard Big Bang model, in the case of flat ($k = 0$) accelerating ($q < 0$) universes. Quantum spacetime has dominated the total temperature in the past and it will again in the future, according to these results. The global coherence condition is not subject to decoherence, rather appears to be only temporarily dominated by the origin of matter in the Universe.

using Euclidean quantum field theory in the fashion of Lee.

From Fig. 6 there is ambiguity in the answer to the question of what is the temperature of the Universe. Normally such a temperature is attributed to the CMBR radiation which at present has a measured value $T_{CMBR} = 2.73$ K. Here we suggest the second point of view that the measured temperature is that of the remnant photon gas from the Big Bang while the temperature of the Universe itself is the quantum one derived from the thermalization theorems. The quantum vacuum state for a scalar field which pervades the nonuniform expanding cosmic space, looks to a comoving observer or detector like a thermal state with temperature (31) and (32) for an open ($k = 0$) accelerating universe[3,4] ($q \sim -1, \Lambda \neq 0$).

As illustrated in Fig. 5, (37) approaches the value $T_{FL} \to T_{dS}$ in (36) or (28) for large values of $y \gg 1$. That is, in the future it approaches a de Sitter space with temperature $T_{dS} \approx 3.96 \times 10^{-30}(\mathcal{H}_o/100)\sqrt{\Omega_\Lambda}K$, where Ω_Λ is the contribution of Λ to the closure parameter Ω. Assuming $\mathcal{H}_o = 70 \ km \ s^{-1}Mpc^{-1}$ and $\Omega_\Lambda = 0.7$, then $T_{FL} \approx 2.32 \times 10^{-30}K$. For

$y \ll 1$ in the past, (37) goes asymptotically as $T \sim 1/t$ like (36) but is scaled as $T_{FL} \sim \frac{\sqrt{2}}{3}t^{-1}$ rather than $T_{dS} \sim \frac{2}{3}t^{-1}$, a result caused by $\sqrt{|-q|}$. Despite the fact that this temperature in (37) is currently extremely low and beyond present measurability, the result is important from a conceptual point of view. The quantum temperature once dominated ($t = 1.46 \times 10^{-45}s$) in the past and it will again in the future ($t \sim 10^{+62}s$). The urge to argue that (36) and (37) are evidence for decoherence and the appearance of the classical world in the matter- and radiation-dominated periods is dispelled. In the future the quantum universe will apparently re-emerge again by virtue of the global coherence condition.

The global vacuum state of the universal scalar field exhibits entanglement. This is a non-trivial result based upon the fact that Lee's thermalization theorem treats interactions between S_i and S_o, and these can be expressed in terms of observables in the CMBR data. Because the approach is essentially based on the global coherence condition, which inherently takes into account the nonlocal features of quantum theory, the result is a first hint for quantum entanglement on the cosmic scale.

Acknowledgments

The authors wish to thank Roger Penrose for several very encouraging comments in the early phase of this work[53]. Also, they are grateful to Kenneth Wilson for preparation of the graphics for the temperatures shown.

References

1. A. Penzias, and R. Wilson, *Ap. J.* **142**, 419 (1965).
2. R. Srianand, P. Petitjean, and C. Ledoux, *Nature* **408**, 931 (2000).
3. S. Perlmutter, et al., *Ap. J.* **517**, 565 (1999).
4. A. G. Riess, et al., *Ap. J.* **536**, 62 (2000).
5. G. Lemaitre, *Nature* **127**, 706 (1931).
6. E. Schrödinger, *Naturwissenschaften* **23**, 807, 823, and 844 (1935); and *Proc. Camb. Phil. Soc.* **31**, 555; **32**, 446 (1935).
7. G. Veneziano, *Phys. Lett.* **B406**, 297 (1997).
8. A. Guth, *Phys. Rev.* **D23**, 347 (1981).
9. G. F. R. Ellis and W. Stoeger, *Class. Quantum Grav.* **5**, 207 (1988).
10. R. Penrose, in *Relativity, Groups and Topology* ed C. DeWitt and B. DeWitt (Gordon and Breach, London, 1964) p 565.
11. W. Rindler, *Mon. Not. Roy. Astron. Soc.* **116**, 662 (1956).
12. W. F. Hornyak, *Nuclear Structure* (Acad. Press, New York, 1975) Sect. III.D.
13. J. P. Jeukenne, A. Lejeune, and C. Mahaux, *Phys. Rep.* **25**, 83 (1976).
14. H. Feshbach *Theoretical Nuclear Physics* (John Wiley, New York, 1992).

38

15. F. Perey and B. Buck, *Nucl. Phys.* **32**, 353 (1962).
16. J. A. Wheeler, *Phys. Rev.* **50**, 643 (1936).
17. J. Bardeen, L. Cooper, and J. Schrieffer, *Phys. Rev.* **108**, 1175 (1957).
18. D. Rogovin and M. Scully, *Phys. Rep.* **25**, 175 (1976).
19. W. Rindler, *Amer. J. Phys.* **34**, 1174 (1966).
20. N. D. Birrell, and P. C. W. Davies, *Quantum Fields in Curved Space* (Cambridge University Press, London, 1982) Sect. 4.5.
21. T. D. Lee, *Nucl. Phys.* **B264**, 437 (1986).
22. T. D. Lee, *Niels Bohr: Physics and the World* ed H. Feshbach, T. Matsui, and A. Oleson (Harwood, New York, 1988) p 183.
23. T. D. Lee, *Prog. Theor. Phys. Suppl.* **85**, 271 (1985).
24. N. N. Bogoliubov, *Nuovo Cim.* **VII**, 794 (1958).
25. R. Haag, H. Narnhofer, and U. Stein, *Comm. Math. Phys.* **94**, 219 (1984).
26. D. Sciama, P. Candelas, and D. Deutsch, *Adv. Phys.* **30**, 327 (1981).
27. M. E. Peskin and D. V. Schroeder, *An Introduction to Quantum Field Theory* (Addison-Wesley, New York, 1995) p 14.
28. J. Bisognano and E. Wichmann, *J. Math. Phys.* **16**, 985 and **17**, 303 (1975).
29. G. L. Sewell, *Phys. Lett.* **79A**, 23 (1980); *Ann. Phys.* **141**, 201 (1982).
30. S. Takagi, *Prog. Theor. Phys. Suppl.* **88**, 1 (1986) Ch. 5.
31. E. Alvarez, *Rev. Mod. Phys.* **61**, 561 (1989).
32. W. G. Unruh, *Phys. Rev.* **D14**, 870 (1976).
33. P. C. W. Davies, *J. Phys. A: Math. Gen.* **8**, 609 (1975).
34. R. Friedberg, T. D. Lee, and Y. Pang, *Nucl. Phys.* **B276**, 549 (1986).
35. S. W. Hawking, *Comm. Math. Phys.* **43**, 199 (1975).
36. G. W. Gibbons and S. W. Hawking, *Phys. Rev.* **D15**, 2738 (1977).
37. T. Wilson and H. Blome, submitted (2002).
38. P. Kustaanheimo, *Proc. Edinburgh Math. Soc.* **9**, 13 (1953).
39. R. C. Tolman, *Relativity, Thermodynamics, and Cosmology* (Clarendon Press, Oxford, 1934) p 239.
40. G. Lemaitre, *Mon. Not. Roy. Astron. Soc.* **90**, 490 (1931).
41. W. de Sitter, *Proc. Roy. Acad. Sci. (Amsterdam)* **20**, 229 (1917).
42. F. Kottler, *Ann. d. Phys.* **56**, 401 (1918).
43. P. G. O. Freund, A. Maheshwari, and E. Schonberg, *Ap. J.* **157**, 857 (1969).
44. S. Fulling and S. Ruijsenaars, *Phys. Rep.* **152**, 135 (1987).
45. J. S. Bell, R. J. Hughes, and J. M. Leinaas, *Z. Phys. C* **28**, 75 (1985).
46. S. Hawking and G. Ellis, *The large-scale structure of space-time* (Cambridge University Press, London 1973) p 125.
47. J. E. Felten and R. Isaacman, *Rev. Mod. Phys.* **58**, 689 (1986).
48. J. Overduin and W. Priester, *Naturwiss.* **88**, 229 (2001); astro-ph/0101484.
49. H. Blome, J. Hoell, and W. Priester, in *Sterne und Weltraum* ed W. Raith, *Lehrbuch der Experimentalphysik* (Walter de Gruyter, Berlin, 2002) Ch. 6.
50. Ya. B. Zel'dovich and I. D. Novikov, *Relativistic Astrophysics*, Vol **2** (University of Chicago Press, Chicago, 1983) p 17 and 137.
51. G. Gamow, *Nature* **162**, 680 (1948).
52. A. D. Chernin, *Nature* **220**, 250 (1968).
53. T. Wilson and H. Blome, *Bull. Amer. Phys. Soc.* **40**, No. 14, 2121 (1995).

THE PHYSICS OF EXTRAGALACTIC JETS FROM MULTIWAVELENGTH OBSERVATIONS

RITA M. SAMBRUNA

George Mason University
Department of Physics and Astronomy, MS 3F3,
4400 University Dr.
Fairfax, VA 22030, USA
E-mail: rms@physics.gmu.edu

I summarize recent progress in the study of kpc-scale jets, focusing on the results from our X-ray and optical survey of radio jets with *Chandra* and *HST*.

1. Introduction and Motivation

Jets are a ubiquitous feature of radio-loud Active Galactic Nuclei (AGN), providing a means to transport energy from the central compact regions to the distant lobes. Recent studies at high angular resolution in the radio band showed that pc-scale radio jets are also present in Seyferts and other radio-quiet sources (e.g., Blundell & Beasley 1998; Brunthaler et al. 2000), thus establishing jets as a common feature of all AGN.

Many important questions are still open in the study of jets. How jets are created near the supermassive black hole which is thought to power AGN, and how they propagate and stay collimated out to kpc scales, are some of the central, and still unsolved, mysteries of jets. A necessary precursor to addressing these and other questions is the knowledge of the physical conditions of jets - the particle distribution and energy, their speeds, the equilibrium between particles and magnetic fields, jet powers and composition.

Clues about jet physical properties are provided by multiwavelength imaging of jets. Until recently, most radio jets were observed at optical and UV with *HST* and other ground-based telescopes. These observations established that the radio-to-UV continuum from jet knots is due to synchrotron. The launch of the *Chandra* X-ray Observatory by NASA in July 1999 opened a new window for the study of jets, thanks to its unprecedented angular resolution and improved sensitivity. And indeed, the first *Chandra* light, the distant quasar PKS0637–752, surprisingly showed the presence of a bright, kpc-scale X-ray jet (Chartas et al. 2000), with only a weak optical counterpart in archival *HST* data (Schwartz et al. 2000). The bright X-ray flux was attributed to inverse Compton (IC) scattering of the Cosmic Microwave Background (CMB) photons by relativistic electrons in the jet, with Lorentz factors $\Gamma \sim 10$ (Tavecchio et al. 2000; Celotti et al. 2001).

The detection of an X-ray jet in PKS 0637–752 raised a new question: how common is optical and X-ray emission from radio jets in AGN? Previous X-ray observations had focused on optical jets. Had we been biased toward synchrotron jets thus missing out on a "new" class of Compton-dominated jets? To answer this question, we designed and performed a survey of radio jets with *Chandra* and *HST* in search of their X-ray and optical counterparts. The first results of the survey were published in Sambruna et al. (2002). Here I give an overview of the work in progress for all the survey jets, which will be reported in Gambill et al. (2002), Scarpa et al. (2002), and Cheung et al. (2002).

2. Sample selection and Observations

The targets of the survey were selected from the radio, without any *a priori* knowledge of their optical and X-ray emission properties. In this sense our survey is unbiased toward detections at shorter wavelengths. However, most sources show one-sided jets, suggesting that beaming is substantial.

The sample was extracted from the list of known radio jets of Bridle & Perley (1984) and Liu & Xie (1992). The selection criteria were chosen in order to match the *Chandra* and *HST* capabilities. Specifically: 1) The radio jet is $\gtrsim 3''$, i.e., long enough to be easily resolved with *Chandra* and *HST*; 2) The radio jet has radio surface brightness $S_{1.4\ GHz} \gtrsim 5\mathrm{mJy/arcsec}^2$ at $> 3''$ from the nucleus, i.e., bright enough to be detected in reasonable *Chandra* and *HST* exposures for average values of the radio-to-X-ray and radio-to-optical spectral indices, $\alpha_{rx} \sim 0.8$ and $\alpha_{ro} \sim 0.8$; and 3) High-resolution ($1''$ or better) published radio maps show that at least one bright ($\gtrsim 5$ mJy) radio knot is present at $> 3''$ from the nucleus, to prevent contamination from the wings of the core PSF. These criteria gave a sample of 17 radio jets, spanning a range of redshifts, core and extended radio powers, and classification of the nuclear activity (13 flat spectrum radio quasars, 3 steep spectrum radio quasars, 1 radio galaxy).

The awarded exposures were 10 ks per target with *Chandra* ACIS-S (occasionally, targets were observed for slightly more or less than 10 ks to accomodate gaps in the *Chandra* schedule) and one orbit per target with *HST*. The *HST* observations were performed with STIS and the `clear` filter. Therefore, both in X-rays and optical the exposures were sufficient to find the X-ray/optical counterpart of the radio jet, but not enough for a detailed study of their morphologies and spectra. Nevertheless, interesting results were obtained.

3. Results

3.1. *Detection rates and Jets morphologies*

Figures 1,2, and 3 show the *Chandra* ACIS-S images. The X-ray images were produced by smoothing the raw *Chandra* data with a Gaussian of width=0.3″ in the energy range 0.4–8 keV, with final resolution of 0.86″ FWHM. Overlaied on the X-ray images are the radio contours from the archival *VLA* data, smoothed to the same (for the detected X-ray jets) resolution as *Chandra*. In 10/17 sources, an X-ray counterpart to the radio jet is apparent, in 1 source (0405–123) there is only an X-ray counterpart to the northern radio hotspot, while in the remaining 3 sources the detection at X-rays of the jet is uncertain due to the low signal-to-noise ratio of the *Chandra* data. Thus, the detection rate of the jets at X-rays is at least 60%.

In the optical (Scarpa et al. 2002) the detection rate of the jets is similar, with 12/17 (70%) sources showing bright optical emission from radio knots. Interestingly, some optical knots do not have X-ray counterparts. Vice-versa, there are X-ray knots which are not detected with *HST*.

Concentrating on a comparison of the radio and X-ray jets in Figure 1–3, a variety of morphologies are apparent. In most cases, the X-rays track the radio one-to-one (e.g., 1354+195, 1150+497); for future reference I will call these jets "class I"[a]. In a few jets of class I (1510–089, 1641+399) the X-ray jets is shorter than the radio.

A jet that stands out because of its peculiar morphology is the jet of 1136–135. Here the X-ray and radio morphologies appear to be "anti-correlated": knot A is moderately bright at X-rays while little or no radio emission is present; the X-ray emission peaks at knot B while the radio

---

[a] Here the term "class" is loosely used as a convenient reference for groups of jets exhibiting similar properties; it does not have the usual meaning of "astrophysical class".

42

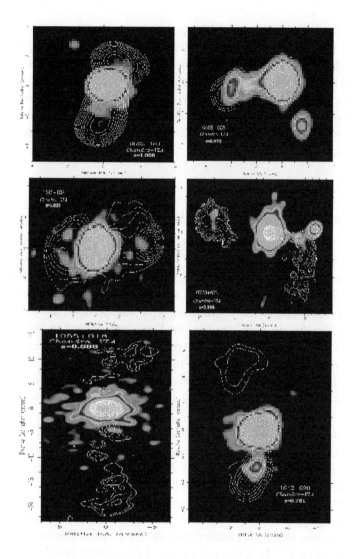

Figure 1. *Chandra* ACIS-S image in the 0.4–8 keV energy range of the newly discovered X-ray jets from our survey (grey image). Overlaied are the radio contours from archival *VLA* data. Both the grey image and the contours are plotted logarithmically, in steps of factor 2. The *Chandra* image is smoothed with a Gaussian of width $\sigma=0.3''$, yielding a resolution of $0.86''$ FWHM. The radio image was restored with a circular beam with FWHM$=0.86''$. The base level for the radio contours is 0.6 mJy/beam.

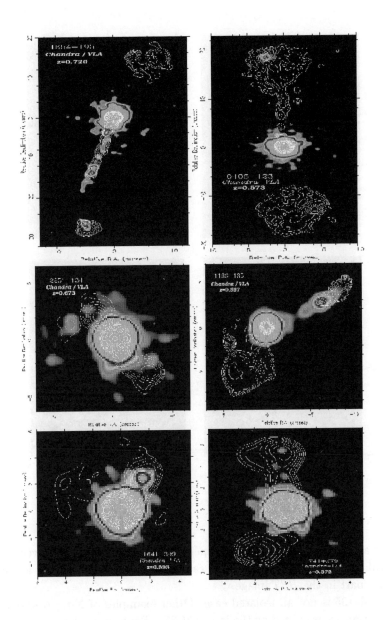

Figure 2.   Same as for Figure 1.

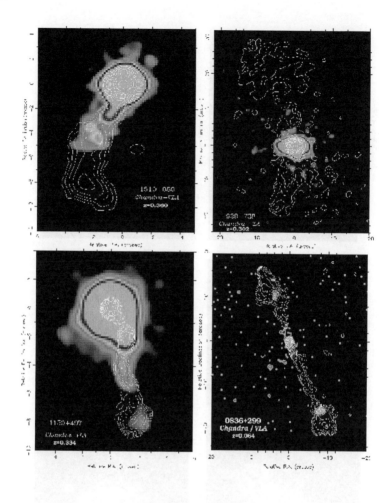

Figure 3. Same as for Figure 1.

peaks at knot C, where only an upper limit to the X-rays is derived. The optical is intermediate with all the radio and X-ray knots having a bright counterpart in the *HST* image.

1136–135 is not an isolated case. Other examples of X-ray/radio anti-correlation are provided by the jets of 3C371 (Pesce et al. 2001) and 3C273 (Sambruna et al. 2001), with similar, but not identical, morphologies to 1136–135. For convenience, I will refer below to this type of jets as "class II".

The multiwavelength jet morphologies offer first general clues as to the origin of the X-ray emission. In a synchrotron plus inverse Compton (IC) scenario (both processes are important in jets as suggested by blazar studies), the same particle population is responsible for emitting both the radio (via synchrotron) and the X-rays (via IC); the jet morphology should thus be very similar at both wavelengths, as observed in class I jets. As the jets are very long (projected lengths $\sim$ 50-100 kpc) and extends outside the host galaxy, the most likley source of seed photons for IC is provided by the Cosmic Microwave Background photons, whose density scales like $(1 + z)^4$ and is amplified in the rest-frame of the jet by a factor Γ^2, with Γ the bulk Lorentz factor of the jet (Tavecchio et al. 2000). Thus it is likely the IC/CMB is the dominant emission process for the X-rays in class I jets.

On the other hand, if the X-ray emission were due to synchrotron, one would expect shorter radiative lifetimes, and thus more compact emission regions, at the shorter wavelengths. The X-ray peaks closer to the core than at radio wavelengths can thus be accounted for in terms of shorter travel distances performed by the high-energy electrons before they dissipate energy through radiative losses. The morphologies of the jets in class II suggest synchrotron is important in at least some or all the knots of these jets.

3.2. Spectral Energy Distributions

Detailed information is provided by the Spectral Energy Distributions (SEDs) of individual knots. These can be derived for a given knot extracting fluxes from the same spatial region around the knot at the various wavelengths.

In interpreting the SED, a critical role is played by the optical flux. Specifically, if the optical emission lies on the extrapolation between the radio and X-ray fluxes or above it, the SED is compatible with a single electron spectrum extending to high energies; instead, if the optical emission falls well below the extrapolation, it argues for different spectral components (and therefore different mechanisms unless two electron populations are hypothesised) below and above the optical range (e.g., synchrotron and IC respectively). Thus, in all knots where IC/CMB dominates the X-ray emission, we expect an up-turn of the spectrum in the X-ray band with respect to the radio-optical extrapolation and thus an optical-to-X-ray index α_{ox} flatter than the radio-to-optical index α_{ro}. Conversely, when synchrotron dominates we expect $\alpha_{ro} \lesssim \alpha_{ox}$, with the inequality holding when radiative losses are important in the X-ray band.

Figure 4 shows the plot of the broad-band indices, α_{ro} versus α_{ox}, defined as the spectral indices between 5GHz and 5500Å, and between 5500Å and 1 keV, respectively (Gambill et al. 2002, in prep.). Only the knots for which a firm detection at all wavelengths is available were used. The dotted line marks the locus of points for which $\alpha_{ro} = \alpha_{ox}$. It can be seen that most knots lie in the region $\alpha_{ro} > \alpha_{ox}$, except for knot A in the jet of 1136–135. This behavior is well consistent with the morphological properties discussed above.

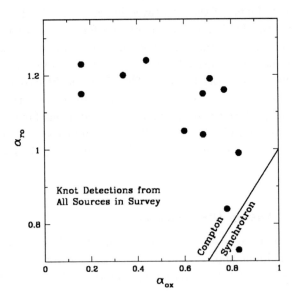

Figure 4. Radio-to-optical index, α_{ro}, versus the optical-to-X-ray index, α_{ox}, for individual knots with firm optical and X-ray detections (Gambill et al. 2002). The dotted line marks the division between knots where inverse Compton dominates over synchrotron for the production of X-rays (see text). Most X-ray knots are unlikely to be due to synchrotron emission from the high-energy tail of the radio-emitting electron distribution.

A more quantitative investigation of the jet physical properties requires a detailed modeling of the SEDs. To date, we have completed the analysis of the SEDs of the first four detected jets (Sambruna et al. 2002), and the work is in progress for the remaining jets of the survey. We note, however, that the sources analyzed in Sambruna et al. (2002) are typical representatives of both class I and II. Briefly, we computed synchrotron and IC/CMB emission models reproducing the SEDs of the most conspicuous knots. However, with only three observed fluxes the models are underconstrained. Following the procedure of Tavecchio et al. (2000), we assume the flux extraction radius is the size of the (spherical) emission region and compute for what values of the magnetic field, B, and of the beaming factor, δ^b, it is possible to reproduce the radio and X-ray fluxes. As discussed in Sambruna et al., in the case of IC/CMB we adopted equipartition as an additional constraint to fix the models univocally. In the cases where X-rays are attributed to direct synchrotron emission, there is no "independent" constraint from the X-ray flux and only the equipartition assumption survives.

Specific models were computed to reproduce the SEDs of X-ray knots A and B of each detected jet with synchrotron plus IC/CMB emission and assuming equipartition. The results are shown in Figure 5 for the jets of 1150+497 (class I) and 1136–135 (class II). The derived magnetic fields are of the order of $B \sim (10 - 40)\mu$Gauss, the Lorentz factors $\Gamma \sim (3 - 5)$, and the Doppler factors $\delta \sim 6 - 7$ (see Table 4 in Sambruna et al. 2002). Thus an important conclusion of the modeling is that jets are still (moderately) relativistic on scales of tens to hundreds of kpc.

While in 1150+497 the X-ray emission from both knots A and B can be ascribed to IC/CMB, in the case of 1136–135 a complex situation emerges. For knot A, where $\alpha_{ro} \lesssim \alpha_{ox}$, we suggest that synchrotron emission is a plausible emission process. Assuming the equipartition magnetic field, the cooling time of the electrons in the knot is $\sim 10^{11}$ s. This is consistent with the light-crossing time for knot A, assuming its radius is $1''$. For knot B in the same jet, clearly $\alpha_{ro} \gtrsim \alpha_{ox}$ and IC/CMB should be the main emission process. The X-rays fading further out and the increasing radio brightness should then be attributed to a deceleration of the the relativistic plasma and to a compression of the magnetic field, as expected at the outer boundary of jets (e.g., Gómez 2001). The 1136–135 jet is clearly an interesting case of a "mixed" jet, and raises the question of what causes different processes to dominate at different locations in the jet. We will address this and other questions in our forthcoming deeper *Chandra* and multicolor *HST* observations of 1136–135 and 1150+497.

---

[b]The beaming factor δ is here defined as $\delta \equiv [\gamma(1 - \beta\cos\theta)]^{-1}$, where β is the bulk velocity of the plasma in units of the speed of light, $\gamma = (1 - \beta^2)^{-1/2}$ the corresponding Lorentz factor, and θ the angle between the velocity vector and the line of sight.

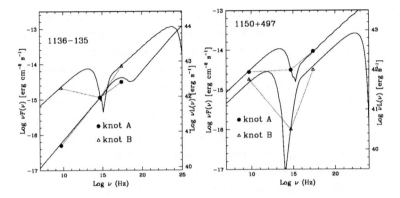

Figure 5. Radio-to-X-ray Spectral Energy Distributions (SEDs) for the brightest X-ray knots in the jets of 1136–135 (left) and 1150+497 (right). In both panels, the left-handed vertical axes are in units of observed flux, νF_ν, while the right-handed axes are in units of luminosity, νL_ν. Typical uncertainties on the fluxes are 33% or larger. The X-ray flux is always above the extrapolation from the radio-to-optical continuum, except for knot A in 1136–135, suggesting a general dominance of inverse Compton for the production of X-rays. In knot A of 1136–135, synchrotron is likely the dominant process. The solid lines are the best-fit models (sum of all components), with the parameters reported in Sambruna et al. 2002.

3.3. Caveats

It is worth remarking a few caveats affecting our analysis. First, the limited signal-to-noise ratio at both X-ray and optical wavelengths gives room to alternative interpretations of the SEDs. Second, a variety of conditions may exist within the relatively large extraction regions we used (1″, dictated by the *Chandra* resolution), for example if the emitting particle distributions are stratified or multiple shocks exist. While higher angular resolutions at X-rays await future generations of space-based telescopes, deeper follow-up X-ray and optical observations of the new jets of this survey with *Chandra* and *HST* can at least remedy the first limitation of our analysis, in providing X-ray and optical continuum spectra for individual knots, a key test for the emission models.

In fact, alternative models have been discussed in the literature for the origin of the X-rays from kpc-scale jets. Harris & Krawczynski (2002) propose that the X-ray emission originates via synchrotron from a population of relativistic electrons separate from the one responsible for the longer wavelengths. Such jet "inhomogeneities" would also be supported by the small (~ 0.2-$0.5''$) positional offsets between the X-rays and radio wavelengths observed in 3C273. Aharonian (2002) proposes synchrotron emission from relativistic protons, requiring much higher magnetic fields and more powerful acceleration processes. Recently, Dermer & Atoyan (2002) discussed a model where the X-rays originate via Thomson losses of the synchrotron electrons on the CMB in an effort to account for the upturn in the optical-to-X-ray continuum observed in 3C273. Indeed, different emission mechanisms predict different slopes in the X-ray band, with steeper spectra (by $\Delta\Gamma \gtrsim 0.5$) expected in the case of synchrotron emission than in the case of IC/CMB.

X-ray and optical spectra of single knots will be essential to discriminate among the various models, as well as detailed maps to measure and quantify more exactly the positional offsets of the radio, optical, and X-ray peaks. Optical observations are necessary to identify the mechanism responsible for the X-ray emission, as the optical band lies at the intersection of the synchrotron and IC components. Finally, an additional important constraint will be provided by future IR observations with *SIRTF*, probing a poorly known region in the SEDs where the synchrotron peak (related to the break energy of the synchrotron electron population) is located.

4. Summary and Conclusions

Extragalactic relativistic jets glow at X-rays on kiloparsec scales. Important constraints on the origin of their intense X-ray emission is provided by multiwavelength imaging and spectroscopy, such as afforded by *Chandra*, *HST*, and the ground-based radio telescopes. Favored candidates for the origin of the X-rays are synchrotron emission from a population of relativistic electrons in the jet, or/and inverse Compton scattering of the Cosmic Microwave Background photons (IC/CMB). These processes imply that the plasma is at motion with relativistic speeds on scales of tens and hundreds of kpc. While alternative models are possible, detailed optical and X-ray spectra of individual jet knots are needed to discriminate among the various possibilities.

Acknowledgments

This work would not have been possible without the help of my collaborators L. Maraschi, F. Tavecchio, C.M. Urry, J.K. Gambill, G. Chartas, C.C. Cheung, and R. Scarpa. I gratefully acknowledge financial support from an NSF CAREER award, and several NASA grants: NAG5-10073, NAG5-10244, GO1-2110A, and grant HST-GO-08881.01-A, from the Space

Telescope Science Institute, which is operated by AURA, Inc., under NASA contract NAS 5-26555.

References

1. Aharonian, F. A., *MNRAS* **332**, 215 (2002).
2. Blundell, K. M. and Beasley, A. J., *MNRAS* **299**, 165 (1998).
3. Bridle, A. H. and Perley, R. A., *ARA&A* **22**, 319 (1984).
4. Brunthaler, A. *et al.*, *A&A* **357**, L45 (2000).
5. Celotti, A., Ghisellini, G., and Chiaberge, M., *MNRAS* **321**, L1 (2001).
6. Chartas, G. *et al.*, *ApJ* **542**, 655 (2000).
7. Cheung, C. C. *et al.*, in prep. (2002).
8. Dermer, C. and Atoyan, A. *ApJ* **568**, L81 (2002).
9. Gambill, J. K. *et al.*, in prep. (2002).
10. Gómez, J. *Ap&SS* **276**, 281 (2001).
11. Harris, D. E. and Krawczynski, H., *ApJ* **565**, 244 (2002).
12. Liu, F. K. and Xie, G. Z., *A&AS* **95**, 249 (1992).
13. Pesce, J. E. *et al.*, *ApJ* **556**, L79 (2001).
14. Sambruna, R. M. *et al.*, *ApJ* **571**, 206 (2002).
15. Sambruna, R. M. *et al.*, *ApJ* **549**, L161 (2001).
16. Scarpa, R. *et al.*, in prep. (2002).
17. Schwartz, D. A. *et al.*, *ApJ* **540**, L69 (2000).
18. Tavecchio, F., Maraschi, L., Sambruna, R. M. and Urry, C. M., *ApJ* **544**, L23 (2000).

SUPERNOVAE

I. J. DANZIGER

INAF - Osservatorio Astronomico di Trieste,
Via Tiepolo 11,
I-34131 Trieste, ITALY
E-mail: danziger@ts.astro.it

We present a description of selected areas of supernova (SN) research with special emphasis on observational results. The subjects cover classification, types, historical SNe, diversity, theoretical understanding, element production, relationship to Gamma Ray Bursts (GRBs).

1. Introduction

Supernovae (SNe) are effectively the end-points in the evolutionary history of stars above a certain mass. They are end-points because the explosion effectively destroys the integrity of the star possibly, though not certainly, leaving a compact object which is doomed to quiet oblivion as it loses energy and cools quiescently. The remainder of the star has been blown into the interstellar medium (ISM) which it enriches in heavy elements produced as a result of stellar evolution and explosive nucleosynthesis. In reality supernovae are the main producers of heavy elements and therefore their role in chemical evolution of galaxies is all important. But it is also worth recalling here that SNe not only inject metals into the ISM but also energy which heats and affects the behaviour of the interstellar gas [1]. These are rather specific reasons for addressing these and related topics below.

2. Types of Supernovae

The currently recognised supernovae are of two types, thermonuclear and core-collapse. This typing results from our understanding of the physical reasons underlying the cause of the explosion. However, of necessity, a classification of SN events when they occur required a classification scheme based on observable properties. In the early phases the characteristic spectra are dominated by PCygni absorption and emission lines. Therefore SNe were classified as Type I or Type II depending on whether Balmer lines of hydrogen were visible at early (photospheric) phases. This classification alone unfortunately does not separate the above 2 types. Type Ia SNe represent the thermonuclear detonation (or deflagration) of an accreting carbon-oxygen (C-O) white dwarf (WD). This is consistent with lack of hydrogen and strong lines of SiII, the criteria for classification.

Type Ib SNe lack hydrogen lines but show lines of HeI and weak or no SiII lines. Type Ic SNe lack hydrogen lines, show no or very little SiII absorption and no helium lines. Both Types Ib and Ic are thought to represent the core-collapse of a massive star whose outer hydrogen envelope was stripped off as a result of winds or binary interaction during previous evolutionary stages. The difference between the 2 types of progenitor may simply reflect the degree to which helium has been stripped. The recent SN1998bw associated with GRB980425 appears to have been an energetic version of a Type Ib/c event. In this case optical lines of HeI were not evident but the strong HeI 10830 line was identified [2].

Type II SNe are so classified because hydrogen lines are evident at early phases. They have been subdivided into at least 4 categories as a result of other observable characteristics. The most frequent or "normal" Type II are subdivided into Type IIL and IIP where L and P result from a linear or plateau shaped light curve. The more recent Type IIn classification is associated with a flat light curve now understood to result from the envelope interaction with circum-stellar material (CSM). Evidence of this interaction also appears in the spectra in the form of low velocity components of emission (and sometimes absorption) lines. A Type IIb classification has been proposed for objects such as SN1993J which initially showed hydrogen lines afterwards fading to invisibility. A natural explanation is that the progenitor star lost most but not all its hydrogen envelope during its evolution prior to the explosion. The small outer layer of hydrogen becomes invisible lacking sufficient energy input following recombination.

It is the presence of material surrounding the exploding star that results in non-thermal radio emission occurring and continuing over various intervals presumably depending on the physical conditions and extent of this material often ascribed to a pre-explosion wind [3]. Radio emission is also generally accompanied by X-ray emission also resulting from shock propagation resulting from the interaction [4]. Thermonuclear SNe have never produced radio or X-ray emission but there are examples of all sub-types of core-collapse SNe where radio and X-ray emission have been detected.

The spectra of these various types of SNe then evolve in distinctive ways as the envelope expands and becomes optically thin giving rise to an almost purely emission line spectrum. Type Ia SNe show strong lines of the following ions - FeII, FeIII, CaII, CoII. Type Ib and Ic SNe at this stage are almost indistinguishable showing lines of OI (very strong), CaII, and FeII. The main spectroscopic difference among Types IIP,L and n involves the presence or absence of lines of OI although in this respect a clear pattern of behaviour is not yet obvious. There are also obvious differences in expansion velocities but how this relates to other characteristics remains a subject for future work. Obviously Type IIn spectra also carry the imprint within the line profiles of interaction with the CSM [5].

All of these characteristics eventually demand a physical explanation. We will see later that also *within* all the various supernovae types there is a diversity of characteristics. With Type Ia SNe whose main sequence progenitors were less massive than 8-10$M_\odot$ we are probably confronting a diversity in progenitor mass and mass of radioactive ^{56}Ni produced, even if only approximately one percent of all white dwarfs explode as thermonuclear SNe. (The remainder are destined to become cold WDs). In the discussion above we have isolated 6 different types of core-collapse SNe. However if we suppose that the various physical quantities that can vary are the progenitor mass, the kinetic energy of the explosion, the radius of the exploding star, the mass of ^{56}Ni, and the presence of CSM and concede just a simple dichotomy in the values of each, we are confronted with the possibility of 32 distinct events. Unravelling the parameters for all such events looks like a formidable but interesting task [6].

3. A Physical Understanding

3.1. *Thermonuclear Supernovae*

The spectroscopic characteristics and their place of occurrence in galaxies where only stars with masses not much greater than 1$M_\odot$ strongly suggests highly evolved progenitors. The most promising candidates are C-O WDs which theoretically can exist in the mass range 0.5-1.4$M_\odot$. Below that mass range helium WDs exist and WDs cannot exist above the Chandresekhar limiting mass. Because of the electron degeneracy obtaining in a C-O WD the relativistic equation of state prevails. This equation gives the relation between the pressure and density without a dependence on temperature. Combined with the hydrostatic condition it leads to an expression for the

Chandrasekhar mass dependent only slightly on composition and physical constants. What also emerges from these considerations is the mass-radius relation for a WD showing that the radius is inversely proportional to some power of the radius. Thus as the mass approaches the Chandrasekhar mass, the radius must shrink towards zero. Near the limiting mass a small increase in mass causes a large increase in density and nuclear compressional heating which leads to a thermal runaway because the equation of state does not provide the normal means of expansion and cooling. This would happen in a C-O WD near the Chandrasekhar limit if it accreted mass from a companion red giant star or coalesced with another WD. If a thermal runaway starts and hydrodynamical cooling fails, there occurs either deflagration or detonation depending on whether the burning flame is subsonic or supersonic. Both types of model have been constructed and both have residual problems confronted with observations. This results in nucleosynthesis with the predominant production of Fe [7].

3.2. Core-collapse Supernovae

A massive star forms on the main sequence when ignition of hydrogen at its center and the consequent heating and pressure increase counterbalances the gravitational infall. When the hydrogen at the center is exhausted and only a shell of hydrogen burning exists the star undergoes core contraction and in the HR diagram moves towards lower temperatures or towards the red supergiant phase. Core contraction causes heating which results in a central temperature sufficient for helium (He) to ignite producing carbon and oxygen. This core ignition results in the star moving again to higher temperatures in the HR diagram. This pattern of core contraction and core ignition repeats as each successive fuel is exhausted. In the case of a star of $25 M_\odot$ its traffic backwards and forwards in the HR diagram ceases because the outer layers of the star do not have time to respond to the machinations of the central core. This is because succeeding stages proceed faster and faster as the mass increases. Thus an unmixed hypothetical star achieves an "onion-ring structure" where the final fusion stage of silicon burning produces an iron core. If unmixed it would consist in successive layers of helium + nitrogen then carbon + oxygen, then neon + oxygen, then oxygen + magnesium, then silicon + sulphur, and finally the iron core. Because the binding energy per baryon reaches a maximum at iron (Fe), heat cannot be extracted by further fusion. This maximum of binding energy occurs at iron because it is at this nuclear mass that the short range attractive nuclear force is exceeded by the longer range repulsive coulomb force. Consequently the iron core which approaches a mass near the Chandrasekhar limit begins to collapse. The nuclear decomposition produces neutrinos which extract heat and speed the infall. In succession the disintegration of Fe produces α particles and neutrons; the α particles photodisintegrate extracting heat to overcome the large binding energy of He; the collapse, now catastrophic, is in free fall under self-gravity. Velocities of 1000 km/sec and densities of 10^{10} gm/cm^3 prevail. Finally protons + electrons combine to form neutrons at near nuclear densities. A neutron star or Black Hole is formed with an accompanying burst of neutrinos. In fact a total of 19 neutrinos was detected over a period of 12.5 seconds by 2 observatories signalling the explosion of SN 1987A. Immediately following this a hydrodynamical rebound of infalling material from the "stiff" neutron star occurs, possibly assisted by neutrino emission. A shock wave propagates outwards expelling the envelope and explosive nucleosynthesis occurs producing elements from oxygen (α-process) to the radioactive elements ^{56}Ni and ^{57}Ni (from explosive Si burning) [8]. The principles are similar for Types Ib, Ic, II although it is thought that Types Ib,c have already lost their hydrogen and possibly helium envelopes as a result of transfer or winds prior to the explosion. How the existence of a progenitor star in a binary system affects the outcome is a more complicated story and surely relevant to some actual situations.

4. Historical Galactic Supernovae

We may summarize briefly what is known of SN events actually observed during historical times when written records were kept. The probable real SNe events for which remnants now exist are the following: AD 185 (C; Type Ia); AD 1006 (A,C,J,E; Type Ia); AD 1054 (C,J,A,E?; Type II, Crab); AD 1572 (E,C,J; Type I?, Tycho); AD 1604 (E; Type I?, Kepler); AD 1670? (E?; Type Ib,c II, Cas A Flamsteed?). Other possible SNe for which remnants may be associated are: AD 386 (C); AD 393 (C); AD 1181 (C,J). The above records originate from: A (Arab lands), C (China), J (Japan), E (Europe). It is not certain that the SN associated with Cas A was sighted by Flamsteed, but the projection backwards in time of the proper motion of filaments provides a reliable date of explosion.

One may use the 4 SNe 185, 1006, 1054 and Cas A, all of which lie within 3 kpc from the Sun and 250 pc from the Galactic plane, to deduce that 4

SNe occurred in a volume of 10^{10} pc^3. This amounts to 1 per 500 years in one tenth of the volume of the Galaxy. Therefore the rate for our Galaxy should be $\sim$ 1 per 50 years. Incompleteness caused by seasonal effects, obscuration, and intrinsic faintness may explain why more SNe have not been observed in recent centuries.

5. Supernova Rates and Places of Occurrence

Thermonuclear SNe (Ia) occur in galaxies of all types, whereas core-collapse SNe do not occur in early-type galaxies (E-S0) but occur in all other galaxies. This already suggests that thermonuclear SNe originate from an older or more evolved population of stars. In addition Type Ib,c SNe are often associated with HII regions and also produce radio emission characteristics suggesting massive progenitors and evolved because of lack of hydrogen.Thus these qualitative characteristics already strongly suggest limits for types of progenitor.

Rates have been established for various types of SNe in various types of galaxies. These rates are constantly up-dated as statistics of SNe occurrence improve. In E-S0 galaxies with a blue luminosity similar to our Galaxy in the nearby universe there occurs about 1 thermonuclear event each 500 years. This and other rates would presumably scale with luminosity of the particular host galaxy. This rate is not significantly different for other types of galaxy with similar luminosities. Core-collapse SNe occur more than twice as frequently as thermonuclear events in spirals and irregulars with a slight tendency for an increased rate towards later types. Type Ib SNe constitute about 15-20 percent of all core-collapse events. Therefore the rates for all SNe in a galaxy similar to our own are about a factor of 2 lower than that deduced from the historical SNe but are consistent, within the uncertainties which also include that of the Hubble constant.

A more detailed analysis of statistics of occurrence reveals that intrinsically faint SNeIa (discussed later) represent <25 percent of all SNIa; SNIIn represent 2-5 percent of all Type II; intrinsically faint Type II (such as SN 1987A) represent 10-30 percent of all Type II [12].

6. Asymmetric Explosions and Polarization of Light

An expanding envelope of a SN where electron scattering of photons predominates should show no net polarization if the envelope is spherically symmetric. Therefore measurements of linear polarization of all types of

SNe at early phases have been increasing recently [9]. In principle accuracies of 0.1-0.2% are, with care, achievable. For SNIa all measures (with one possible exception) show p<0.2%. For Type II the situation is less clear and values at the 1% level have been reported. There are several major obstacles to a clear interpretation of real observed polarization at this level. These include knowing how to correct for interstellar polarization caused by the interstellar medium of our Galaxy and the host galaxy. In addition the net polarization emanating from an asymmetric object depends on the angle of viewing relative to any axis of symmetry. Net polarization could also result from an unequal surface brightness of the envelope resulting from clumping rather than a large scale asymmetry of the envelope. Polarization measurements have become more important with the increasing number of claims that γ-ray bursts originate from collapse of massive stars where 3-D symmetry may not be preserved owing to disk formation and a preferred axis of jet emission.

7. Energies

Although we can say that the kinetic energy released by Type Ia, Type Ib and Type II SNe all approximate 10^{51} ergs, there is reasonably clear evidence of significant variations around these values. There are different sources of information leading to these approximate estimates. 1. This source results from a study of supernova remnants of the various types and the radio and X-ray emission combined with a theoretical understanding of remnant evolution. 2. In the case of Type Ia SNe where the disintegration of a C-O WD is supposed, the incineration of a $^{12}C/^{16}O$ mixture requires that 0.8×10^{51} ergs are produced to which must be added 0.5×10^{51} ergs, the binding energy of the WD. Modelling of light curves and spectra of SNIa suggest a range of $0.7-1.7 \times 10^{51}$ ergs reflecting the dispersion in the observed spectral characteristics of this class. For Type IIP, hydrocode matching of the light curves together with a knowledge of velocity at the photosphere gives $0.5-2 \times 10^{51}$ ergs depending on the SN. Consequently the values for Type IIL and IIn are uncertain because of the paucity of light curve models for such objects. For some classes of Type Ib,c SNe such as SN 1998bw (GRB980425) modelling and velocities indicate an order-of-magnitude higher kinetic energy [10,11]. For both types of SNe the radiative energy released is much less, being $\ll 10^{49}$ ergs for SNIa where radiation is trapped and converted to KE accelerating the envelope; whereas it is of this order for SNII.

8. Light Curves

The luminosities of all types of SNe and the temporal variation over extended phases provide valuable insights into the nature of the explosion and its subsequent evolution. This information, coupled with quantitative spectroscopy, at early (photospheric) and late (nebular) phases forms the basis of the following discussion. Bolometric light curves are particularly valuable as they embrace the total radiative energy budget of the SN. They are still sparse, with the best observed object being SN 1987A because it was so bright. In general the main lack of data lies with the IR region of the spectrum where normally only a few observations exist for the best observed, and virtually nothing for the remainder. In the absence of dust formation, which is known to occur in some SNe, the IR part of the spectrum while not negligible, does not dominate the radiative output at most phases. It will be seen however that IR spectra carry much valuable information and deserve particular attention and effort.

Some very general statements about the diversity of light curves of all types of SNe may be made. The maximum luminosity and shape of the early light curve around maximum light are functions of the mass of the ejecta, the radius of the exploding star, the mass of radioactive ^{56}Ni produced, and the kinetic energy of the explosion. These parameters are in turn related to the evolutionary status and mass of the progenitor star. The shape and luminosity of the light curves in the nebular phase will be determined by the initial mass of ^{56}Ni produced which has transformed by β-decay into radioactive ^{56}Co whose decay into stable ^{56}Fe producing γ-rays and positrons deposited in the envelope provides the energy source. The shape is also dictated by the fraction of γ-rays which are actually absorbed and thermalized by the envelope. Other radioactive species play a secondary role in the early nebular phases. The shape of the light curve can also be modified by the formation of dust in the envelope which has now been observed to occur in 2 Type II SNe SN 1987A and SN 1999em approximately 500 days after the explosion. This modification occurs because shorter wavelength radiation is absorbed by the dust and re-emitted thermally in the IR according to the dust temperature.

8.1. *Thermonuclear or Type Ia SNe*

For many years Type Ia SNe had been thought to be a family of very homogeneous objects. The accumulation of much larger better observed samples reveals that this is not the case, although the variations of different observed properties appear to be correlated. For example the total range in absolute luminosity at maximum light so far reported for Type Ia is about 2 magnitudes in V. This absolute magnitude correlates strongly with the rate of decay of the light curve immediately after maximum light [13]. These parameters also correlate with expansion velocities of the photosphere and also with those of the envelope at nebular phases. These variations are linked via the idea that the underlying cause is the mass of radioactive ^{56}Ni. This is supported by the fact that although the exponential decay of the light curve at later phases is faster than that of ^{56}Co, owing to increasing γ-ray escape from the thin envelope, the luminosities at late phases also correlate in the expected way. Nevertheless the reason for these variations emerging from what was assumed to be a single mechanism of WD instability at the Chandrasekhar limit is still being examined. Some clues come from the following statistics of observed events. a. Type Ia SNe in E and SO galaxies have lower photospheric velocities. b. Type Ia in spiral galaxies tend to be brighter than those in E and SO galaxies. c. The occurrence frequency/unit IR luminosity is higher in late spirals than in E/SO galaxies. These characteristics are qualitatively consistent with the possibility that Type Ia progenitors occupy a significant range in mass.

8.2. *Core-collapse or Type Ib,c and II*

There is a range of observed properties in the family of Type Ib,c. Absolute luminosity at maximum is the most obvious and this plus the differences in the luminosities on the exponential "tail" point to differences in the mass of radioactive ^{56}Ni produced. The limited modelling available also points to this in a quantitative way. The light curves also tend to decay faster than the ^{56}Co decay rate, presumably for the same reason as in the case of Type Ia. For Type II we have already mentioned that the IIn class with a flat light curve is caused by the ejecta-wind interaction whereby kinetic energy is converted into radiative energy. The differences between Types IIP and IIL may, to first order be understood as follows. The size and duration of the plateau phase is a function of the ejecta mass, where larger ejecta masses cause longer plateaus - a result of longer radiative diffusion times.

The luminosity at maximum can be a function of the energy but is also a strong function of the radius of the star at the point of explosion. Therefore IIP SNe would have massive ejecta $> 10 M_\odot$, while IIL might have ejecta masses of 1-2$M_\odot$ [14]. SN 1987A is a good example of dependence of maximum luminosity on radius and not energy or mass of ^{56}Ni. It has a relatively large, independently determined, energy and mass of ^{56}Ni, but it was intrinsically faint coming from the explosion of a B supergiant with a known small radius. Clear differences are also apparent within the families of Types IIP and IIL at late phases although results are not systematized owing to limited statistics. Luminosities in the late phases tend to follow a decay line in V close to that given by ^{56}Co decay. However there is, at a given phase, an observed range in absolute luminosity of about a factor of 40, indicative of a similar range in the mass of ^{56}Co. Relating these established facts to the properties of a progenitor star and its evolution is a task in its infancy.

9. Distance Determinations

9.1. Type Ia

The use of Type Ia as standard candles to determine distances and cosmolgical parameters has attracted much attention in recent years. One can exemplify the use of SNe as well as the need for accuracy as follows. At a redshift z=0.5, a standard source would have an apparent difference of 0.3 mag. depending on whether we were in an empty ($q_o = 0$) or a closed ($q_o = 0.5$) universe. Similarly at z=0.3, a standard source would differ in brightness by 0.3 mag. depending on whether the cosmological constant Ω is 1 or 0. This demonstrates the basis for current work on Type Ia SNe at redshifts of z=0.5 - 1 and beyond. It also shows the accuracy required in order that a standard candle technique be effective. Fortunately, as alluded to above, although real SNeIa have a large spread in absolute magnitude at maximum, this strongly correlates with rate of decline of the light curve. Measuring this rate of decline allows one to know what the actual absolute luminosity of any particular SN is and thus recover the accuracy of the method.

9.2. Type II

Previous discussion emphasized that the family of Type II SNe could not be used as standard candles, not only because of the large range of absolute magnitudes but because of an uncorrelated diversity of observed parameters describing their behaviour. However the method of "Expanding Photospheres" can be usefully employed though not yet at cosmological distances. One envisages in the photospheric phase that when the envelope expands ballistically, a change in the radius and temperature gives rise to a change in brightness. We may write 2 equations to describe this situation. The observed luminosity

$$f = r^2/D^2 \cdot A^2 \cdot \pi \cdot B(T) \tag{1}$$

where r = radius of envelope, D = distance, B = Planck function, A = correction factor computed and applied because the radiation field is dilute (~ 0.5 at higher temperatures and increasing to ~ 1 at lower temperatures [15]). The ballistic expansion condition allows:

$$r = r_o + v \cdot t \tag{2}$$

where r_o = initial radius (negligible), v = the velocity at time t after explosion. Observationally determined are f, B, and v. B comes from the best fit of a black body curve to photometry or spectrophotometry [16,17] ; v comes from measuring velocities of weak lines formed near the photosphere; A has been calculated from models. One solves the 2 equations for the 2 unknowns r and D. A series of observations during the photospheric phase ensures higher accuracy. Note the significant point that this distance is independent of any earlier calibration of steps in the distance determination ladder. The method is limited to SNe bright enough to realise spectra of sufficient quality to enable accurate velocity determinations.

10. Element Production and Abundances

10.1. Methods

There are 4 main methods for determining abundances from observations of SNe. 1. Photospheric absorption line spectroscopy which is the analog of abundance analyses of the Sun and stars. It can be used in the optical and infrared (IR) regions and gives only relative abundances for material at or above the photosphere. 2. Nebular spectroscopy involves analysis of emission lines which is a powerful method for determining the total mass of an ion when the envelope is optically thin and densities are below the critical densities for the lines being observed. In practice a knowledge of the ionization states is necessary as well as the temperature. 3. Bolometric light curves which are sustained by radioactive β-decay through deposition of γ-rays in the envelope provide accurate masses of those radioactive elements because the ratio of deposited/escaped γ-rays is expected to be a well determined quantity. 4. Gamma-ray emission (+ associated X-rays from Comptonization) from radioactive elements can be used to measure masses of these elements. As in the previous method a knowledge of deposition vs. escape is required through a model.

10.2. Photospheric Line Analysis

Unlike the spectral analysis of a stellar photosphere where turbulent velocities are small, in a SN envelope one is dealing with a large differential velocity of systematic outwards expansion. This has necessitated the development of spectrum synthesis codes incorporating Monte Carlo solution of the transfer equation based on the Schuster-Scwarzschild approximation and radiation transport based on the Sobolev approximation. In analysing an observed spectrum one typically inputs the time of explosion, the photospheric velocity and an estimate of the luminosity; and by matching one hopes to recover relative abundances, mass of ^{56}Ni and possibly the mass of exploding star.

This method has been effectively applied to the early spectra of SN 1987A where strong lines of BaII were identified and whose velocity profile suggested this element was confined in velocity and therefore in radial distribution more than other elements [18]. Information on other elements has also been derived.

The technique has been applied with some success to early spectra of Type Ia SNe. While the quality of fits to observed spectra are such that precise models of exploding stars cannot yet be discriminated, some understanding of observed differences have been usefully delineated. SN 1991T the most luminous of the family of SNIa showed early photospheric spectra that , to the discriminating eye, were noticeably different from spectra of less luminous objects at the same phase. Modelling showed that this difference was due to a higher temperature producing a higher state of ionization in Fe for example rather than to an abundance effect. The higher temperature is obviously related to the higher energy, and velocities produced by the explosion and the higher mass of ^{56}Ni apparently produced [19]. On the other hand the presence of a distinct absorption feature identified with a triplet of OI7772 in SN 1991bg, the intrinsically faintest SNIa, and not nearly so obvious in brighter objects raises the possibility of unburnt oxygen in the outer parts of the envelope emerging from a less energetic explosion [20].

10.3. Nebular Spectral Line Analysis

In the expanding envelope where a Monte Carlo approach is applicable, the following assumptions are made. A non-LTE treatment of the rate equations in a nebula of uniform density obtains. This latter assumption is surely too simple. Heating is calculated from energy deposition of γ-rays resulting from β-decay of ^{56}Co into ^{56}Fe after which thermal and statistical equilibrium is assumed. In order to match the observed spectra one inputs the mass, composition, time after explosion and outer velocity of the expanding sphere.

As for the photospheric phase of SNeIa, here the method has been effectively used to show that there is a range of mass of Fe apparently produced with mass depending on the intrinsic brightness of the SN. The following range, $< 0.1M_\odot \rightarrow > 1M_\odot$ is indicated with SN 1991bg and SN 1991T at the lower and upper limits. A detailed comparison of observed and model spectra suggests that there is still significant emission unaccounted for [20]. An analysis of the nebular spectrum of SN 1987A has resulted in first-order abundance estimates for 8 different elements [21]. Of these the abundance of CoII is well determined because the fine structure lines of the ground state transitions occurring at 10.5μ could be observed from the ground. Since these lines originate from very low excitation levels the strength is insensitive to temperature, yielding (after the envelope became optically thin in these lines) the most direct measure of the mass of ^{56}Co which was also in good agreement with other independent methods. Because both ^{56}Co and ^{57}Co β-decay with significantly different half-lives the study of the temporal variation of these lines has allowed one to disentangle the different masses of these 2 species to high accuracy. Unfortunately the important mass of oxygen is much less well determined from the observed [OI]6300,64 lines. This is because the upper level for these transitions is high enough that an accurate temperature is necessary but not observationally available. Consequently one is limited to a range $0.2M_\odot$ - $3M_\odot$ for OI by this method. Less direct methods suggest 1.5-2M$_\odot$. The measurement of the flux of these lines in other Type II SNe at the same phase combined with a knowledge of the energy input from the mass of radioactive material allows one to estimate oxygen masses for these objects. This is why an accurate measure for SN 1987A is so important.

10.4. Abundances from Light Curves

The shape and absolute luminosity of light curves of both types of SNe may and have been used to derive masses of the radioactive material that is the dominant source of energy supporting the radiative emission. In fact 45 years ago radioactive ^{254}Cf with a half-life of 55 days was proposed as the energy source for Type Ia SNe such as SN 1937C because its light curve changed at the same rate. The idea was a good one but could not take account of the fact that γ-ray deposition decreases with time in such a SN envelope with a resultant apparent decay rate faster than the 77-day half life of the actual energy source, namely ^{56}Co. One estimates for example that at 200 days only 0.5 percent of γ-rays are deposited in the envelope. The following radioactive species and decays following explosive silicon burning are expected.

$$^{56}Ni \rightarrow {}^{56}Co \rightarrow {}^{56}Fe$$

$$^{57}Ni \rightarrow {}^{57}Co \rightarrow {}^{57}Fe$$

$$^{44}Ti \rightarrow {}^{44}Sc \rightarrow {}^{44}Ca$$

The deposited γ-rays produce non-thermal electrons which are thermalized and cause ionic heating and excitation. Positrons which are thermalized will annihilate into photon pairs. The following half-lives obtain:

$$^{56}Ni - 6.1\,days\,(5\,lines)\,;\ {}^{56}Co - 77\,days\,(8\,lines)$$

$$^{57}Ni - 36\,hours\,(1\,line)\,;\ {}^{57}Co - 274\,days\,(2\,lines)$$

$$^{44}Ti - 60.3\,years\,(2\,lines)\,;\ {}^{44}Sc - 3.7\,hours\,(2\,lines)$$

From all these radioactive decays there are 20 γ-ray lines spanning the energy range 69 KeV - 1.8 MeV which have some chance of being detected if rapid follow-up takes place. So far the only clear evidence from the light curves for these various radioactive decays has been the ^{56}Co and possibly the ^{57}Co decay. This is because of the suitable half-lives and mass of material involved.

10.4.1. Type Ia

Modelling of light curves of Type Ia SNe from before maximum light and extending to 500 days has led to a conclusion similar to that from the nebular spectra viz. there is a range of mass of ^{56}Co produced from 0.1 to $1M_\odot$ which correlates again with the absolute magnitude at maximum. This general conclusion seems independent of whether deflagration or detonation models are invoked [14,22].

In summary for SNeIa it seems that absolute luminosity at maximum, expansion velocities at both photospheric and nebular phases, rate of decline of light curves all correlate with masses of radioactive ^{56}Co deduced either from nebular spectra or from light curves.

10.4.2. Type II

The quasi-bolometric light curves of core-collapse SNe are now yielding accurate masses of ^{56}Co by comparison with the light curve of SN 1987A. Accurate photometry yielding bolometric light curves for this SNe which, combined with modelling and demonstrated consistency with spectroscopic and γ-ray measurements, clearly points to $0.075M_\odot$ of ^{56}Co having been produced. The light curve is also consistent with a $^{57}Co/^{56}Co$ mass about 1.5 times the solar system value of the $^{57}Fe/^{56}Fe$ ratio, a result in good agreement again with spectroscopy and γ-ray measurements [23].

Since the V light curve for many SNeII follows the ^{56}Co decay line and since Type IIP also seem to follow a similar color evolution, it has proven convenient and accurate to compare the luminosity level of of the V light curve with that of SN 1987A in order to obtain a mass of ^{56}Co relative to that in SN 1987A. In the future it may prove possible to use relative Hα luminosities in a similar way. Observationally one must be sure that the radioactively generated light curves are not disturbed by other contributions to the observed radiative luminosity. For the ^{56}Co decay the period 150 - 800 days is reliable provided there are no light echoes and no interaction of the expanding envelope with the surrounding CSM. Echoes, which result from radiation scattered into the line-of-sight from interstellar dust lying outside the line-of-sight, have now been detected from Type Ia and Type II

SNe. Energy injected into the envelope from a compact object at the center could be visible at various times depending on its nature and the mass of radioactive material. Such an effect has not so far been recognised in the well observed SN 1987A nor in other SNe. After 800 days other longer-lived radioactive decays such as that of ^{57}Co begin to become important as do the effects of recombination radiation which require modelling as they do not leave a unique unambiguous signature [24].

This method applied to Type II SNe has yielded ^{56}Co masses in the range 0.075 - 0.002$M_\odot$ for SN 1987A and SN 1997D respectively [25].

10.5. Gamma-Ray Observations

All 8 lines from the ^{56}Co decay mentioned above have been detected in SN 1987A when data taken over the interval 160 - 460 days is added [26]. These are the γ-rays which escaped the envelope. An analysis of these together with the X-ray emission resulting from Comptonisation of the harder radiation through a reasonable model yields a mass of ^{56}Co very similar to the 2 independently determined values discussed above. An analysis of the 2 lines from decay of ^{57}Co also detected in SN 1987A near day 1600 [27] yields a mass of this isotope in good agreement with the value deduced from IR spectroscopy and mentioned above. The two strongest lines (847 and 1238 KeV) from ^{56}Co decay have been detected from the intrinsically bright Type Ia SN 1991T. The value of the mass of ^{56}Co depends on the distance to SN 1991T and a value near the Chandrasekhar limit requires a distance of 13 Mpc. A larger distance therefore seems to be excluded if uncertainties in the measurement are small. Finally a 5 σ detection of the 2 γ-ray lines near 1150 KeV from decay of ^{44}Ti has been made in the young SN remnant Cas A yielding a mass of this isotope in the range 1 - 2.4$\cdot 10^{-4}M_\odot$. We do not have a measure of the total mass of ^{56}Ni produced in this core-collapse event to test whether the Ti mass is consistent with our knowledge of explosive silicon burning, but recent X-ray observations of Cas A show that Fe certainly exits at high velocities as evidenced by Fe X-ray line emission in the outskirts of this remnant [28].

11. The Gamma-Ray Burst Connection

Because of an apparently close association in position and time of occurrence GRB980425 has been identified with the SN explosion identified as SN 1998bw [29]. It remains until now the most convincing argument that at least some GRBs originate in SN explosions. SN 1998bw occurred in a face-on low luminosity spiral galaxy where it is superimposed on HII regions in a spiral arm. This SN has some almost unique distiguishing characteristics which tend to reinforce the idea of an association. It was very bright at maximum light though not now the intrinsically brightest known SN. It was also a strong radio emitter reaching a maximum some 50 days or so (depending on frequency) after outburst. It was also an X-ray emitter although the precise knowledge of the character of this emission is hampered by the presence of other X-ray sources in the field of view, and possibly even some as yet unresolved sources very close to the SN. Early optical spectra were judged peculiar, after which it was realised that the peculiarity was largely due to a very high expansion velocity ($\sim$ 30000 km/sec) even 8 days after the explosion. A more subtle anomaly is that these very high velocities decrease to the normal Type Ib,c velocities approaching the nebular phase. The apparent absence of optical lines of HeI suggested a Type Ic but subsequent identification of intrinsically stronger HeI lines in IR spectra leaves it dangling as Type Ib,c with consequent uncertainty concerning the progenitor. Models which make use of the spectra, light curves and velocities suggest a core-collapse event in the evolved He-C-O core of a once massive ($\sim$ 40 $M_\odot$) main sequence star which lost mass in a wind or binary interaction. The mass which exploded was arguably in the range 10 - 15 $M_\odot$. That an explosion energy of $> 10^{52}$ ergs was involved seems indicated by the models fitting the light curves and spectra [10,11]. Such objects have acquired the name "hypernova" although they may not conform to what the original proposal had in mind [30]. Linear polarization measurements of $\sim$ 0.4 % were not very helpful in discriminating asymmetry because the ISM of the host galaxy may be contributing all or most of this amount. Some types of collapse model require or allow an interval of time between the SN collapse and the GRB owing to the formation of an accretion disk dictating a preferred axis of symmetry for jet formation coming from infall of material [31]. This type of model can relax the constraint on the coincidence in time between SN outburst and GRB although by how much is not yet confidently known. In fact the cosmologically distant GRB 011211 is used as evidence for this type of delay since the detection of X-ray emission lines of various ions in its afterglow suggests that the γ-rays in the burst heat the already expanding envelope to X-ray temperatures [32]. Other claims for cosmologically distant GRBs associated with SN events use the shape of the afterglow light curve as evidence. In these cases instead of an exponential decay, a bump in the light curve is observed as if additional radiation were

68

coming from a SN near maximum light. Unfortunately for these events the data points are sufficently sparse that, with their associated errors, it is not possible to show convincingly that the SN explosion had preceded the GRB [33,34,35]. In the nearby universe there have occurred other SNe which have been associated with GRB events, but with considerably less certainty than SN 1998bw, this because of greater uncertainties in timing and position. Some, such as SN 1997ef [36] bear some resemblance to SN 1998bw, i.e. Type Ib,c but with lower energies. Others are Type II events such as SN 1997cy [37] with, for example, very high optical luminosities. In spite of the increasing number of suggested associations between GRBs and SNe, there is not, at the time of writing, a consensus that all GRBs are associated with SN events.

Acknowledgments

This work was supported by the Italian Space Agency (ASI), and by the Italian Ministry of Education, University , and Research (MIUR Grant COFIN2000). The papers referenced result from the lectures and do not do justice to the entire body of recent work on SNe. Some aspects of this presentation derive from a continuing enjoyable collaboration with Nikolai Chugai.

References

1. S. Recchi eta al., *MNRAS* **322**, 800 (2001)
2. F. Patat et al., *ApJ* **555**, 900 (2001)
3. R. A. Chevalier and C. Fransson, *ApJ* **420**, 268 (1994)
4. K. W. Weiler et al., *ARA&A* **40**, 387 (2002)
5. N. N. Chugai and I. J. Danziger, *MNRAS* **268**, 173 (1994)
6. A. V. Filippenko, *ARA&A* **35**, 309 (1997)
7. K. Nomoto et al., *Nuc. Phys A* **621**, 467 (1997)
8. S. E. Woosley and T. A. Weaver, *ApJS* **101**, 181 (1995)
9. D. C. Leonard et al., *PASP* **113**, 920 (2001)
10. K. Iwamoto et al., *Nature* **395**, 672 (1998)
11. S. E. Woosley, et al., *ApJ* **516**, 788 (1999)
12. E. Cappellaro, et al., *A&A* **351**, 459 (1999)
13. M. M. Phillips, *ApJ* **413**, L105 (1993)
14. D. Arnett, *Supernovae and Nucleosynthesis:* Princeton Uni. Press (1996)
15. B. Schmidt et al., *ApJ* **395**, 366 (1992)
16. M. Hamuy et al., *ApJ* **558**, 615 (2001)
17. A. Elmhamdi et al., *MNRAS* **in press** (2002)
18. P. Mazzali et al., *A&A* **297**, 509 (1995)
19. P. Mazzali et al., *A&A* **303**, 118 (1995)

20. P. Mazzali et al., *MNRAS* **284**, 151 (1997)
21. I. J. Danziger, *Supernovae: The Tenth Santa Cruz Workshop in Astr. and Astrophys.* ed. S. E. Woosley, Springer-Verlag, 69 (1991)
22. E. Cappellaro, et al., *A&A* **328**, 203 (1997)
23. P. Bouchet and I.J.Danziger, *A&A* **273**, 451 (1993)
24. C. Fransson and C. Kozma, *ApJ* **408**, L25 (1993)
25. N. N. Chugai and V. P. Utrobin, *A&A* **354**, 557 (2000)
26. R. Diehl and F. X. Timmes, *PASP* **110**, 637 (1998)
27. J. D. Kurfess et al., *ApJ* **399**, L37 (1992)
28. J. P. Hughes et al., *ApJ* **528**, L109 (2000)
29. T. J. Galama, et al., *Nature* **395**, 670 (1998)
30. B. Paczyński, *ApJ* **494**, L45 (1998)
31. A. I. MacFayden and S. E. Woosley, *ApJ* **524**, 262 (1999)
32. J. N. Reeves, et al., *Nature* **416**, 512 (2002)
33. J. S. Bloom, et al., *Nature* **401**, 453 (1999)
34. T. J. Galama, et al., *ApJ* **536**, 185 (2000)
35. J. S. Bloom, et al., *ApJ* **572**, L45 (2002)
36. K. Iwamoto, et al., *ApJ* **534**, 660 (2000)
37. M. Turatto, et al., *ApJ* **534**, L57 (2000)

GAMMA-RAY AND NEUTRINO SIGNATURES OF COSMIC RAY ACCELERATION BY PULSARS

WŁODEK BEDNAREK

Department of Experimental Physics, University of Łódź,
90-236 Łódź, ul. Pomorska 149/153, Poland

Since their discovery, pulsars were considered as one of the most promising candidate sources of cosmic rays because in the extreme cases (milisecond periods) they can in principle generate potentials able to accelerate particles up to $\sim 10^{20}$ eV. In fact observations of nonthermal X-ray emission and also high energy γ-ray emission from nebulae around the youngest pulsars in Galaxy (e.g. Crab Nebula, Vela Nebula, PSR 1706-44) support that expectations. However, in spite of many efforts, the question of how and to what energies particles are accelerated by pulsars remains still unanswered. In this paper we review various models of particle acceleration by pulsars and estimates of pulsar's contribution to the cosmic ray content in the Galaxy. Recent predictions for the fluxes of γ-rays, neutrinos and neutrons, produced during the process of acceleration of particles and their propagation in the medium surrounding pulsars are discussed.

1. Introduction

It is widely believed that cosmic rays, at least with energies below 10^{15} eV, are accelerated by the shock waves of supernovae which exploded in our Galaxy. In fact, the nonthermal X-ray emission and TeV γ-ray emission have been reported from a few shell type supernova remnants (e.g. SN 1006 - Tanimori et al. 1998; Cas A - Aharonian et al. 2001; SNR RX J1713.7-3946 - Muraishi et al. 2000). This emission is usually interpreted in terms of leptonic models. Recent claim on the evidence of hadron acceleration in SNR RX J1713.7-3946 (Enomoto et al. 2002) is not supported by calculations of the γ-ray production by hadrons (e.g. Reimer & Pohl 2002).

Also the pulsars and their nebulae have been considered for a long time as possible sources of cosmic rays, even with energies higher than in the case of the shell type supernova remnants ($> 10^{15}$ eV). The non-thermal radio, optical, and X-ray emission, observed from about 10 pulsar wind nebulae (Chevalier 2000; Camilo et al. 2002), and detection of TeV γ-rays from 3 nebulae (around pulsars B0531 (Crab) - Weekes et al. 1989; B0833-

45 (Vela) - Yoshikoshi et al. 1997; and PSR 1706-44 - Kifune et al. 1995; Chadwick et al. 1998) idicate that relativistic particles have to be present in these objects. However these results are also usually interpreted in terms of leptonic models for the high energy radiation leaving the question of hadron acceleration by these objects still open.

In this paper we review models of hadron acceleration to energies $> 10^{14}$ eV in the pulsar wind nebulae and consider contribution of such sources to the cosmic rays in the Galaxy. We concentrate on the neutral signatures of hadron acceleration and discuss their possible detection by the next generation of the γ-ray and neutrino telescopes.

2. Mechanisms of particle acceleration

Pulsars are attractive accelerators of cosmic rays since they can generate huge potential differences in their rotating magnetospheres. If the magnetosphere of neutron star rotates as a rigid body then the observer in the lab system sees the electric field,

$$E = -v \times B/c, \tag{1}$$

where $v = \Omega \times r$ is the velocity of rotation and Ω is the angular velocity, r is the distance from the axis of rotation, B is the magnetic field strength, and c is the velocity of light. The potential drop across the polar cap region of the pulsar is

$$U = \Omega^2 B_{\mathrm{NS}} R_{\mathrm{NS}}^3 / 2c^2 \tag{2}$$

where B_{NS} is the surface magnetic field of the star, and R_{NS} is its radius. For example the potential drop across the polar cap of the Crab pulsar ($P = 0.033$ s, and $B = 4 \times 10^{12}$ G) is $\sim 3 \times 10^{16}$ eV. However, these huge electric fields should be immediately balanced by the plasma extracted from the neutron star or created in cascades occuring in the magnetosphere (Goldreich & Julian 1969). Therefore only a part of the full available potential may not be compensated by the plasma present in the pulsar magnetosphere. The appearence of such acceleration regions, called "gaps", is postulated in different parts of the pulsar magnetosphere (for recent review on the pulsar electrodynamics in the context of the acceleration of particles see Rudak 2001).

It is believed that the surface of a neutron star is composed of heavy nuclei from the iron group. These nuclei are binded into the surface. However their binding energy is not well known. The common opinion is that these nuclei can be stripped out of the surface of the neutron star by the very strong electric field or as a result of bombardment and heating by the products of cascades occuring in the pulsar magnetosphere. If this is the case then the iron nuclei can gain energy in the inner pulsar magnetosphere, in the region of the pulsar wind, or in the nebula surrounding the pulsar. Below we review these different possibilities for acceleration of heavy nuclei in more detail.

2.1. *Formation of a neutron star*

The violent processes during the formation of a neutron star might create good conditions for acceleration of particles. For example, de Gouveia Dal Pino & Lazarian (2000) proposed that particles can be accelerated in the magnetic reconnection sites just above the magnetosphere of newborn milisecond pulsars, originated by the accretion induced collapse of the white dwarf. The model is atractive since it avoids the problems with the propagation of particles in the massive envelope surrounding the neutron star if it is formated during the core collapse of the massive star. It may be difficult to localize such sources of cosmic rays due to the possible lack of neutral radiation accompanying the process of particle acceleration.

2.2. *Inner magnetosphere*

In some regions of the magnetosphere the amount of $e^{\pm}$ plasma is insufficient to short out induced electric field. Some models of such acceleration regions, called 'gaps', predict quite large potential drops. For example, Arons & Scharlemann (1979) has shown that a gap should appear along field lines close to the magnetic pole and at the critical line which bend toward the rotation axis. Because of the overall curvature of these magnetic field lines, photon created at the specific line leaves it before being absorbed and creates new $e^{\pm}$ pair which is bounded to another line. This leads to the formation of a "slot gap" in the pulsar magnetosphere (Arons 1983). In such gaps a huge potential drops can be generated, e.g. for the 5 ms pulsar with the surface magnetic field of 10^{12} G, the potential drop along the gap can reach the value $\sim 10^{17}$ V.

The regions of particle acceleration can also appear along the magnetic field lines between the null surface and the light cylinder radius which curve towards the equator (e.g. Cheng, Ho & Ruderman 1986). In these "outer gaps" the maximum potential drops for the Crab type pulsars are (Cheng et al. 1990)

$$\Phi \approx 1.25 \times 10^{16} B_{12} P_{\mathrm{s}}^{4/3} \ \mathrm{V}, \tag{3}$$

where $B = 10^{12}B_{12}$ G is the surface magnetic field of the pulsar, and P_s is its period in seconds. The iron nuclei, accelerated by relatively young and strongly magnetized pulsars, can move through the potentials which in principle can accelerete them up to a few 10^{18} eV.

However these nuclei lose also energy during the acceleration process. In principle, two processes may become important, the curvature energy losses and the photo-desintegration energy losses. The curvature energy losses become important when the nuclei move along the curved magnetic field lines. They lose energy at a rate

$$\frac{dE}{dt} = \frac{2(Ze)^2 c}{3R_c^2}\gamma^4,$$ (4)

where Ze is the charge of the particle, c is the velocity of light, R_c is the radius of curvature which is usually approximated by $R_c \approx (R_{NS}R_{LC})^{1/2}$, $R_{LC} = cP/2\pi$ is the light cylinder radius, and γ is the Lorentz factor of the nuclei. For example, in the outer gap of the Crab pulsar, the radius of curvature of the magnetic field is $R_c \approx 10^7$ cm, and the average electric field is $\sim 5 \times 10^6$ V cm^{-1}. The acceleration of iron nuclei is balanced by the curvature energy losses for the Lorentz factor of nuclei $\sim 3.5 \times 10^6$ (corresponding energy $\sim 2 \times 10^{17}$ eV).

The photodesintegration of heavy nuclei in collisions with the radiation from the neutron star surface, or produced in the inner magnetosphere can become also important process for their energy losses. It has been shown (Protheroe et al. 1998) that the iron nuclei can lose even several nuclei in collisions with the thermal radiation emitted by; (1) the whole surface of the neutron star at early time after neutron star formation when the temperature of the neutron star is above 5×10^6 K; (2) the polar cup region of the neutron star heated to temperatures above 10^7 K. The nuclei can be also efficiently photodesintegrated in collisions with the nonthermal photons produced in the electromagnetic cascades developing in gaps (Bednarek & Protheroe (1997). These processes will be discussed further in detail. The buyproducts of photodesintegration of nuclei, i.e. relativistic neutrons, are injected into the pulsar surrounding. Neutrons, not confined by the magnetic field present inside the expending envelope, decay at large distances from the neutron star supplying an attractive mechanism for energy transport from the pulsar surrounding.

2.3. Light cylinder region

The rotating magnetic field of the pulsar can in principle generate huge potential drops close to the light cylinder radius since the Goldreich & Julian (1969) density of plasma, necessary to saturate induced electric field, tends there to infinity. Beskin & Rafikov (2000) argue that $e^\pm$ plasma can be efficiently accelerated in this region provided that the magnetic energy density at the light cylinder dominates over the plasma energy density, i.e. the magnetization parameter $\sigma \gg 1$ (Michel 1969), and the plasma density at this region is much higher than the Goldreich & Julian density, i.e. the multiplication parameter of the plasma in the inner magnetosphere is $\lambda \gg 1$. If the longitudinal electric current is smaller than the Goldreich & Julian one (Goldreich & Julian density times velocity of light) then induced electric field near the light cylinder is $|\mathbf{E}| = |\mathbf{B}|$. This model can be tested if the acceleration of particles occurs in a strong radiation field. Recently Bogovalov & Aharonian (2000) has calculated the γ-ray flux produced by $e^\pm$ plasma, accelerated close to the light cylinder, which comptonize nonthermal radiation from the pulsar inner magnetosphere and thermal radiation from the pulsar surface. The comparison of calculations with the observations of the γ-ray flux from the Crab Nebula excludes the possibility of formation of relativistic $e^\pm$ plasma within 5 light radii from the pulsar. This causes serious problems for the Beskin et al. model.

2.4. Pulsar wind zone

Just after discovery of pulsars, Gunn & Ostriker (1969) suggested that particles can be accelerated to very high energies in the pulsar wind zone. The authors argue that rotating neutron star with the oblique magnetic field generates low frequency electromagnetic waves with the pulsar frequency. Because of low frequency and very large amplitude, these waves can capture particles and accelerate them to relativistic energies in times very short compared with the period of the wave (the mechanism called magnetic slingshot). However the maximum energies reached by particles depend on the presence of plasma above the light cylinder radius of the pulsar in which such waves prepagate. As we discussed in previous subsection the region above the light cylinder should be fulfilled with a dense $e^\pm$ plasma created in the pulsar inner magnetosphere. In such conditions the waves cannot propagate (Kennel & Pellat 1976; Asseo et al. 1978), and the magnetized $e^\pm$ plasma should be analysed in the magnetohydrodynamic (MHD) approximation.

Observations of young plerions (e.g. the Crab Nebula) indicate that the value of the magnetization parameter of the plasma (the ratio of the magnetic energy of the wind to the kinetic energy of particles in the wind) is quite high $\sigma \sim 10^{4-5}$ close to the light cylinder (e.g. Rees & Gunn 1974, Coroniti 1990). However, at the termination shock of the wind the value of the magnetization parameter drops to $\sigma \sim 10^{-3}$ for the Crab Nebula (Kennel & Coroniti 1984, De Jager & Harding 1992). Therefore, it looks that somewhere in the pulsar wind most of the rotational energy of the pulsar is converted into relativistic particles (Rees & Gunn 1974, Kennel & Coroniti 1984). However how the wind dissipates energy remains a mystery. It has been argued that the ideal, ultrarelativistic MHD wind is not able to convert the Poynting flux into particles (Chiueh, Li & Begelman 1998, Bogovalov & Tsinganos 1999). This difficulty can be overcomed due to the wind rapid expansion in a magnetic nozzle (Chiueh, Li & Begelman 1998) or nonideal MHD effects in a two-fluid plasma (Melatos & Melrose 1996). Recent analysis of the propagation of the MHD wind (Contopoulos & Kazanas 2001), which base on the exact solution of the axisymmentric magnetosphere (Contopoulos, Kazanas & Fendt 1999), shows that the Lorentz factor of outflowing plasma increases linearly with distance from the light cylinder up to the moment when the flow collimates drastically towards the direction of the axis of symmetry.

Another possible explanation for the conversion of the Poynting flux into the particles energy is the acceleration in the reconnection regions of the oppositely directed magnetic fields in the wind (Michel 1982, Coroniti 1990). Recent investigation of this process by Lyubarsky & Kirk (2001) shows that since the wind accelerates in the course of reconnection, the conversion of Poynting flux to particles has to occur on longer timescale and the process becomes not efficient for pulsars with the Crab pulsar parameters, although it is still efficient for the milisecond pulsars.

2.5. Pulsar nebula

Rees & Gunn (1974) pointed out that pulsar wind terminates in a standing reverse shock due to the interaction of a wind with a supernova envelope. The details of such model were constracted in a steady state, spherically symmetric situation by Kennel & Coroniti (1984a) and in the time dependent picture by Emmering & Chevalier (1987). The standing shock deccelerates and heats the $e^{\pm}$ wind. This shock can accelerate leptons and hadrons by the first order Fermi acceleration mechanism (Gaisser et al. 1987;1989, Berezinsky & Ginzburg 1987, Harding et al. 1991), up to energies $\sim 10^{16}$ eV in the case of the Crab pulsar. Ultrarelativistic leptons interact with the magnetic field and generate observed radiation of the nebula around the pulsar. The details of such model applied to the Crab nebula have been discussed by Kennel & Coroniti (1984b) and De Jager & Harding (1992). Another way of acceleration of leptons in the pulsar nebula has been proposed by (Gallant & Arons 1994). If the pulsar wind contains significant amount of relativistic heavy ions (e.g. iron nuclei), which dominate the energy density of the wind, then the nuclei can transfer their energy to positrons at the shock as a result of rezonant scattering (Hoshino et al. 1992, Gallant et al. 1992).

The pulsar nebula may also work as an re-accelerator of cosmic rays (Bell 1992) if the magnetic field is much weaker close to the polar regions than at the equator as postulated by the Coroniti (1990) model. Already energetic cosmic rays, e.g. accelerated by the supernova shock wave, can enter the pulsar nebula at the poles and drift inside the nebula to exit near the equator, thus gaining energy due to the potential difference between the pole and the equator. During the single entrance individual cosmic rays can increase their energies by two orders of magnitude. However such process occurs if other effects do not prevent particle's penetration of the nebula at the poles, like e.g. expension of the wind along the polar axis, or magnetic reconnection at the polar axis.

3. Contribution of pulsars to the cosmic rays

The mechanism of particle acceleration by the long scale electromagnetic waves in the pulsar wind zone (Ostriker & Gunn 1969) has been proposed as responsible for the acceleration of cosmic rays to the highest observed energies (Gunn & Ostriker 1969) and at the knee region (Karakuła, Osborne & Wdowczyk 1974). These estimations assume that particles are accelerated by Galactic population of observed radio pulsars to the maximum possible energies allowed by the pulsar electrodynamics (Karakuła et al. 1974) or by the population of pulsars with the maximum allowed periods determined by the stability condition of a rotating neutron star (Gunn & Ostriker 1969). Karakuła et al. (1974) concluded that the change of mass composition from light (at the region knee) to heavy (above the knee) should occur in the cosmic ray spectrum if protons and iron nuclei are accelerated by the pulsar at this same moment. This general prediction is consistent with the recent measurements of the mass composition above 10^{15} eV (e.g. Glasmacher et al. 1999, Fowler et al. 2001).

The problem of pulsar contribution to the cosmic ray spectrum has been recently refreshed in the context of the origin of the highest energy cosmic rays. Blasi, Epstein & Olinto (2000) have suggested that the iron nuclei could be accelerated to $\sim 10^{20}$ eV by the neutron stars in our Galaxy which periods are shorter than ~ 10 ms and their surface magnetic fields are in the range $10^{12} - 10^{14}$ G. In this simple model the iron nuclei are accelerated to the maximum energies allowed by the pulsar electrodynamics. They are injected with the rate which scales with the maximum possible rate described by the Goldreich & Julian (1969) density of charged particles close to the pulsar light cylinder radius. The authors assume that nuclei with such extreme energies are not photo-desintegrated during propagation in the pulsar magnetosphere and not confined by the expending supernova envelope. These two assumptions are crucial for this model and should be studied in detail.

Recently detailed model for the contribution of the galactic population of pulsars to the observed cosmic ray spectrum above 10^{15} eV has been developed by Bednarek & Protheroe (2002). It is assumed that the iron nuclei are accelerated in the outer gaps of pulsars (Cheng, Ho & Ruderman 1986), and then suffer partial desintegration in the non-thermal radiation fields of the outer gap and in collisions with the matter of the expending envelope of the supernova. The spectra and expected mass composition of particles escaping from the supernova remnant are computed taking into account the observed population of the radio pulsars and the adiabatic and interaction energy losses of particles in the expending nebula. Moreover, this model include the diffusion of particles in the supernova nebula and in the Galaxy. It is predicted that heavy nuclei, accelerated directly by pulsars, and the light nuclei, from their desintegrations, should contribute to the cosmic ray spectrum at the knee region. The contribution of heavy nuclei significantly increases the average value of $< \ln A >$ with increasing energy as suggested by recent observations (Glasmacher et al. 1999, Fowler et al. 2001).

The contribution of pulsars to the cosmic rays in the energy range $10^{15} - 10^{19}$ eV has been also considered in the paper by Giller & Lipski (2002). As in the work by Blasi et al. 2000), the authors do not specify the model for particle injection by pulsars but only simply assume that all rotational energy of a pulsar is transfered to hadrons (protons or iron nuclei) with the maximum possible energies allowed by the pulsar electrodynamics. The consistency with the observed cosmic ray spectrum is obtained by assuming that pulsars are born with relatively long periods which distribution is described by the gamma function with $s = 3.86$ and the average period $< P_0 > = 0.5$ s. The effects of propagation and adiabatic energy losses of accelerated particles in the expending supernova envelope have not been taken into account in derivation of the spectrum. The other features of cosmic rays, e.g. the mass composition of cosmic rays, are not discussed in this work.

4. Signatures of cosmic ray acceleration

Young pulsar wind nebulae, which accelerate cosmic rays, should become sources of high energy γ-rays and neutrinos resulting from interactions of hadrons with the matter of the supernova envelope, e.g. Berezinsky & Prilutsky (1978, Sato (1977). It has been shown in this early works that supernovae, containing energetic pulsar, should become efficient sources of neutrinos during first year after explosion. They should be detected by the planned at that time the DUMAND experiment. A few months after explosion, supernovae with pulsars should produce also significant fluxes of γ-rays. More recent calculations of neutral emission generated in such supernova-pulsar scenario had been stimulated by the explosion of supernova SN1987A in LMC (e.g. Gaisser et al. 1987; Berezinsky & Ginzburg 1987; Nakamura et al. 1987; Gaisser & Stanev 1987; Berezinsky et al. 1988). These authors usually assume that hadrons are injected with the power law spectrum in some not well specified mechanism for particle acceleration. The predicted fluxes of high energy neutrinos and γ-rays > 1 TeV from SN 1987A should be detectable if the created pulsar accelerates hadrons with the power $\sim 10^{39}$ erg s^{-1}.

Further in this section we describe recent scenarios for the neutral emission from supernovae remnants containing pulsars. We apply specific models for particle acceleration in such sources which were developed for the explanation of the γ-ray emission from pulsars and pulsar nebulae, i.e. acceleration in the inner magnetosphere and in the pulsar wind zone. We consider the radiation processes during the early stage of pulsar supernova interaction, when the supernova envelope is still optically thick, and at the later stage, when hadrons accumulated in the pulsar nebula interact with relatively rare medium. At the end, the energetic pulsar in the high density medium, e.g. in a large molecular cloud, is discussed.

4.1. Prompt gamma-ray and neutrino emission

Let us estimate the fluxes of neutrinos produced in the early phase of the pulsar - supernova envelope interaction assuming two different models for injection of relativistic heavy nuclei, i.e. acceleration in the inner pulsar magnetosphere at the slot gap (Arons & Scharlemann 1979; Arons 1983), and acceleration in the wind zone as discussed by Blasi et al. (2000) and Beall & Bednarek (2002).

The energetic pulsars are probably formed during explosions of type Ib/c supernovae, whose progenitors are Wolf-Rayet type stars (for the evolution models of such stars and their explosions see Woosley et al. 1993). The iron core collapses to a very hot proto-neutron star which cools to the neutron star during about $t_{NS} \approx 5 - 10$ s from the collapse. The rest of the mass of a presupernova (the envelope) is expelled with the velocity at the inner radius of the order of $v_1 = 3 \times 10^8$ cm s^{-1}. The velocities of matter in the envelope and the density profiles can be approximated by (for details see Beall & Bednarek 2002), $v(R) = v_1(R/R_1)^b$, ($b = 0.5$ and $R_1 = 3 \times 10^8$ cm) and $n(R) = n_1(R/R_1)^{-a}$ (the density at R_1 is $n_1 = 1.2 \times 10^{31}$ cm^{-3}, and the parameter $a = 2.4$). The initial column density decreases with time, t, due to the expansion of the envelope according to $\rho(t) = \int_{R_1}^{R_2} n(R)(R/R + v(R)t)^2 dR$, where $R_2 = 3 \times 10^{10}$ cm is the outer radius of the envelope at the moment of explosion, and $v(R)$ and $n(R)$ are described above. The volume above the pulsar and below the expanding envelope is filled with thermal radiation which is not able to escape because of the high optical depth of the envelope. The temperature of this radiation drops with time during the expansion of the envelope according to $T(t) = T_0(R_1/R_1 + v_1(t_{NS} + t))^{3/4}$. At early stage, the pulsar can lose energy on emission of electromagnetic and gravitational radiation. In the first model we assumed that both processes are important (Protheroe et al. 1998) and in the second model we include only electromagnetic energy losses (Beall & Bednarek 2002).

The standard models of the neutron star cooling predict that its surface temperature decreases to $\sim 10^7$ K at a few days after neutron star formation and later cools to $\sim 4 \times 10^6$ K in about one year. However the polar cup temperature remains $\sim 10^7$ K or higher, due to the heating by electrons and γ-rays from cascades in the pulsar magnetosphere. If heavy nuclei, e.g. iron, are extracted from the neutron star surface and accelerated in the pulsar slot gap, then they should suffer multiple photodesintegrations, injecting neutrons which escape freely (Protheroe et al. 1998). We calculate the spectra of injected neutrons for some specific parameters of the pulsar (see Figs. 6 and 7 in Protheroe et al. 1998), normalizing the number of iron nuclei injected by the pulsar to the pulsar rotational energy loss rate (factor ξ). The neutrons move balistically through the expending supernova envelope and produce high energy neutrinos and γ-rays. The spectra of neutrinos are calculated from

$$F_\nu(E_\nu) \approx [(1 - e^{-\tau_{pp}})/\Omega_b d^2] \int \dot{N}_n(E_n)P_{n\nu}(E_\nu, E_n)dE_n, \qquad (5)$$

where d is the distance to the pulsar, $P_{n\nu}(E_\nu, E_n)$ is the number of neutrinos produced via decay of pions in multiple necleon-nucleon interactions of a nucleon with energy E_n, $\dot{N}_n(E_n)$ is the production rate of neutrons by Fe nuclei, τ_{pp} is the optical depth of the supernova shell to nucleon-nucleon collisions, and Ω_b is the solid angle in which Fe nuclei are accelerated and neutrons are injected. In contrary to neutrinos, significant fraction of produced γ-rays is absorbed in collisions with the matter of the supernova during the first few months. Taking into account these effects the γ-ray flux can be calculated from

$$F_\gamma(E_\gamma) \approx [(e^{-\tau_{pp}} - e^{-\tau_{\gamma p}})/(\tau_{\gamma p}/\tau pp - 1)/\Omega_b d^2] \times$$
$$\int \dot{N}_n(E_n)P_{n\nu}(E_\nu, E_n)dE_n, \qquad (6)$$

where $\tau_{\gamma p} \approx 0.7\tau_{pp}$ is the optical depth of the shell for $e^{\pm}$ pair production.

Using the above formulae we calculated the fluxes and spectra of neutrinos and γ-rays above 100 MeV and 1 TeV (Protheroe et al. 1998). The neutrino spectra from the source at the distance 10 kpc are above the atmospheric neutrino background within $10°$ around the source during first one year after supernova explosion if the parameter describing the model is $\xi\Omega_b^{-1} = 1$ sr^{-1}. The γ-ray flux, peaking at about 2 months after the explosion, should be detectable with the future GLAST detector provided that $\Omega_b\xi^{-1} < 0.07$ sr (for the pulsar with the initial period 5 ms) and $\Omega_b\xi^{-1} < 3 \times 10^{-3}$ sr (10 ms pulsar), and by the next generation of the Cherenkov telescopes provided that $\Omega_b\xi^{-1} < 3.8$ sr (10 ms pulsar) and $\Omega_b\xi^{-1} < 75$ sr (5 ms pulsar).

Now let us estimate the flux of neutrinos expected in the similar general scenario but with different model for acceleration of heavy nuclei by the pulsar. In these calculations it is assumed that the spectrum of iron nuclei accelerated in the pulsar wind is given by (Beall & Bednarek 2002)

$$\frac{dN}{dEdt} = \frac{2\pi c\chi r_{\rm LC}^2 n_{\rm GJ}(r_{\rm LC})(E_{\rm Fe}E^2)^{-1/3}}{3\left[(E_{\rm Fe}/E)^{2/3} - 1\right]^{1/2}}$$

$$\cong \frac{3 \times 10^{30}\chi(B_{12}P_{\rm ms}^{-2}E^{-1})^{2/3}}{\left[(E_{\rm Fe}/E)^{2/3} - 1\right]^{1/2}} \frac{\rm Fe}{\rm s\ GeV}. \tag{7}$$

where χ is the efficiency of particle acceleration normalized to the Goldreich & Julian density close to the light cylinder radius (Goldreich & Julian 1969), and the maximum energies of nuclei are given by Blasi et al. (2000), $E_{Fe} = B^2(r_{\rm LC})/8\pi n_{\rm GJ}(r_{\rm LC}) \approx 1.8\times 10^{11}B_{12}P_{ms}^{-2}$ GeV, where $r_{\rm LC} = cP/2\pi$ is the light cylinder radius.

When the envelope in opaque for the radiation the nuclei are accelerated in the dense thermal radiation filling the cavity below expending supernova envelope. These nuclei will interact with strong thermal radiation field inside the supernova cavity, suffering at first multiple photo-disintegration of nucleons. The secondary nucleons lose energy mainly via pion production. The pions then decay into high energy neutrinos if their decay distance scale $\lambda_\pi \approx 780\gamma_\pi$ cm, is shorter than their characteristic energy loss mean free path. We show (Beall & Bednarek 2002) that pions decay before losing energy only for temperatures of radiation $T \leq 3 \times 10^6$ K. The temperature of the radiation inside the envelope drops to $T \leq 3 \times 10^6$ K at about $t_{\rm dec} \sim 10^4$ s after the supernova explosion. When the optical depth through the expanding envelope drops below $\sim 10^3$, the radiation is not further confined in the region below the envelope and its temperature drops rapidly. This happens at the time $t_{\rm conf} \sim 2 \times 10^6$ s after the explosion. At later times, the relativistic iron nuclei do not desintegrate in the radiation field but interact directly with the matter of the envelope whose density is already low enough so that pions produced by that interaction are able to decay into neutrinos and muons.

We now compute the differential spectra of muon neutrinos produced in the interaction of nuclei: (1) with the radiation field below the envelope during the period $1 \times 10^4 - 2 \times 10^6$ s after the supernova explosion; and (2) with the matter of the envelope during the period from $2 \times 10^6 - 3 \times 10^7$s after the explosion, assuming that the nucleons cool to the lowest energies allowed by the column densities of photons and matter, respectively. For a supernova inside our Galaxy at a distance $D = 10$ kpc, we estimate the expected flux of muon neutrinos produced in these two processes. If EHE CRs are produced by pulsars within our Galaxy, than the observed flux

Table 1. Expected number of neutrinos from a supernova at
the distance of 10 kpc

	$P_{ms} = 3, B_{12} = 4$	$P_{ms} = 10, B_{12} = 100$
N-$\gamma \rightarrow \nu_\mu$ (H)	0.8	96
Fe-M$\rightarrow \nu_\mu$ (H)	8.7	720
N-$\gamma \rightarrow \nu_\mu$ (N)	0.2	20
Fe-M$\rightarrow \nu_\mu$ (N)	3.4	256

of particles allows to constrain some free parameters of considered model. By comparing the observed flux of cosmic rays at $\sim 10^{20}$ eV with model estimations of the flux of iron nuclei, Blasi et al. (2000) finds that following condition should be fulfilled $\chi \epsilon Q / \tau_2 R_1^2 B_{13} \approx 4 \times 10^{-6}$, ϵ is the fraction of pulsars which have the parameters required for particle acceleration to 10^{20} eV, Q is the trapping factor of particles within the Galactic Halo, $\tau = 100\tau_2$ yr is the rate of neutron star production, $R = 10R_1$ kpc is the radius of the Galactic Halo, and $B_{13} = 0.1B_{12}$ is the pulsar surface magnetic field. For plausible parameters: $\tau_2 = 1$, $R_1 = 3$, $Q \sim 1$, and $\epsilon = 0.1$, we obtain the limit on the particle acceleration efficiency $\xi \approx 10^{-4}$ (for the pulsar with $P_{ms} = 3$ and $B_{12} = 4$), and $\xi \approx 4 \times 10^{-3}$ ($P_{ms} = 10$ and $B_{12} = 100$). The results of calculations for the pulsars with these two sets of parameters and the density factors estimated above, are shown in Table 1 for the case of neutrinos arriving from directions close to the horizon, i.e. not absorbed by the Earth (H), and for neutrinos which arrive moving upward from the nadir direction and are partially absorbed (N). These calculations show that some neutrinos might be observed in the 1 km^2 neutrino detector. Neutrinos from a Crab-type pulsar located at the distance of ~ 2 kpc might be also observable by such detector during the first year after pulsar formation if the particle acceleration efficiency is $\xi > 3 \times 10^{-3}$ (Beall & Bednarek 2002).

If the considered model works, then the whole population of pulsars created in the Universe should contribute to the extragalactic neutrino background. This could be detectable, because in such a case we do not need to be lucky to find the pulsar within the Galaxy during such an early phase. In another paper (Bednarek 2001) we estimate the extragalactic neutrino background from the population of pulsars with parameters similar to those of classical radio pulsars formed in the Universe.

4.2. Delayed gamma-ray and neutrino emission

Possible contribution of γ-rays from hadronic processes to the spectrum of pulsar wind nebulae has been considered by a few authors (e.g. Berezinsky & Prilutsky 1978; Cheng et al. (1990); Atoyan & Aharonian 1996). For example Cheng et al. (1990) calculated the high energy γ-ray fluxes assuming that relativistic protons accelerated in the Crab pulsar outer gap interact with the matter inside the Nebula. They noticed that this process may contribute in the TeV γ-ray range. In this subsection we describe calculations of the neutral emission produced by hadrons injected by: the pulsar in the interstellar medium, taking as an example the case of the Crab nebula (Bednarek & Protheroe 1997), and the pulsar immersed in a high density medium (Bednarek 2002).

4.2.1. Pulsar in the interstellar medium

Let us consider (Bednarek & Protheroe 1997) a young pulsar inside its wind powered nebula and investigate the consequences of acceleration of heavy nuclei (e.g. iron nuclei) by the pulsar. As an example we concentrate on the Crab Nebula (pulsar) which is a well established γ-ray source. The iron nuclei, extracted from the neutron star surface and accelerated in the pulsar's magnetosphere, photodisintegrate in collisions with soft photons from the pulsar's outer gap, injecting energetic neutrons which can decay either inside or outside the Crab Nebula. The protons from neutron decay inside the nebula are trapped by the Crab Nebula magnetic field, and accumulate inside the nebula producing gamma-rays and neutrinos in collisions with the matter in the nebula. Neutrons decaying outside the Crab Nebula contribute to the Galactic cosmic rays.

In order to calculate the expected fluxes of gamma-rays and neutrinos, it is assumed that Fe nuclei can escape from the polar cap surface of the Crab pulsar, and move along magnetic field lines to enter the outer gap where they can be accelerated as in the model of Cheng et al. (1986). The nuclei interact with photons either producing secondary $e^{\pm}$ pairs (with negligible loss of energy) or extracting a nucleon. In order to obtain the energy spectrum of neutrons extracted from Fe nuclei, $N_n(\gamma_n)$, we simulate their acceleration and propagation through the outer gap using a Monte Carlo method. To obtain the rate of injection of neutrons per unit energy we multiply $N_n(\gamma_n)$ by the number of Fe nuclei injected per unit time, $\dot{N}_{\text{Fe}}$, which can be simply related to the total power output of the pulsar $L_{\text{Crab}}(B, P)$, $\dot{N}_{\text{Fe}} = \xi L_{\text{Crab}}(B, P)/Z\Phi(B, P)$, where ξ

is the parameter describing the fraction of the total power taken by relativistic nuclei accelerated in the outer gap, $Z = 26$ is the atomic number of Fe, B is the surface magnetic field, P is the pulsar's period, and $\Phi(B, P) \approx 5 \times 10^{16}(B/4 \times 10^{12}\text{G})(P/1\,\text{s})^{4/3}$ V, is the potential difference across the outer gap (Cheng et al. 1986).

Soon after the supernova explosion, when the nebula was relatively small, nearly all energetic neutrons would be expected to decay outside the nebula. However, at early times we must take account of collisions with matter. The optical depth may be estimated from $\tau_{nH} \approx \sigma_{pp}n_H r \approx 8.6 \times 10^{14}M_1 v_8^{-2}t^{-2}$, where $M = M_1 M_\odot$ is the mass ejected during the Crab supernova explosion in units of solar masses, $r = vt$, $n_H = M/(4/3\pi r^3 m_p)$ is the number density of target nuclei, and $v = 10^8 v_8$ cm s^{-1} is the expansion velocity of the nebula. We note that $\tau_{nH} = 1$ at $t \approx 0.93M_1^{1/2}v_8^{-1}$ yr.

The number and spectrum of relativistic protons from neutron decay at the present time is determinated by the injection rate of neutrons into the Crab Nebula integrated over time since the pulsar's birth. We estimate the evolution of the Crab pulsar's period from birth to the present time taking account of magnetic dipole radiation energy losses and gravitational energy losses for an ellipticity of 3×10^{-4} (Shapiro & Teukolsky 1983). Two initial periods of the pulsar are considered: 5 ms, and 10 ms.

The spectrum of protons from neutron decay outside the Crab Nebula is given by

$$N_p^{\text{out}}(\gamma_p, t_{\text{CN}}) = \int_0^{t_{\text{CN}}} dt \dot{N}_{\text{Fe}}(t)N_n(\gamma_p, t)e^{-\tau_{nH}(t)}e^{-vt_{\text{CN}}/c\gamma_p\tau_n} \qquad (8)$$

where $\tau_n \approx 900$ seconds is the neutron decay time, and we make the approximation that the Lorentz factor of protons is equal to that of parent neutrons, $\gamma_p \approx \gamma_n$.

The equation for the spectrum of protons injected inside the Crab Nebula is more complicated because we must take account of proton-proton collisions and adiabatic energy losses due to the expansion of the nebula. The Lorentz factor of these protons at time t after the explosion such that their present Lorentz factor is γ_p is given by $\gamma_p(t) \approx \gamma_p(t + t_{\text{CN}})/2tK^{\tau_{pp}(t)}$, where $\tau_{pp}(t)$ is the optical depth for collision of protons with matter between t and t_{CN}, and is given by $\tau_{pp}(t) \approx 1.3 \times 10^{17}M_1 v_8^{-3}(t^{-2} - t_{\text{CN}}^{-2})$, and $K \approx 0.5$ is the inelasticity coefficient in proton–proton collisions.

Since we are interested in protons interacting inside the nebula at the present time, we must also include those neutrons which decayed at locations outside the nebula at time t but which will be inside the nebula at

Table 2. Model parameters for the Crab supernova and limits on $\xi\mu$.

Model	I	II	III	IV
P_0 (ms)	5	10	10	10
$r_{\rm CN}$ (pc)	1	2	2	1
M_1	3	3	10	3
$(\xi\mu)_\gamma$ at 10 TeV	0.63	6.9	2.0	1.0
$(\xi\mu)_\nu$ at 10 TeV	0.22	1.8	0.54	0.29

time $t_{\rm CN}$. Taking account of all these effects, we arrive at the formula below
for the proton spectrum inside the nebula at time $t_{\rm CN}$,

$$N_p^{\rm in}(\gamma_p, t_{\rm CN}) = \gamma_p^{-1} \int_0^{t_{\rm CN}} \mathrm{d}t \dot{N}_{\rm Fe}(t) e^{-\tau_n H(t)}$$

$$\times \left[N_n(\gamma_p(t), t)\gamma_p(t)[1 - \exp(-vt/c\gamma_p(t)\tau_n)] \right.$$

$$\left. + \int_t^{t_{\rm CN}} \mathrm{d}t' N_n(\gamma_p(t'), t)\gamma_p(t') \frac{v\exp(-vt'/c\gamma_p(t')\tau_n)}{c\gamma_p(t')\tau_n} \right]. \qquad (9)$$

The first term gives the contribution from neutrons decaying initially inside
the nebula while second term gives the contribution from neutrons decaying
at points initially outside the nebula which will be inside the nebula at time
$t_{\rm CN}$. We compute the expected γ-ray spectra produced by particles inside
the Crab Nebula for four different models, taking various initial pulsar periods, present sizes and masses of the Crab Nebula. The fluxes may possibly
be enhanced if protons are efficiently trapped by the dense filaments as
suggested by Atoyan and Aharonian (1996). Therefore we introduce an effective density experienced by the protons inside nebula, $n_H^{\rm eff} = \mu n_H$, where
n_H is defined above, and the parameter μ takes into account the possible
effects of proton trapping by the filaments. The photon fluxes expected at
the Earth are shown in Fig. 1 for $\xi\mu = 1$, and are compared with observations of the Crab Nebula above 0.2 TeV. Results are shown for models I
to IV having P_0, $r_{\rm CN}$ and M_1 as specified in Table 2, and for a distance to
the Crab Nebula of 1830 pc. Comparison with the Whipple observations at
10 TeV allows us to place constraints on the free parameters of the model,
and upper limits on $\xi\mu$ which are given in Table 2 as $(\xi\mu)_\gamma$.

The neutrino spectrum produced in collisions of protons with matter
inside the Crab nebula for the models considered above is above the atmospheric neutrino background flux within $1°$ of the source direction. Neutrino
detectors with good angular resolution should be able to detect neutrinos at

10 TeV from the Crab Nebula if $\xi\mu$ is greater than $(\xi\mu)_\nu$ given in Table 2. Note that in all cases $(\xi\mu)_\nu < (\xi\mu)_\gamma$ and so the possibility of ν detection is allowed by the existing γ-ray observations.

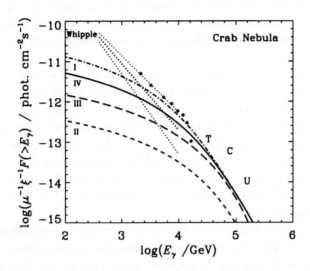

Figure 1. Spectra of γ-rays from interactions of protons with matter inside the Crab Nebula for the different proton spectra and for the different models of the nebula (I – IV) considered in the text. Observations: Whipple Observatory (dotted line and error box); THEMISTOCLE (+); and CANGAROO (solid line and error box). Upper limits from various experiments: T - Tibet, H - HEGRA, C - CYGNUS, and U - CASA-MIA.

4.2.2. Pulsar in the high density medium

We consider the young pulsar formated in the high density medium typical for the Galactic Center. This scenario is motivated by the existence of an extended excess of cosmic rays over a narrow energy range $10^{17.9} - 10^{18.3}$ eV from directions close to the Galactic Center (GC) reported recently by AGASA collaboration (Hayashida et al. 1999). The GC excess was confirmed in the analysis of the SUGAR data (Bellido et al. 2001).

The Galactic Center region (inner ~ 50 pc) is rich in many massive stellar clusters with a few to more than 100 OB stars (Morris & Serabyn 1996). These stars should soon explode as supernovae. We consider the possibility that the excess of cosmic rays from the direction of the Galactic Center, is caused by a young, very fast pulsar in the high density medium. As in

88

previous discussions we assume that the pulsar can accelerate iron nuclei to energies $\sim 10^{20}$ eV. These nuclei leave the supernova envelope without energy losses and diffuse through the dense central region of the Galaxy. Some of them collide with the background matter creating neutrons (from desintegration of Fe), neutrinos and gamma-rays (in inelastic collisions). We suggest that neutrons produced at a specific time after the pulsar formation are responsible for the observed excess of cosmic rays at $\sim 10^{18}$ eV.

The particles accelerated close to the pulsar can escape into the surrounding and diffuse in the magnetic field of the cloud, suffering collisions with the matter from time to time. In our further considerations we discuss, as an example, two media typical of the GC region in which the pulsar may be immersed. The first one is a huge molecular cloud with the radius $R_c = 10$ pc, the density $n_c = 10^3$ cm^{-3}, and the magnetic field $B_c = 10^{-4}$ G (the total mass $\sim 10^5 M_\odot$), and the second one is an extended high density region inside the GC with $R_c = 50$ pc, $n_c = 10^2$ cm^{-3}, and $B_c = 3 \times 10^{-5}$ G (the total mass $\sim 10^6 M_\odot$).

We estimated that these nuclei can escape through the supernova envelope after ~ 1 yr after the supernova explosion for typical parameters of the supernova, i.e the mass of the envelope in the case of type Ib/c supernovae $M_{env} = 3 M_\odot$, and the expansion velocity of the envelope at the inner radius is $v_{env} = 3 \times 10^8$ cm s^{-1} (Beall & Bednarek 2002). The iron nuclei diffuse in the magnetic field of the high density medium in the GC region, i.e. huge molecular clouds. Some of them interact producing neutrons, neutrinos, and γ-rays. In order to obtain the equilibrium spectrum of iron nuclei inside the cloud, we have to integrate over the activity time of the pulsar since its parameters evolve in time due to the pulsar's energy losses. As before we assume that the pulsar loses energy on electromagnetic waves. The equilibrium spectrum of iron nuclei at a specific observation time, t_{obs}, is calculated from

$$\frac{dN}{dE} = \int_{t_0}^{t_{obs}} \frac{dN}{dEdt} K e^{-c(t_{obs}-t)/\lambda} dt, \quad (10)$$

where $dN/dEdt$ is the spectrum of iron nuclei, see Eq. (7), $t_0 = 1$ yr, K gives the part of nuclei produced at the time 't' which do not escape from the cloud due to the diffusion and are still present inside the cloud at the time t_{obs}. λ is the mean free path for collision of the iron nuclei with the matter of the cloud. The value of K is estimated from $K = (R_c/D_{dif})^3$, where $D_{dif} = (r_L ct/3)^{1/2}$ is the diffusion distance of iron nuclei in the

Table 3. Expected gamma-ray fluxes and numbers of neutrino events from the supernova at the Galactic Center.

Model	$N_\gamma(> 1\ \mathrm{TeV})$	$N_\gamma(> 10\ \mathrm{TeV})$	N_ν^{a}	N_ν^{na}
(I)	4.3×10^{-12}	2.2×10^{-12}	23	30
(II)	8.7×10^{-13}	6.6×10^{-13}	11	16
(III)	2.5×10^{-13}	1.7×10^{-13}	5.3	8.8

magnetic field of the cloud, and r_{L} is the Larmor radius of the iron nuclei with energy E. For the case, $D_{\mathrm{dif}} \leq R_{\mathrm{c}}$, we take $K = 1$.

The part of iron nuclei confined within the molecular cloud, interact with a relatively dense medium suffering desintegrations and pion energy losses. The pions decay into neutrinos and γ-rays. We calculate the differential spectra of neutrons (from desintegrations of the iron nuclei), muon neutrinos, and γ-rays (from inelastic collisions of iron) on the Earth taking into account the decay of neutrons and absorption of γ-rays in MBR on the path from the GC $D_{\mathrm{GC}} \approx 8.5$ kpc.

If this excess of CRs is caused by neutrons produced in the pulsar model discussed here, then the expected flux of neutrons can be compared with the observed by the SUGAR experiment. Basing on this normalization we predict absolute fluxes of neutrinos and gamma-rays on Earth. This procedure allows us to derive the free parameter of our model (i.e. the efficiency ξ of iron acceleration by the pulsar) and limit the age of the pulsar for other fixed parameters, $P, B, R_{\mathrm{c}}, n_{\mathrm{c}}, B_{\mathrm{c}}$, which are in fact constrained by the observations. We consider five different sets of parameters describing our scenario: model (I) R $= 10$ pc, $n = 10^3$ cm^{-3}, $B_{\mathrm{c}} = 10^{-4}$ G, $t_{\mathrm{obs}} = 10^4$ yr; (II) $t_{\mathrm{obs}} = 3 \times 10^3$ yr and other parameters as above; (III) R $= 50$ pc, $n = 10^2$ cm^{-3}, $B_{\mathrm{c}} = 3 \times 10^{-5}$ G, $t_{\mathrm{obs}} = 10^4$ yr. In all these models we assume that the pulsar is born with $B = 6 \times 10^{12}$ G and $P_0 = 2$ ms. Normalizing the predicted neutron flux to the observed excess of CR particles we derive the value of the parameter ξ which has to be $\xi \approx 1$ (model I), 0.18 (II), and 0.3 (III).

Using the above estimates for ξ we can now predict the expected fluxes of γ-rays and muon neutrinos and antineutrinos in the case of every model. In Table 3 the γ-ray fluxes above 1 TeV and 10 TeV in units cm^{-2} s^{-1} are reported. The γ-ray fluxes in the energy range 1-10 TeV produced in models, (I) $\sim 2.1 \times 10^{-12}$ cm^{-2} s^{-1}, and (II) $\sim 2.1 \times 10^{-13}$ cm^{-2} s^{-1}, and probably also in (IV) $\sim 8 \times 10^{-14}$ cm^{-2} s^{-1}, should be observed by the future systems of Cherenkov telescopes. We estimate the number of muon neutrino events during one year in the IceCube detector (see Table 3). The case of neutrinos coming to the neutrino detector from directions close to the horizon, i.e. not absorbed by the Earth (N_ν^{na}), and neutrinos which arrive moving upward from the nadir direction, i.e. partially absorbed by the Earth (N_ν^a is distinguished. It is clear that the IceCube detector should detect a few up to several neutrinos per year from the Galactic Center region provided that the excess of cosmic rays at $\sim 10^{18}$ eV from the GC region is caused by neutrons.

5. Conclusion

Young pulsars are at present one of the best candidates responsible for acceleration of cosmic rays up to the energies $\sim 10^{18}$ eV, and maybe even up to the highest energies observed. However, in spite of growing interest to this problem in last years, the details of the acceleration mechanism and propagation of accelerated hadrons in the pulsar surrounding (supernova remnant) remains unclear. Recent estimations of the contribution of hadrons, accelerated in terms of pulsar scenarios, to the cosmic ray content in the Galaxy, made under rather optimistic assumptions, confirm the old expectations that pulsars can be responsible for the bulk of the cosmic rays between the knee and the ankle (Bednarek & Protheroe 2002; Giller & Lipski 2002). Extreme pulsars, with milisecond periods and/or superstrong surface magnetic field, might also accelerate hadrons up to $\sim 10^{20}$ (Blasi, Epstein & Olinto 2000; de Gouveia Dal Pino & Lazarian 2000).

Observations of non-thermal radio to X-ray emission and γ-ray emission from supernovae and pulsar wind nebulae do not show at present clear evidences of acceleration of hadrons. There is a hope that next generation of γ-ray observatories and large scale neutrino detectors give the answer to this problem since recent calculations of neutral emission from sources containing young pulsars suggest that:

- Neutrino and gamma-ray emission should be detectable from the supernova within our Galaxy during the first year after explosion if the pulsar with reasonable parameters is formed (initial period 5-10 ms and surface magnetic field $\sim 10^{12}$ G);
- Neutrinos should be observed by the 1 km^2 detectors from the Crab type nebulae if the heavy nuclei have been accelerated efficiently by the pulsar;
- The neutrons produced in the interaction of the energetic pulsar with the high density medium, typical for the Galactic Center region, can be responsible for the excess of cosmic rays reported

by the AGASA and SUGER groups. Such pulsar produces also neutrinos and TeV γ-rays which should be observed by the next generation of the neutrino and Cherenkov telescopes.

We have to mention that discussed fluxes of neutral particles, produced in scenarios which involve very young pulsars, base on the assumption that pulsars are formed with the surface magnetic field strengths derived from the observations of the classical radio pulsars. This may not be the case. For example if the milisecond pulsar was created during the famous supernova SN 1987A (Middleditch et al. 2000), then the limits on the pulsar power output requires the surface magnetic field for this pulsar $< 5 \times 10^{10}$ G (Nagataki & Sato 2001). Thus the acceleration efficiencies and maximum energies of particles can be significantly lower than considered in the reviewed papers.

Acknowledgements

I would like to thank the Organizers of the School for the invitation and their support. The work presented in this paper has been also partially supported by the KBN grant 5P03D 02521 and the University of Łódź grant No. 505/275.

References

1. Aharonian, F.A., et al., *A&A* **370**, 112 (2001).
2. Atoyan, A.M., Aharonian, F.A., *MNRAS* **278**, 525 (1996).
3. Arons, J., *ApJ* **266**, 215 (1983).
4. Arons, J., Scharlemann, E.T., *ApJ* **231**, 854 (1979).
5. Asseo, E., Kennel, C.F., Pellat, R., *A&A* **65**, 401 (1978).
6. Beall, J.H., Bednarek, W., *ApJ* **569**, 343 (2002).
7. Bednarek, W., *A&A Lett.* **378**, 49 (2001).
8. Bednarek, W., *MNRAS* **331**, 483 (2002).
9. Bednarek, W., Protheroe, R.J., *PRL* **79**, 2616 (1997).
10. Bednarek, W., Protheroe, R.J., *Astropart.Phys.* **16**, 397 (2002).
11. Bell, A.R., *MNRAS* **257**, 493 (1992).
12. Bellido, J.A., Clay, R.W., Dawson, B.R., Johnston-Hollit, M., *Astropart.Phys.* **15**, 167 (2001).
13. Berezinsky, V.S., Castagnoli, C., Navarra, G., *A&A* **203**, 317 (1988).
14. Berezinsky, V.S., Ginzburg, V.L., *Nature* **329**, 807 (1987).
15. Berezinsky, V.S., Prilutsky, O.F., *A&A* **66**, 325 (1978).
16. Beskin, V.S., Rafikov, R.R., *MNRAS* **313**, 433 (2000).
17. Blasi, P., Epstein, R.I., Olinto, A.V., *ApJ* **533**, 123 (2000).
18. Bogovalov, S.V., Aharonian, F.A., *MNRAS* **313**, 504 (2000).

92

19. Bogovalov, S.V., Tsinganos, K., *MNRAS* **305**, 211 (1999).
20. Camilo, F. et al., *ApJ*, in press (2002).
21. Chadwick, P.M. et al., *Astropart.Phys.* **9**, 131 (1998).
22. Cheng, K.S., Ho, C., Ruderman, M., *ApJ* **300**, 500; **300**, 522 (1986).
23. Cheng, K.S et al., *J.Phys. G.* **16**, 1115 (1990).
24. Chevalier, R.A., *ApJ* **539**, L45 (2000).
25. Chieueh, T., Li, Y.-Z., Begelman, M.C., *ApJ* **505**, 835 (1998).
26. Contopoulos, I., Kazanas, D., *ApJ* **566**, 336 (2002).
27. Contopoulos, I., Kazanas, D., Fendt, C., *ApJ* **511**, 351 (1999).
28. Coroniti, F.V., *ApJ* **349**, 538 (1990).
29. De Gouveia Dal Pino, E.M., Lazarian, A., *ApJ* **560**, 358 (2001).
30. De Jager, O.C., Harding, A.K., *ApJ* **396**, 161 (1992).
31. Emmering, R.T., Chevalier, R.A., *ApJ* **321**, 334 (1987).
32. Enomoto, R. et al., *Nature* **416**, 823 (2002).
33. Fowler, J.W., *Astropart.Phys.* **15**, 49 (2001).
34. Gallant, Y.A., et al., *ApJ* **391**, 73 (1992).
35. Gallant, Y.A., Arons, J., *ApJ* **435**, 230 (1994).
36. Gaisser, T.K., Harding, A.K., Stanev, T., *Nature* **329**, 314 (1987).
37. Gaisser, T.K., Harding, A.K., Stanev, T., *ApJ* **345**, 423 (1989).
38. Gaisser, T.K., Stanev, T., *PRL* **58**, 1695 (1987).
39. Giller, M., Lipski, M., *J.Phys. G* **28**, 1275 (2002).
40. Glasmacher, M.A.K. et al., *Astropart.Phys.* **12**, 1 (1999).
41. Goldreich, P., Julian, W.H., *ApJ* **157**, 869 (1969).
42. Gunn, J., Ostriker, J., *PRL* **22**, 728 (1969).
43. Harding, A.K. et al., *ApJ* **378**, 163 (1991).
44. Hayashida, N. et al., *Astropart.Phys.* **10**, 303 (1999).
45. Hoshino, M. et al., *ApJ* **390**, 454 (1992).
46. Karakuła, S., Osborne, J.L., Wdowczyk, J., *J.Phys. A* **7**, 437 (1974).
47. Kennel, C.F., Coroniti, F.V., *ApJ* **283**, 694 (1984a).
48. Kennel, C.F., Coroniti, F.V., *ApJ* **283**, 710 (1984b).
49. Kennel, C.F., Pellat, R., *J.Plasma Phys.* **15**, 335 (1976).
50. Kifune, T. et al., *ApJ* **438**, L91 (1995).
51. Lyubarsky, Y., Kirk, J.G., *ApJ* **547**, 437 (2001).
52. Melatos, A., Melrose, D.B., *MNRAS* **279**, 1168 (1996).
53. Middleditch, J. et al., *New Astronomy* **5**, 243 (2000).
54. Michel, F.C., *ApJ* **158**, 727 (1969).
55. Michel, F.C., *Rev.Mod.Phys.* **54**, 1 (1982).
56. Morris, M., Serabyn, E., *ARA&A* **34**, 645 (1996).
57. Muraishi, H. et al., *A&A* **354**, L57 (2000).
58. Nagataki, S., Sato, K., *Prog. Theor. Phys.* **105**, 429 (2001).
59. Nakamura, T., Yamada, Y., Sato, H., *Prog. Theor. Phys.* **78**, 1065 (1987).
60. Ostriker, J.P., Gunn, J.E., *ApJ* **157**, 1395 (1969).
61. Protheroe, R.J., Bednarek, W., Luo, Q., *Astropart. Phys.* **9**, 1 (1998).
62. Rees, M.J., Gunn, J.E., *MNRAS* **167**, 1 (1974).
63. Reimer, O., Pohl, M., *A&A Lett.*, **390**, 43 (2002).
64. Rudak, B., *Lecture Notes in Physics*, in press (2002)

65. Shapiro, S.L., Teukolsky, S.L., *Black Holes, White Dwarfs and neutron Stars*, (New York: Wiley, 1983) (1983).
66. Sato, H., *Prog. Theor. Phys.* **58**, 549 (1977).
67. Tanimori, T. et al., *ApJ*, **497**, L25 (1998).
68. Weekes, T.C. et al., *ApJ* **342**, 379 (1989). (2000).
69. Woolsey, S.E., Langer, N., Weaver, T.A., *ApJ* **411**, 823 (1993).
70. Yoshikoshi, T. et al., *ApJ* **487**, L65 (1997).

GAMMA RAYS FROM PSR B1259-63/BE BINARY SYSTEM

A. SIERPOWSKA & W. BEDNAREK

Department of Experimental Physics, University of Łódź,
90-236 Łódź, ul. Pomorska 149/153, Poland

We assume that the pulsar in PSR B1259-63/Be binary system injects electrons (and positrons) which lose energy on: (1) comptonization of thermal radiation from the massive companion, and (2) synchrotron process. The γ-rays are produced by these leptons during their propagation in the pulsar wind region and after their isotropisation by the shock region formated in the interaction of the pulsar and stellar winds. From normalization of the synchrotron spectrum, produced in the shock region, to the X-ray flux observed from this binary system near the periastron, we show under which conditions the γ-ray emission should be detected by the planned GLAST experiment.

1. Introduction

The binary system PSR B1259-63/Be star (SS2883) contains young radio pulsar, $P = 47.76$ ms, on a highly eccentric orbit. The massive star, with the effective temperature $T_{eff} = 2.6 \times 10^4 K$ and radius $R_\star = 4.2 \times 10^{11}$ cm, creates strong target for the electrons at the periastron distance $23R_\star$. The system was detected at hard X-rays by the OSSE [1]. Only the upper limits are available at MeV energies or above [2]. The γ-ray emission from PSR B1259-63/Be system has been calculated assuming that electrons: (1) are accelerated in the shock region between the stars [3,4]; (2) move in the pulsar wind zone [5,6]. We calculate the X-ray and γ-ray spectra from electrons which propagate radially in the wind zone and after their isotropization in the shock between the pulsar and the star winds.

2. High energy processes inside a binary system

The power injected into the surroundings by the pulsar PSR B1259-63 ($L_{rot} = 8 \times 10^{35}$ erg s^{-1}) is high enough that the matter from the massive companion can not accrete. However, in collisions of the pulsar and stellar winds a double shock structure seperated by the contact discontinuity is formed. The geometry of this structure can be described by the parameter

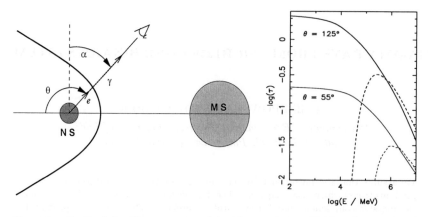

Figure 1. On the left - The geometry of the binary system. The location of the shock region is marked by the thick full curve. Leptons are injected in the wind zone and lose a part of their energy on ICS process. These ones, which reach the shock region, are isotropised and produce synchrotron and inverse Compton radiation at large solid angle. The observer is located at the angle θ measured from the direction defined by the centres of the stars. On the right - The optical depths for electrons on ICS process (full curves) and on absorption of gamma-ray photons (dashed curves) are shown as a function of electron energy for two locations of the observer $\theta = 55°$ and $125°$.

$\eta = (L_{rot}/c)/(\dot{M}v)$, which is the ratio of momentum carried by the pulsar wind and the momentum of the stellar wind determined by the mass loss rate $\dot{M}$, and the velocity of the wind v. The distance from the pulsar to the shock is given by $\rho = \sqrt{\eta}D/(1+\sqrt{\eta})$, where D is the seperation between the stars. The system is viewed from the Earth at the angle θ which changes drastically close to the periastron between $55°$ and $125°$ (see Fig. 1). The value of η for this binary is estimated on $\sim 0.01 - 0.02$, applying a typical mass loss rate for Be stars 10^{-7} $M_\odot$ yr^{-1} [7] and the wind velocity 2×10^8 cm s^{-1} [8]. Significantly lower values η are also likely due to the collimation of the stellar wind in the equatorial plane.

We assume that relativistic electrons (positrons) are injected by the pulsar into the wind zone. In order to check if these electrons can efficiently interact with radiation of the massive star, we calculate the optical depth for electrons moving rectilinearly from the pulsar and the optical depth for $e^\pm$ pair creation by the γ-ray photon. Fig. 1 show that the collisions of electrons with soft photons can be quite frequent, especially in the Thomson regime (the optical depth up to a few for $\theta = 125°$), but the probability of photon conversion into the $e^\pm$ pair is much lower. Therefore, contrary to other massive binaries (e.g. Cyg X-3, LSI 61°303 [9]) the effects of cascading in the radiation of the massive star are not important in the case of PSR

B1259-63/Be star (SS2883).

3. Gamma-rays from inverse Compton Scattering

We consider three different models for injection of leptons by the pulsar: (1) monoenergetic injection from the inner magnetosphere into the wind zone with energies 10^6 MeV; (2) injection from the inner pulsar magnetosphere with the spectrum given by Hibshman & Arons [10]; and (3) as in (2) but with additional acceleration in the pulsar wind zone as postulated by Contopoulos & Kazanas [11]. These leptons move at first radially in the wind zone and then are isotropized in the double shock structure created in the interaction of the pulsar and stellar winds.

3.1. The wind region

Leptons injected from the inner pulsar magnetosphere move radially in the wind with negligible synchrotron energy losses. However they comptonize radiation from the massive star. We have calculated the inverse Compton γ-ray spectra for the mentioned above initial distributions of leptons and for two locations of the pulsar on the orbit, defined by the angles $\theta = 55°$ and $125°$ (see full histograms in Fig. 2). The results are shown for two values of $\eta = 0.01$ and 0.001 (thin and thick lines). Note that there is no efficient γ-ray production for the angle $\theta = 55°$, when the observer is on the same side of the massive star as the pulsar, since the angles of interaction between leptons and soft photons are large. For $\theta = 125°$, the γ-ray spectrum extends between MeV up to ~ 1 TeV for the case of lepton injection with the spectrum. The intensity is higher for $\eta = 0.01$ than $\eta = 0.001$, since in the first case the distance on which leptons propagate in the wind zone is longer.

3.2. The shock region

Leptons, after losing (or gaining) energy during propagation in the wind zone, fall onto the shock in the pulsar wind. They are isotropised there losing energy on the synchrotron and inverse Compton processes.

We calculate the equilibrium spectrum of electrons in the shock region, N_e, by assuming that they lose energy on synchrotron and IC processes and can also escape from this region with the characteristic escape time T_{esc}. Since the synchrotron process dominate in most of the considered cases, N_e can be obtained by solving the continuity equation under stady state

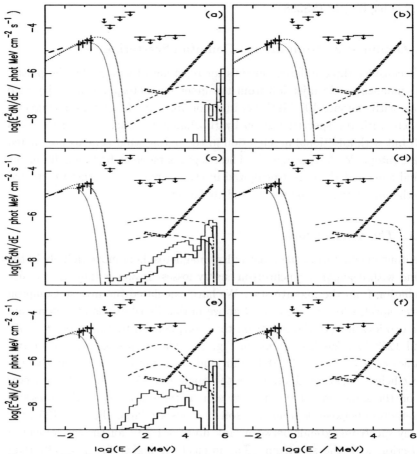

Figure 2. The photon spectra from synchrotron (dotted lines) and ICS (dashed lines) processes are calculated for two values of $\eta = 0.001$ (thicker lines) and $\eta = 0.01$ (thiner lines), and three different models for injection of electrons by the pulsar: monoenergetic electrons (upper figures), electrons with the spectrum of Hibschman & Arons (middle figures), and electrons with the spectrum and additional acceleration in the pulsar wind zone (bottom figures). The left and right figures are for the observation angles $\theta = 125°$ and $55°$, respectively. The synchrotron and ICS spectra from electrons, propagating in the shock, are marked by the dotted and dashed lines, respectively. The full histograms show the ICS spectra produced by leptons in the pulsar wind. The experimental results are from ASCA (thick dashed line), OSSE (error boxes), and the upper limits from COMPTEL and EGRET. The sensitivity limit of the GLAST detector is marked by double dot-dashed lines.

conditions,

$$\frac{\partial}{\partial E}\left(\frac{dE}{dt} \cdot N_{\mathrm{e}}\right) = \frac{N_{\mathrm{e}}}{T_{\mathrm{esc}}} + Q_{\mathrm{e}}, \qquad (1)$$

where Q_e $[e^{\pm} \text{ s}^{-1}]$ is the injection spectrum of electrons by the pulsar which arrive to the shock region. We use the solution of Eq. 1

$$N_e(E) = \left(\frac{dE}{dt}\right)^{-1} \int_E^{E_{\max}} Q_e(E')dE', \qquad (2)$$

which is valid for electrons with energies fulfiling the condition $T_{\text{cool}} \ll T_{\text{esc}}$. The lower limit on T_{esc} can be estimated by L/v_{esc}, where L is the distance between the pulsar and the shock. The escape velocity of $e^{\pm}$ pairs after the wind shock, v_{esc}, is assumed equal to one third of the velocity of light. The cooling time of $e^{\pm}$ pairs is estimated using the energy loss formulae on synchrotron and ICS processes in the Thomson regime, $T_{\text{cool}} = [(A_{\text{syn}} + A_{\text{ICS}})E_e B^2]^{-1}$. A_{syn} and A_{ICS} are the coefficients in formulae for the synchrotron and the ICS energy losses $(dE/dt = AE^2)$. Note that in considered model the synchrotron losses usually dominate over ICS losses. The condition, $T_{\text{cool}} \ll T_{\text{esc}}$, allows us to estimate the minimum energy of $e^{\pm}$ pairs, $E_{\min} \approx c/(3AL)$, to which they can cool efficiently before escaping. For example, for the value of $\eta = 0.001$, the magnetic field at the shock region in the direction towards the observer located at the angle $\theta = 125°$ is equal to $B_{\text{sh}} \approx 21.5$ G, provided that the parameter describing the magnetization of the wind $\sigma = L_{\text{mag}}/L_{\text{rot}} = (L_{\text{rot}} - L_{e^{\pm}})/L_{\text{rot}} \approx 1$ at the shock distance from the pulsar, where L_{mag} and $L_{e^{\pm}}$ is the energy carried in the wind in the form of the magnetic field and $e^{\pm}$ pairs, respectively. In fact if the electrons cool eficiently then $L_{e^{\pm}} \approx L_X \sim 10^{34}$ erg s^{-1} which is much less than $L_{\text{rot}} = 8 \times 10^{35}$ erg s^{-1}. For this value of B_{sh}, we obtain $E_{\min} \approx 3$ GeV, applying $L = 2.5 \times 10^{12}$ cm, and $A_{\text{ICS}} \ll A_{\text{syn}} \approx 1.2 \times 10^{-6} \text{MeV}^{-1}\text{s}^{-1}$. Therefore ICS γ-ray spectra, calculated from the equilibrium spectrum of electrons, can be valid for photon energies above $<\varepsilon> \gamma_{\min}^2 \approx 100$ MeV, where $<\varepsilon>$ is a typical energy of soft photons from the massive star and $\gamma_{\min} = E_{\min}/m_e$.

We calculate the synchrotron and γ-ray spectra produced by $e^{\pm}$ pairs with energies above $E_{\min}$. The results for three considered electron injection models are shown in Fig. 2, after normalization of the synchrotron spectrum to the X-ray spectrum observed by the ASCA [12] and the OSSE detectors[1]. From the fitting to the X-ray spectrum we find that only for $\eta = 0.001$ the spectrum can extend to the OSSE energy range. The γ-ray spectra from the shock region should be detected by the planned GLAST experiment for the case of $e^{\pm}$ injection from the pulsar magnetosphere with the spectrum given by Hibschman & Arons [10], if additional reacceleration of pairs in the pulsar wind zone occurs. Without reacceleration, the predicted fluxes are on the level of GLAST detectability. In fact the γ-ray emission from the wind zone might be also detected by the GLAST if the collimation of these photons is much stronger than the collimation of photons produced in the shock region. Unfortunatelly, we have no information on the angular distribution of $e^{\pm}$ pairs injected from the pulsar into the wind zone which makes the absolute determination of photon fluxes produced in the wind zone impossibile.

4. Conclusion

We estimated the γ-ray flux from the PSR B1259-63/Be binary system at energies above ~ 100 MeV for three models of electron (positron) injection by the pulsar. It is found that this source should be observed by the GLAST detector in the case of injection of electrons from the inner pulsar magnetosphere with the spectrum of Hibschman & Arons [10] if additional reacceleration process takes place in the pulsar wind zone. The system might be only marginally detected if the reacceleration does not operate, and it should not be detected if monoenergetic electrons with energy 10^6 MeV are injected by the pulsar as considered in some previous works [5,6].

Acknowledgements

We would like to thank the Organizers of the School for the invitation and their support. This work has been also supported by the grants: KBN 5P03D02521 and UŁ No. 505/275.

References

1. Grove, J.E. et al., *ApJ* **447**, L113 (1995).
2. Tavani, M. et al., *A& A* **120**, 221 (1996).
3. Tavani, M., Arons, J., *ApJ* **477**, 439 (1997).
4. Kirk, J.G. et al., *Astropart.Phys.* **10**, 31 (1999).
5. Ball, L., Kirk, J.G., *Astropart.Phys.* **12**, 335 (2000).
6. Ball, L., Dodd, J., *Pub.Astr.Soc.Aust.* **18**, 98 (2000).
7. Waters, L.B.F.M. et al., *A& A* **198**, 200 (1988).
8. Johnston, S. et al., *MNRAS* **326**, 643 (2001).
9. Bednarek, W., *A& A* **322**, 523 (1997); *A& A* **362**, 646 (2000).
10. Hibschman, J.A., Arons, J., *ApJ* **560**, 871 (2001).
11. Contopoulos, I., Kazanas, D., *ApJ* **566**, 336 (2002).
12. Kaspi, V.M. et al., *ApJ* **453**, 424 (1995).

YOUNG COMPACT OBJECTS IN THE SOLAR VICINITY

S. B. POPOV

Sternberg Astronomical Institute
Universitetski pr. 13, Moscow 119992, Russia
E-mail: polar@sai.msu.ru
University of Padova
via Marzolo 8, Padova 35131, Italy

M. E. PROKHOROV

Sternberg Astronomical Institute
Universitetski pr. 13, Moscow 119992, Russia

M. COLPI

University of Milano-Bicocca
Piazza della Scienza 3, Milano 20126, Italy

R. TUROLLA

University of Padova
via Marzolo 8, Padova 35131, Italy

A. TREVES

University dell'Insubria
via Vallegio 11, Como 22100, Italy

We present Log N – Log S distribution for close young isolated neutron stars. On the basis of Log N – Log S distribution it is shown that seven ROSAT isolated neutron stars (if they are young cooling objects) are genetically related to the Gould Belt. We predict, that there are about few tens unidentified close young isolated neutron stars in the ROSAT All-Sky Survey. In the aftermath of relatively close recent supernova explosions (1 kpc around the Sun, few Myrs ago), a few black holes might have been formed, according to the local initial mass function. We thus discuss the possibility of determining approximate positions of close-by isolated black holes using data on runaway stars and simple calculations of binary evolution and disruption.

1. Introduction

Neutron stars (NSs) and black holes (BHs) are among the most interesting astrophysical objects. Usually NSs are observed as radio pulsars or as accreting objects in close binaries. Stellar mass BHs normally are observable if they also accrete matter from companions. Here we will focus on isolated NSs (which can show no radio pulsar activity) and BHs.

An isolated NS can be relatively bright (in soft X-rays especially) due to its thermal emission during the first Myr of its life, when it is still hot ($T > 10^6$ K). Such objects are observed in the Solar vicinity and in supernova (SN) remnants[1].

An isolated BH can be found due to accretion of the interstellar medium (NSs can also be found in such a way, see Treves et al.[2]), or due to the microlensing effect[3].

In this paper we will construct Log N — Log S distribution for young close isolated NSs and discuss how one can estimate an approximate positions of close young isolated BHs.

2. Close young neutron stars

In this section we discuss young close isolated NSs. Partly we are based on the results published in Popov et al.[4] and partly we present our recent results[5].

2.1. Log N — Log S distribution

Main components of our model are[5]: spatial distribution of NS progenitors, NS formation rate, NS cooling histories, and model of interstellar absorption (i.e. interstellar medium (ISM) distribution). In addition we calculate dynamical evolution of NSs in the galactic potential. In brief our model can be described in the following way: NSs are born in the Galactic plane and in the Gould Belt; at birth they receive a kick velocity; then we follow the evolution of NSs in the Galactic potential; while a NS moves in the Galaxy we calculate (with some time step) ROSAT count rate basing on cooling curves and assumed model of interstellar absorption.

We assume that NSs are born in the Galactic disk and in the Gould Belt. NSs are considered to be born with a constant rate. 20 NSs per Myr come from the Gould Belt, and 250 NSs per Myr are uniformly distributed in the Galactic plane with a limiting distance from the Sun 3 kpc. The Gould Belt is modeled as a disk with 500 pc radius and inclination to the

Galactic plane equal to 18°. Center of the disk is situated at 100 pc from the Sun in the Galactic anticenter direction. In the center of the disk there is a hole with a radius 150 pc, where NSs are not born (see a review by Pöppel[6] for detailed description of the Gould Belt, and Torra et al.[7] for a shorter one).

To calculate cooling of NSs we use the data obtained by Sankt-Petersburg group (see recent data in Kaminker et al.[8], and a review in Yakovlev et al.[9]). We use cooling curves for NSs of masses from $1.1\,M_\odot$ to $1.8\,M_\odot$ with a step $0.1\,M_\odot$. Mass spectrum is assumed to be flat ("realistic" spectrum with sharp maximum around 1.35-1.4 solar masses gives nearly the same result). Curves take into account all processes of neutrino emission. Equation of state used in Kaminker et al.[8] was introduced in Prakash et al.[10]. It is *Model I* of Prakash et al. for symmetry energy and compression modulus of saturated nuclear matter, K, is equal to 240 MeV. Maximum NS mass in that model is $1.977\,M_\odot$. Neutron superfluidity in the crust and core is ignored, as far as it does not influenced the final results significantly. Calculations for each NS are truncated when its temperature falls down to 100 000 K; it corresponds to a NS's age 4.25 Myrs for the lightest NSs ($M = 1.1\,M_\odot$) and less for more massive objects.

As far as we expect our NSs to emit most of their luminosity at soft energies $\sim 20 - 200$ eV (which corresponds to temperatures about 10^5–10^6 K) we have to take into account interstellar absorption. Absorption is a very important feature of our model. Any attempts to estimate the amount of observable cooling isolated NSs using unabsorbed flux *greatly* overestimate this number.

Main results are presented in the Fig. 1. We present a curve for NSs born both in the Gould Belt and the Galactic disk and a curve only for the disk to show the relative importance of the Gould Belt. All curves are referring for the whole sky.

As can be seen the Gould Belt *alone* can explain all observed points. Absorption and flat geometry of NSs distribution naturally explain very flat (< -1) Log N – Log S.

Our calculations show, that there are about few dozens of unidentified close isolated NSs in ROSAT All-Sky Survey (> 0.015 cts s^{-1}) depending on parameters of the model, also there can be few unidentified ROSAT isolated NSs (RINSs) with fluxes > 0.1 cts s^{-1} at low Galactic latitudes (see also Schwope et al.[11]). Most of objects should be observed at $\sim \pm 20°$ from the Galactic plane towards the directions of lower absorption. Some of them can have counterparts among unidentified gamma-ray sources (also connected with the Gould Belt, see Grenier[12]). Identification of these objects can be important for choosing a correct cooling model and for determination of mass spectrum of NSs.

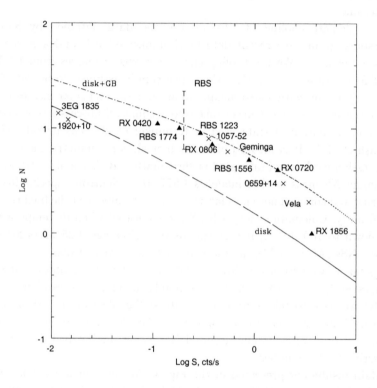

Figure 1. All-sky Log N - Log S distribution. Black triangles — seven RINSs. Crosses — Geminga, "three musketeers", 1929+10 and 3EG J1835. We also show the ROSAT Bright Sources (RBS) limit (Schwope et al.[11]). Upper curve: NSs born in the Gould Belt and in the Galactic disk ($r_{disk} = 3$ kpc, total birth rate 270 Myr^{-1}). Lower curve: only NSs born at the Galactic disk ($r_{disk} = 3$ kpc, birth rate 250 Myr^{-1}).

2.2. *Census of close young isolated neutron stars of different types*

At the present moment it is known about 20 NSs satisfying the following criteria: age less than 4.25 Myrs, distance from the Sun less than 1 kpc (see the Table). It includes: the "magnificent seventh" (seven RINSs), Geminga and the Geminga-like object 3EG J1835, "three musketeers" (Vela, 0656+14,

1055-52), 1929+10 and seven young radio pulsars, which are not detected in X-rays.

We see, that among local young isolated NSs we have normal radio pulsars, pulsars beams of which do not pass the Earth (Geminga and probably 3EG J1835), and RINSs. Also we can expect, that around us (inside 1 kpc) there are about one hundred of isolated NSs younger than $\sim$4 Myr (some of NSs born inside 1 kpc of course can leave this volume in several Myrs). These NSs are not detected as radio pulsars, but tens of them can be identified in ROSAT data as dim sources (others are too old to be hot enough). The beaming effect can be responsible only for part of these undetected (as radio pulsars) young NSs (about 50-70% of young pulsars are not visible from Earth[13]), and most of RINSs should be really radio silent (it is difficult to construct a model in which one observes X-ray pulsations, pure black body spectrum and no signs of radio emission from a close off-beam pulsar, also one has to take into account long periods of four RINSs). It gives strong support to the arguments by Gotthelf and Vasisht[14], that "at least half of the observed young neutron stars follow an evolutionary path quite distinct from that of the Crab pulsar".

2.3. Alignment

An interesting feature of RINSs population is an existence of periods about 5-20 seconds (typical for magnetars) for four objects and their absence for the rest three.

Vasily Beskin (2001, private communication) suggested, that it can happen due to alignment of magnetic and spin axis (see for example Tauris, Manchester[13] for a recent discussion on alignment). Alignment is a process which also leads to a "period freezing" and low pulsation fraction (see Haberl, Zavlin[18] for the data on pulsation fraction in RINSs).

However in the case of coolers alignment should operate on short timescale — the timescale of NS cooling ($\sim$ 1 Myr). As far as for radio pulsars timescale of alignment is about 10 Myrs or longercitetm, we think, that it is unlikely that this mechanism is responsible for RINSs distribution over pulsation fraction, otherwise one has to assume that RINSs form a population separate from normal radio pulsars. For example, if we assume, that alignment timescale is $\propto (\Omega_0^2 \cos^2\alpha_0\, B_0^2)^{-1}$ (it comes from magnetodipole spin-down and condition $\Omega_0 \cos\alpha_0 = \Omega \cos\alpha$; also there can be other assumptions about alignment process, but at the moment there is no generally accepted model), then we have to assume that RINSs have dif-

Table 1. Local ($r < 1$ kpc) population of young (age < 4.25 Myrs) isolated neutron stars.

Object name	Period, s	CR[a], cts/s	$\dot{P}$ /10^{-15}	Dist., kpc	Age[b], Myrs	Ref.
RX J1856.5-3754	—	3.64	—	0.117[e]	~ 0.5	2,15
RX J0720.4-3125	8.37	1.69	$\sim 30 - 60$	—	—	2,16
RX J1308.6+2127	5.15	0.29	$\sim 10^4$	—	—	2,17
RX J1605.3+3249	—	0.88	—	—	—	2
RX J0806.4-4123	11.37	0.38	—	—	—	2,18
RX J0420.0-5022	22.7	0.11	—	—	—	2
RX J2143.7+0654	—	0.18	—	—	—	19
PSR B0633+17	0.237	0.54[d]	10.97	0.16[e]	0.34	20
3EG J1835+5918	—	0.015	—	—	—	21
PSR B0833-45	0.089	3.4[d]	124.88	0.294[e]	0.01	20,22,23
PSR B0656+14	0.385	1.92[d]	55.01	0.762[f]	0.11	20,23
PSR B1055-52	0.197	0.35[d]	5.83	$\sim 1^c$	0.54	20,23
PSR B1929+10	0.227	0.012[d]	1.16	0.33[e]	3.1	20,23
PSR J0056+4756	0.472	—	3.57	0.998[f]	2.1	23
PSR J0454+5543	0.341	—	2.37	0.793[f]	2.3	23
PSR J1918+1541	0.371	—	2.54	0.684[f]	2.3	23
PSR J2048-1616	1.962	—	10.96	0.639[f]	2.8	23
PSR J1848-1952	4.308	—	23.31	0.956[f]	2.9	23
PSR J0837+0610	1.274	—	6.8	0.722[f]	3.0	23
PSR J1908+0734	0.212	—	0.82	0.584[f]	4.1	23

[a] ROSAT count rate
[b]) Ages for pulsars are estimated as $P/(2\dot{P})$,
for RX J1856 the estimate of its age comes from kinematical considerations.
[c]) Distance to PSR B1055-52 is uncertain (~ 0.9-1.5 kpc)
[d]) Total count rate (black body + non-thermal)
[e]) Distances determined through parallactic measurements
[f]) Distances determined with dispersion measure

ferent (from radio pulsars) distribution in B_0 or/and α_0, and the observed situation is just a selection effect. For example they can be a sample selected by relatively high surface temperatures, which for young NSs means by low masses.

2.4. Possible correlation between initial magnetic fields and masses of neutron stars

If any part of RINSs (with or without observed pulsations) is explained by high magnetic field, then one has to explain why the fraction of magnetars is so high ($\sim 50\%$). It can be explained as a selection effect if NSs with higher magnetic fields are hotter. It can be so, if they have (on average) lower masses (below some value, which is about 1.3 $M_\odot$ but depends on the equation of state, direct URCA process are not allowed, and NS cooling is much slower).

The possible correlation can be explained if more massive NSs get their additional masses in the process of fall-back. In that case their magnetic field can be significantly suppressed[24], so more massive NSs should have lower initial magnetic fields.

Also there is a question if the matter is swept out, or it is accumulated close to the magnetospheric boundary, and then accreted by the compact object. But numerical estimates[25,26] suggest, that significant part of matter is swept from the propelling NS.

Vice versa, strong initial magnetic field together with fast rotation can prevent strong fall-back (it is especially possible if magneto-rotational mechanism of supernova explosion is valid, see Bisnovatyi-Kogan[27] and Prokhorov, Postnov[28]), and may be it can also lead to long initial spin periods.

Different crust thickness and temperature of NSs of different masses can give additional correlations between mass and magnetic field, the same can be said about different properties at cores of NSs of different masses (for example, deconfinement in central parts of massive NSs[29]).

Correlations should be different in different scenarios of SN explosion. Study of such correlations can give additional possibilities to select a correct scenario.

To summarize, we can expect correlations between magnetic fields and masses of NSs, and observations of RINSs can be of great importance here. Future determination of RINSs parallaxies and proper motions can help to reconstruct their kinematical history and derive their ages. It can give a clue to their mass determination basing on cooling curves (see Kaminker et al.[8]). Our results suggest, that the fraction of low massive NSs (M <1.3-1.4 $M_\odot$) can not be small, but on the other hand there should be a room for massive NSs, because other wise we overpredict the number of bright objects.

3. Close young isolated black holes

In this section we are based on the results published in Popov et al.[4] and Prokhorov, Popov[30].

SNae explosions produce not only NSs, but also BHs. Usually it is accepted that BHs are one order of magnitude less abundant than NSs. This estimate comes from the critical mass for BH formation. If this mass is about 35 $M_\odot$ then the fraction of BHs is about 10%. Having dozens of SNae in the close solar vicinity during the last 10 Myr we can expect several BHs to have formed during the same period in the solar neighborhood.

In 700 pc around the Sun 56 runaway stars are known[31]. Only few of them can result from star-star interactions. Others are products of SNae explosions in binary systems. If the above considerations are correct we can expect about 5 BHs formed in about 50 disrupted binaries.

Close massive runaway stars give us a chance to calculate an approximate positions of close-by young isolated BHs. Among runaway stars we can distinguish the most massive: λ Cep, ζ Pup, HIP 38518 and ξ Per (see Hoogerwerf et al.[31]). Their masses are larger than $\sim 33\ M_\odot$. It means, that the companion (actually the *primary* in the original binary) was even more massive on the main sequence stage. So, the most likely product of the explosion of such a massive star should be a BH.

If the present velocities of runaway stars are known, one can estimate their ages and places of birth. This has been done by Hoogerwerf et al.[31]. To calculate the present position of a BH we have to know the binary parameters, i.e., the masses of stars before the explosion, the BH mass, the eccentricity of the orbit before the explosion, the orbit orientation, and finally the kick velocity of the BH. Some parameters can be inferred from the observation of the secondary star. We can assume a zero kick velovity for BHs and zero orbital eccentricity. Other parameters should be varied within assumed ranges (see details in Prokhorov, Popov[30].

We calculated approximate positions of isolated BHs for four systems mentioned above and estimated erroe boxes where these BHs can be found. For ξ Per and ζ Pup we obtained not very big error boxes inside each of which only one unidentified EGRET source is known. We suggest, that these objects can be young isolated BHs. For two other systems (HIP 38518 and λ Cep) error boxes of the present day BH localization are very large (even with our assumption of zero kick velocity).

4. Conclusions

We conclude that the seven radio-quiet ROSAT isolated NSs can be connected with recent SNae explosions, which produced nearby runaway stars and peculiar features in the local ISM including the Local Bubble. Relatively high local spatial density of young NSs is due to large number of massive progenitors in the Belt. As far as there are few young radio pulsars in the Solar vicinity many (about 1/2 or more) of isolated NSs should be radio quiet. We find that in ROSAT All-Sky Survey it can be about few tens of unidentified RINSs. Also there can be few unidentified RINSs at fluxes > 0.1 cts s^{-1} at low Galactic latitudes.

Massive runaway stars can be used to estimate approximate positions of young close-by isolated BHs. We do the correspondent calculations and estimate present day positions of four BHs. For two of them error boxes are not too large.

5. Acknowledgments

We want to thank D.G. Yakovlev for the data on cooling curves and comments on them, and V.S. Beskin, Matteo Chieregato, Andrea Possenti and Luca Zampieri for discussions.

The work of S.P. was supported by the RFBR grant 02-02-06663. The work of M.P. — by RFBR 01-15-99310.

S.P. thanks Universities dell'Insubria, Milano-Bicocca, Padova and the Organizers of the School for hospitality.

References

1. W. Becker and G. Pavlov, in: "The Century of Space Science", eds. J. Bleeker, J. Geiss and M. Huber, Kluwer Academic Publishers (in press) (astro-ph/0208356) (2002).
2. A. Treves, R. Turolla, S. Zane and M. Colpi, *PASP* **112**, 297 (2000).
3. E. Agol, M. Kamionkowski, L.V.E. Koopmans and R.D. Blandford, *ApJ* **576**, L131 (2002).
4. S.B. Popov, M.E. Prokhorov, M. Colpi, A. Treves and R. Turolla, *Gravitation & Cosmology* (in press) (astro-ph/0201030) (2002).
5. S.B. Popov, M. Colpi, M.E. Prokhorov, A. Treves and R. Turolla, in preparation.
6. W. Pöppel, *Fund. Cosm. Phys.* **18**, 1 (1997).
7. J. Torra, D. Fernández and F. Figueras, *A&A* **359**, 82 (2000).
8. A.D. Kaminker, D.G. Yakovlev and Y.O. Gnedin, *A&A* **383**, 1076 (2002).
9. D.G. Yakovlev, K.P. Levenfish and Yu.A. Shibanov, *Physics-Uspekhi* **42**, 737 (1999).

10. M. Prakash, T.L. Ainsworth and J.M. Lattimer, *Phys. Rev. Lett.* **61**, 2518 (1988).

11. A.D. Schwope, G. Hasinger, R. Schwarz, F. Haberl and M. Schmidt, *A&A* **341**, L51 (1999).

12. I.A. Grenier, *A&A* **364**, L93 (2000).

13. T.M. Tauris and R.N. Manchester, *MNRAS* **298**, 625 (1998).

14. E.V. Gotthelf and G. Vasisht, in: Proceedings of IAU Coll. 177 "Pulsar astronomy - 2000 and beyond", eds. M. Kramer, N. Wex, and N. Wielibinski, ASP Conf. Series **202**, 699 (2000).

15. D.L. Kaplan, M.H. van Kerkwijk and J. Anderson, *ApJ* **571**, 447 (2002).

16. S. Zane, F. Haberl, M. Cropper, V. Zavlin, D. Lumb, S. Sembay and C. Motch, *MNRAS* **334**, 345 (2002).

17. V. Hambaryan, G. Hasinger, A.D. Schwope and N.S. Schulz, *A&A* **381**, 98 (2001).

18. F. Haberl and V.E. Zavlin, *A&A* **391**, 571 (2002).

19. L. Zampieri et al., *A&A* **378**, L5 (2001).

20. W. Becker and J. Trümper, *A&A* **326**, 682 (1997).

21. N. Mirabal and J.P. Halpern, *ApJ* **547**, L137 (2001).

22. G.G. Pavlov, V.E. Zavlin, D. Sanwal, V. Burwitz, and G.P. Garmire, *ApJ* **552**, L129 (2001).

23. ATNF Pulsar Catalogue, http://wwwatnf.atnf.csiro.au/research/pulsar/catalogue/

24. D. Page, U. Geppert and T. Zannias, *Mem. Soc. Astr. It.* **69**, 1037

25. Yu.M. Toropin, O.D. Toropina, V.V. Savelyev, M.M. Romanova, V.M. Chechetkin and R.V.E. Lovelace, *ApJ* **517**, 906 (1999).

26. M.M. Romanova, O.D. Toropina, Yu.M. Toropin and R.V.E. Lovelace, *ApJ* (in press) (2003).

27. G.S. Bisnovatyi-Kogan, *AZh* **14**, 652 (1970).

28. M.E. Prokhorov and K.A. Postnov, *Odessa Astr. Publ.* **14**, 78 (astro-ph/0110176) (2001).

29. A. Sedrakian and D. Blaschke, *Astrofizika* **45**, 203 (hep-ph/0205107) (2002).

30. M.E. Prokhorov and S.B. Popov, *Astronomy Letters* **28**, 542 (2002).

31. R. Hoogerwerf, J.H.J. de Bruijne and P.T. de Zeeuw, *A&A* **365**, 49 (2001).

COSMIC RAYS

COSMIC RAY DIFFUSION IN THE DYNAMIC MILKY WAY: MODEL, MEASUREMENT AND TERRESTRIAL EFFECTS

NIR J. SHAVIV

Racah Institute of Physics, Hebrew University
Jerusalem, 91904, Israel
E-mail: shaviv@phys.huji.ac.il

We study the problem of cosmic ray diffusion in the Milky Way while considering the galactic spiral arm dynamics. Once this new ingredient is added to cosmic ray diffusion models, we find that the cosmic ray flux reaching the solar system should increase periodically with each passage of a spiral arm. We continue with studying the meteoritic exposure ages and find that a record of past crossings of the arms can be extracted. We then briefly review recent evidence which links cosmic rays to climatic change on Earth. Given the suspected link, we argue that spiral arm passages are responsible for the periodic appearance of ice age epochs on Earth. This hypothesis is supported with a clear correlation between ice age epochs and the meteoritic record and also between longer term activity in the Milky Way and glacial activity on Earth. More speculatively, the last such passage may have been partially responsible for the demise of the dinosaurs.

1. Introduction

Most cosmic rays (CRs), with the possible exception of extremely high energies, are believed to originate from supernova (SN) remnants[1,2]. This is also supported by direct observational evidence[3]. Moreover, most SNe in spiral galaxies like our own are those which originate from massive stars, thus, they predominantly reside in the spiral arms, where most massive stars are born and shortly thereafter explode as SNe[4]. Indeed, high contrasts in the non-thermal radio emission are observed between the spiral arms and disks of external galaxies. Assuming equipartition between the CR energy density and the magnetic field, a CR energy density contrast can be inferred. In some cases, a lower limit of 5 can be placed for this ratio[5]. Thus, while modeling the diffusion of cosmic rays in the galaxy, we should take into consideration the non-axisymmetric nature of the cosmic ray source.

We first construct a diffusion model which considers the spiral arms as the primary source of cosmic rays. We then continue with a study of the exposure ages of meteors which hide within them the history of the cosmic ray flux. As we shall soon see, this record registered the past half dozen spiral arm passages.

Quite unrelated, or so it may seem at first, there are indications that solar activity is responsible for at least some climatic variability on time scales ranging from days to millennia[6-15]. This link appears to be related to our topic of cosmic rays, since circumstantial evidence indicates that the observed solar-climate link could be though solar wind modulation of the cosmic rays flux (CRF)[13-15]. This is not unreasonable considering that the CRF governs the tropospheric ionization rate[16]. In particular, if was found that the low latitude cloud cover variations are in sync with the variable CRF reaching Earth, while the inverse of both signals lag by typically half a year after the solar activity[17].

Thus, if indeed climatic effects arise from *extrinsic CRF* variability induced by solar wind modulation, then also the much larger *intrinsic* variations in the CRF reaching the solar system should be a source of climatic effects. In particular, low altitude clouds have a net cooling effect, such that we should expect a colder climate while we are in the cosmic ray wake of spiral arms. This was shown to be supported by various data[18], and we bring here a more elaborate description. A complete unabridged version can be found in Shaviv[19] with the exception of the possible relation between cosmic rays and the demise of the dinosaurs, which appears here for the first time.

2. Diffusion in a Dynamic Galaxy

To estimate the variable CRF expected while the solar system orbits the galaxy, we construct a simple diffusion model which considers that the CR sources reside in the Galactic spiral arms. We expand the basic CR diffusion models (e.g., ref. [2]) to include a source distribution located in the Galactic spiral arms. Namely, we replace a homogeneous disk with an arm geometry as given by Taylor & Cordes[20], and solve the time dependent diffusion problem (see fig. 1). To take into account the "Orion spur"[21], in which the Sun currently resides, we add an arm "segment" at our present location. Since the density of HII regions in this spur is roughly half of the density in the real nearby arms[21], we assume it to have half the typical CR sources as the main arms. We integrate the CR sources assuming a diffusion coefficient of $D = 10^{28} \text{cm}^2/\text{sec}$, which is a typical value obtained in diffusion models for the CRs[2,22,23]. We also assume a halo half-width of 2kpc, which again is a typical value obtained in diffusion models[2], but more importantly, we reproduce with it the ^{10}Be survival fraction[24]. Thus, the only free parameter in the model is the angular velocity $\Omega_\odot - \Omega_p$ around the Galaxy of the solar system *relative* to the Spiral arm pattern speed, which is later adopted using observations. Results of the model are depicted in fig. 2. For the nominal values chosen in our diffusion model and the particular pattern speed which will soon be shown to fit various data, the expected CRF changes from about 25% of the current day CRF to about 135%. Moreover, the average CRF obtained in units of today's CRF is 76%. This is consistent with measurements showing that the average CRF over the period 150-700 Myr before present (BP), was about 28% lower than the current day CRF[25].

Interestingly, the temporal behavior is both skewed and lagging after the spiral arm passages. The lag arises because the spiral arms are defined through the free electron distribution. However the CRs are emitted from SNe which on average occur roughly 15 Myr after the average ionizing photons are emitted. The skewness arises because it takes time for the CRs to diffuse after they are emitted. As a result, before the region of a given star reaches an arm, the CR density is low since no CRs were recently injected in that region and the sole flux is of CRs that succeed to diffuse to the region from large distances. After the region crosses the spiral arm, the CR density is larger since locally there was a recent injection of new CRs which only slowly disperse. This typically introduces a 10 Myr lag in the flux, totaling about 25 Myr with the SN delay. This lag is actually observed in the synchrotron emission from M51, which shows a peaked emission trailing the spiral arms[1].

The spiral pattern speed of the Milky Way has not yet been reasonably determined through astronomical observations. Nevertheless, a survey of the literature[19] reveals that almost all observational determinations cluster either around $\Omega_\odot - \Omega_p \approx 9$ to 13 (km s^{-1})/kpc or around $\Omega_\odot - \Omega_p \approx 2$ to 5 (km s^{-1})/kpc. In fact, one analysis[26] revealed that both $\Omega_\odot - \Omega_p = 5$ or 11.5 (km s^{-1})/kpc fit the data. However, if the spiral arms are a density wave[27], as is commonly believed[28], then the observations of the 4-arm spiral structure in HI outside the Galactic solar orbit[29] severely constrain the pattern speed to $\Omega_\odot - \Omega_p \gtrsim 9.1 \pm 2.4$ (km s^{-1})/kpc, since the four arm density wave spiral cannot extend beyond the outer 4 to 1 Lindblad resonance[19]. We therefore expect the spiral pattern speed obtained to coincide with one of the two aforementioned ranges, with a strong theoretical argumentation favoring the first range.

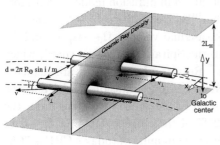

Figure 1. The components of the diffusion model constructed to estimate the Cosmic Ray flux variation. We assume for simplicity that the CR sources reside in Gaussian cross-sectioned spiral arms and that these are cylinders to first approximation. This is permissible since the pitch angle i of the spirals is small. The diffusion takes place in a slab of half width l_H, beyond which the diffusion coefficient is effectively infinite.

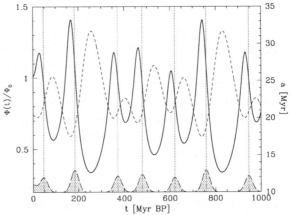

Figure 2. The cosmic-ray flux variability and age as a function of time for $D = 10^{28}$ cm^2/s and $l_H = 2$ kpc. The solid line is the cosmic-ray flux, the dashed line is the age of the cosmic rays as measured using the Be isotope ratio. The shaded regions at the bottom depict the location, relative amplitude (i.e., it is not normalized) and width of the spiral arms as defined through the free electron density in the Taylor and Cordes model. The peaks in the flux are lagging behind the spiral arm crosses due to the SN-HII lag. Moreover, the flux distribution is skewed towards later times.

3. The cosmic ray flux variability from meteorites

To validate the above prediction, that the CRF varied periodically, we require a direct "historic" record from which the actual time dependence of the CRF can be extracted. To find this record, we take a compilation of 74 Iron meteorites which were $^{41}K/^{40}K$ exposure dated[30]. CRF exposure dating (which measures the duration a given meteorite was exposed to CRs) assumes that the CRF history was constant, such that a linear change in the integrated flux corresponds to a linear change in age. However, if the CRF is variable, the apparent exposure age will be distorted. Long periods during which the CRF is low would correspond to slow increases in the exposure age. Consequently, Fe meteorites with real ages within this low CRF period would cluster together since they will not have significantly different integrated exposures. Periods with higher CRFs will have the opposite effect and spread apart the exposure ages of meteorites. To avoid real clustering in the data (due to one parent body generating many meteorites), we remove all occurrences of Fe meteorites of the same classification that are separated by less than 100 Myr and replace them by the average. This leaves us with 42 meteorites. A graphical description of the method appears in fig. 3.

From inspection of fig. 4, it appears that the meteorites cluster with a period of 143 ± 10 Myr, or equivalently, $|\Omega_\odot - \Omega_p| = 11.0 \pm 0.8$ (km s^{-1})/kpc, which falls within the preferred range for the spiral arm pattern speed. If we fold the CR exposure ages over this period, we obtain the histogram in fig. 4. A K-S test yields a probability of 1.2% for generating this non-uniform signal from a uniform distribution. Moreover, fig. 4 also describes the prediction from the CR diffusion model. We see that the clustering is not in phase with the spiral arm crossing, but is with the correct phase and shape predicted by the CR model using the above pattern speed. A K-S test yields a 90% probability for generating it from the CR model distribution. Thus, we safely conclude that spiral arm passages modulate the CRF with a ~ 143 Myr period.

4. Do Cosmic rays affects the climate?

In 1959, Ney[16] suggested that the Galactic CR flux (CRF) reaching Earth could be affecting the climate since the CRF governs the ionization of the lower atmosphere, which in turn could be affecting cloud condensation. This, Ney postulated, could explain the observed climate variability synchronized with the solar cycle through the known modulation of the CRF by the solar wind. In 1991, Tinsley and Deen[13] brought first evidence in support. They showed that the Forbush events during which the CRF suddenly drops and gradually increases correlate with the Northern hemisphere Vorticity Area Index during winter. A much clearer and direct link was subsequently found in the form of an intriguing correlation between cloud cover and the CRF reaching Earth[15]. It was later shown to be cor-

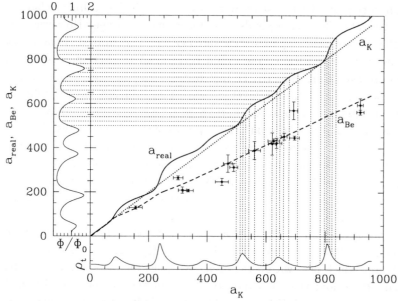

Figure 3. Theoretical comparison between different exposure ages of Iron meteorites and their real age. Plotted as a function of the Potassium exposure age (a_K) are the real age (a_{real}, in solid line) and a non-Potassium exposure age (a_{Be}, such as using $^{10}B/^{21}Ne$ dating, with a dashed line), and a_K (using a dotted line, with a unit slope). Also plotted are the predicted CRF relative to the present flux (Φ/Φ_0) as a function of a_{real}, and ρ_t–the (unnormalized) expected number of Potassium exposure ages per unit time, as a function of a_K. A histogram of a_K should be proportional to ρ_t. The horizontal and vertical dotted lines describe how ρ_t is related to the relation between a_{real} and a_K— equally spaced intervals in real time are translated into variable intervals in a_K, thereby forming clusters or gaps in a_K. The graph of a_{Be} vs. a_K demonstrates that comparing the different exposure ages is useful to extract recent flux changes (which determine the slope of the graph). On the other hand, the graph of ρ_t demonstrates that a histogram of a_K is useful to extract the cyclic variations in the CRF, but not for secular or recent ones. The points with the error bars are about two dozen meteorites which where have both Be and K exposure dating.

related with the low altitude cloud cover (LACC) in particular, which is known to reduce the average global temperature[31,17].

The apparent CRF-LACC link was found through CRF modulation induced by the variable solar wind[15]. If this link is indeed genuine, then long term changes in the CRF induced by spiral arm crossing are too expected to episodically increase the average LACC, thereby reducing the average global temperature and triggering an ice-age epoch (IAE). We shall assume this link to be bona fide and study its consequences, though one should

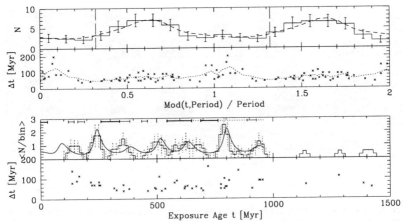

Figure 4. Histogram of the Iron meteorites' exposure ages. The lowest panel marks the a_K ages on the x-axis and the quoted age error on the y-axis. Even by eye, the ages appear to cluster periodically. The second panel is a 1:2:1 averaged histogram of meteorites with a quoted age determination error smaller than 100 Myr, showing more clearly the clustering peaks. Altogether, there are 6 peaks from 210 to 930 Myr BP. The period that best fits the data is 143 ± 10 Myr. The third panel is similar to the first one, with the exception that the data is folded over the periodicity found. It therefore emphasizes the periodicity. A Kolmogorov-Smirnov test shows that a homogeneous distribution could generate such a non-homogeneous distribution in only 1.2% of a sample of random realization. Namely, the signal appears to be real. This is further supported with the behavior of the exposure age errors, which supply an additional consistency check. If the distribution is intrinsically inhomogeneous, the points that fill in the gaps should on average have a larger measurement error (as it is 'easier' for these points to wonder into those gaps accidentally, thus forming a bias). This effect is portrayed by the dotted line in the panel, which plots the average error as a function of phase—as expected, the points within the trough have a larger error on average.

bear in mind that this issue still highly debated.

The apparent effect is on LACC (< 3.2 km), and therefore arises from relatively high energy CRs ($\gtrsim 10$ GeV/nucleon). This CRF can reach equatorial latitudes, in agreement with observations showing a CRF-LACC correlation also near the equator[17]. Thus, when estimating the CRF-LACC forcing, the relevant flux is that of CRs that reach low magnetic latitude stations and that has a high energy cut-off. The flux measured at the University of Chicago Neutron Monitor Stations in Haleakala, Hawaii and Huancayo, Peru is probably a fair measurement of the flux affecting the LACC. Both stations are at an altitude of about 3 km and relatively close to the magnetic equator (rigidity cutoff of 12.9 GeV). The relative change in the CRF for the period 1982-1987 at Haleakala and Huancayo[32] is about

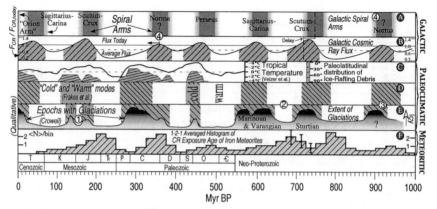

Figure 5. Earth's recent history. The top panel describes past crossings of the Galactic spiral arms assuming a relative pattern speed of $\Omega_p - \Omega_\odot = -11.0$ (km s^{-1})/kpc(which best fits the IAEs). Note that the Norma arm's location is actually a logarithmic spiral extrapolation from its observations at somewhat smaller Galactic radii. The second panel describes the Galactic CRF reaching the solar system using the CR diffusion model, in units of the current day CRF. An important feature is that the flux distribution around each spiral arm is lagging behind spiral arm crossings. This can be seen with the hatched regions in the second panel, which qualitatively show when IAEs are predicted to occur if the critical CRF needed to trigger them is the average CRF. Two arrows point to the middle of the spiral crossing and to the expected mid-glaciation point. The third panel qualitatively describes the epochs during which Earth experienced ice-ages. By fine-tuning the actual pattern speed of the arms (relative to our motion) to best fit the IAEs, a compelling correlation arises between the two. The correlation does not have to be absolute since additional factors may affect the climate (e.g., continental structure, atmospheric composition, etc.). The bottom panel is a 1-2-1 smoothed histogram of the exposure ages of Fe/Ni meteors. The ages should cluster around epochs with a lower CRF flux.

7%, while the relative change in the LACC[17] is about 6%. Namely, to *first approximation*, there is apparently a roughly linear relation between the relevant CRF and LACC.

Next, a 1% *relative* change in the global LACC corresponds[17] to a net effective reduction of the solar flux of $\Delta F_\odot \sim -0.17$ W/m^2. The relation between radiation driving and global temperature change is poorly known. Typical values[33] are $\Delta T/\Delta F_\odot = 0.7 - 1.0°$K/(W m^{-2}). We take a nominal value of $0.85°$K/(W m^{-2}).

Thus, changing the CRF by $\pm 1\%$ would correspond to a nominal change of $\mp 0.14°$K. For the nominal values chosen in our diffusion model, the expected CRF changes from about 25% of the current day CRF to about 135%. This corresponds to a temperature change of about $+10°$K to $-5°$K, relative to today's temperature. This range is sufficient to markedly help or hinder Earth from entering an IAE.

5. Ice Age Epochs and Spiral Passages

Extensive summaries of IAEs on Earth can be found in Crowell[34] and Frakes et al.[35]. Those of the past Eon are summarized in fig. 5. The nature of some of the IAEs is well understood while others are sketchy in detail. The main uncertainties are noted in fig. 5. For example, it is unclear to what extent can the milder mid-Mesozoic glaciations be placed on the same footing as other IAEs, nor is it clear to what extent can the period around 700 Myr BP be called a warm period since glaciations were present, though probably not to the same extent as the periods before or after. Thus, Crowell[34] concludes that the evidence is insufficient to claim a periodicity. On the other hand, Williams[36] claimed that a periodicity may be present. This was significantly elaborated upon by Frakes et al.[35].

Comparison between the CRF and the glaciations in the past 1 Gyr shows a compelling correlation (fig. 5). To quantify this correlation, we perform a χ^2 analysis. *To be conservative*, we do so with the Crowell data which is less regular. Also, we do not consider the possible IAE around 900 Myr, though it does correlate with a spiral arm crossing. For a given pattern speed, we predict the location of the spiral arms using the model. We find that a minimum is obtained for $\Omega_\odot - \Omega_p = 10.9 \pm 0.25$ (km s^{-1})/kpc, with $\chi^2_{min} = 1.1$ per degree of freedom (of which there are 5=6-1). We also repeat the analysis when we neglect the lag and again when we assume that the spiral arms are separated by 90° (as opposed to the somewhat asymmetric location obtained by Taylor and Cordes[20]). Both assumptions degrade the fit ($\chi^2_{min} = 2.9$ with no lag, and $\chi^2_{min} = 2.1$ with a symmetric arm location). Thus, the latter analysis assures that IAEs are more likely to be related to the spiral arms and not a more periodic phenomena, while the former helps assure that the CRs are more likely to be the cause, since they are predicted (and observed) to be lagged.

The previous analysis shows that to within the limitation of the uncertainties in the IAEs, the predictions of the CR diffusion model and the actual occurrences of IAE are consistent. To understand the significance of the result, we should also ask the question what is the probability that a random distribution of IAEs could generate a χ^2 result which is as small as previously obtained. To do so, glaciation epochs where randomly chosen. To mimic the effect that nearby glaciations might appear as one epoch, we bunch together glaciations that are separated by less than 60 Myrs (which is roughly the smallest separation between observed glaciations epochs). The fraction of random configurations that surpass the χ^2 obtained for the best fit found before is of order 0.1% for *any* pattern speed. (If glaciations are not bunched, the fraction is about 100 times smaller, while it is about 5 times larger if the criterion for bunching is a separation of 100 Myrs or less). The fraction becomes roughly 6×10^{-5} (or a 4-σ fluctuation), to coincidentally fit the actual period seen in the Iron meteorites.

6. Star Formation Rate and Long Term Glacial Activity

Another interesting correlation between predicted CRF variability and glacial activity on Earth appears on a much longer time scale. Before 1 Gyr BP, there are no indications for any IAEs, except for periods around 2 - 2.5 Gyr BP (Huronian) and 3 Gyr BP (late Archean)[34]. This too has a good explanation within the picture presented. Different estimates to the Star formation rate (SFR) in the Milky Way (and therefore also to the CR production) point to a peak around 300 Myr BP, a significant dip between 1 and 2 Gyr BP (about a third of today's SFR) and a most significant peak at 2-3 Gyr BP (about twice as today's SFR)[37,38]. This would imply that at 300 Myr BP, a more prominent IAE should have occurred—explaining the large extent of the Carboniferous-Permian IAE. Between 1 and 2 Gyr BP, there should have been no glaciations and indeed none were seen. Last, IAEs should have also occurred 2 to 3 Gyr BP, which explains the Huronian and late-Archean IAEs. This can also be seen in fig. 6.

7. And the Dinosaurs?

Given the above scenario, cosmic rays may even be related to the disappearance of the dinosaurs. Indeed, it is most likely that the last of the dinosaurs saw the light of day during the K/T event, some 66 Myr ago, when a bolide hit the Yucatan peninsula (e.g., [39]). However, a careful study of fossils in N. America actually suggests that the number of dinosaur genera decreased by about a factor of 3 in the 10-15 Myr preceding the K/T event[40,41], and in Europe, the last dinosaurs appear to have disappeared 1-3 Myr before the K/T[42]. There are no widely accepted reason as to why this has happened. However, some point to the fact that the global climate cooled by typically 5-10°C in the 10-20 Myr preceding the K/T event[43,40,44] and that this cooling could have been an environmental stress that was too large for the dinosaurs to endure[43,40]. *If* this hypothesis is correct, namely, that most of the dinosaurs became extinct because the climate cooled quickly,

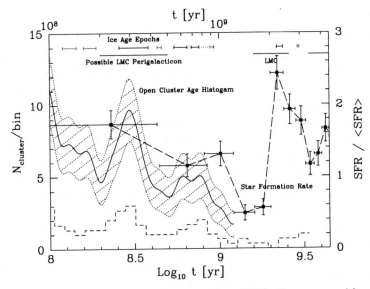

Figure 6. The history of the Star Formation Rate (SFR). The squares with error bars are the SFR calculated using chromospheric ages of nearby stars (Rocha Pinto et al.). These data are corrected for different selection biases and are binned into 0.4 Gyr bins. The line and hatched region describe a 1-2-1 average of the histogram of the ages of nearby open clusters using the Lotkin et al. catalog, and the expected 1-σ error bars. (These data are not corrected for selection effects). Since the clusters in the catalog are spread to cover two nearby spiral arms, the signal arising from the passage of spiral arms is smeared, such that the graph depicts a more global SFR activity (i.e., in our Galactic 'quadrant'). On longer time scales (1.5 Gyrs and more), the Galactic stirring is efficient enough for the data to reflect the SFR in the whole disk. The dashed histogram underneath is the same as the histogram above it, though only with clusters having a better age determination ($w > 1.0$, as defined in Lotkin et al.). There is a clear minimum in the SFR between 1 and 2 Gyr BP, and there are two prominent peaks around 0.3 and 2.2 Gyr BP. Interestingly, the LMC perigalacticon should have occurred sometime between 0.2 and 0.5 Gyr BP in the last passage, and between 1.6 and 2.6 Gyr before present in the previous passage. This would explain the peaks in activity seen. This is corroborated with evidence of a very high SFR in the LMC about 2 Gyrs BP and a dip at 0.7 - 2 Gyr BP. Also depicted are the periods during which glaciations were seen on Earth: The late Archean (3 Gyr) and mid-Proterozoic (2.2-2.4 Gyr BP) which correlate with the previous LMC perigalacticon passage (Gardiner 1994, Lin 1995) and the consequent SFR peak in the MW and LMC. The lack of glaciations in the interval 1 - 2 Gyr before present correlates with a clear minimum in activity in the MW (and LMC). Also, the particularly long Carboniferous-Permian glaciation, correlates with with the SFR peak at 300 Myr BP and the last LMC perigalacticon. The late Neo-Proterozoic ice ages correlate with a less clear SFR peak around 500-900 Myr BP.

then most of the extinction can be directly related to the fact that the solar system entered the Sagittarius-Carinae spiral arm during that period.

8. Summary

To conclude, we first considered that most CR sources reside in the Galactic spiral arms, and incorporated this fact into a cosmic ray diffusion model. Unsurprisingly, this model predicts a variable CRF. By analyzing the exposure ages in Iron meteorites, it was found that the cosmic ray flux history can be reconstructed. It was found to vary periodically, and it nicely agrees with the observations of the Galactic spiral arm pattern speed.

Next, if recent evidence linking the CRF to low altitude cloud cover on Earth is real, typical variations of $O(10°K)$ are predicted from the variable CRF. Each spiral arm crossing, the average global temperature should reduce enough to trigger an IAE. The record of IAEs on Earth is fully consistent with the predicted and observed CRF variation—both in period and in phase. Next, the fit is also found to be better when the predicted lag in the mid-point of the IAEs after each crossing is included and when the actual asymmetric location of the arms is taken into account. Moreover, a random mechanism to generate the IAEs is excluded.

On a more speculative note, there is a curious correlation between the global cooling experienced on Earth at the end of the Cretaceous and the disappearance of the dinosaurs. If the the latter is related to climate change, it could be attributed to the solar system entering a spiral arm.

References

1. M. S. Longair, *High Energy Astrophysics*, 2nd ed., vol. 2 (Cambridge Univ. Press, Cambridge, 1994)
2. V. S. Berezinskiĭ, S. V. Bulanov, V. A. Dogiel, V. L. Ginzburg, V. S. Ptuskin, *Astrophysics of Cosmic Rays*, (North-Holland, Amsterdam, 1990)
3. N. Duric, In *Proceedings 232. WE-Heraeus Seminar, 22-25 May 2000, Bad Honnef, Germany. Edited by Elly M. Berkhuijsen, Rainer Beck, and Rene A. M. Walterbos. Shaker, Aachen, 2000*, p. 179 (2000).
4. P. M. Dragicevich, D. G. Blair, and R. R. Burman, *Mon. Not. Roy. Astr. Soc.*, **302**, 693 (1999).
5. N. Duric, *Astrophys. J.*, **304**, 96 (1986).
6. G. Bond et al., *Science*, **294**, 2130 (2001).
7. U. Neff, S. J. Burns, A. Mangnini, M. Mudelsee, D. Fleitmann, and A. Matter, *Nature*, **411**, 290 (2001).
8. D. A. Hodell, M. Brenner, J. H. Curtis, and T. Guilderson, *Science*, **292**, 1367 (2001).
9. E. Friis-Christensen and K. Lassen, *Science*, **254**, 698 (1991).
10. W. H. Soon, E. S. Posmentier, and S. L. Baliunas, *Astrophys. J.*, **472**, 891 (1996).
11. W. H. Soon, E. S. Posmentier, and S. L. Baliunas, *Annales Geophysicae*, **18**,

583 (2000).

12. J. Beer, W. Mende, and R. Stellmacher, *Quat. Sci. Rev.*, **19**, 403 (2000).

13. B. A. Tinsley and G. W. Deen, *J. Geophys. Res.*, **12**, 22283 (1991).

14. L. Y. Egorova, V. Ya. Vovk, and O. A. Troshichev, *J. Atmos. Solar-Terr. Phys.*, **62**, 955 (2000).

15. H. Svensmark, *Phys. Rev. Lett.*, **81**, 5027 (1998).

16. E. P. Ney, *Nature*, **183**, 451 (1959).

17. N. Marsh and H. Svensmark, *Sp. Sci. Rev.*, **94**, 215 (2000).

18. Shaviv, N. J., *Phys. Rev. Lett.*, **89**, 051102 (2002).

19. Shaviv, N. J., *New Astron.*, in press (2002).

20. J. H. Taylor and J. M. Cordes, *Astrophys. J.*, **411**, 674 (1993).

21. Y. M. Georgelin and Y. P. Georgelin, *Astron. Astrophy.*, **49**, 57 (1976).

22. W. R. Webber and A. Soutoul, *Astrophys. J.*, **506**, 335 (1998).

23. U. Lisenfeld, P. Alexander, G. G. Pooley, and T. Wilding, *Mon. Not. Roy. Astr. Soc.*, **281**, 301 (1996).

24. A. Lukasiak, P. Ferrando, F. B. McDonald, and W. R. Webber, *Astrophys. J.*, **423**, 426 (1994).

25. B. Lavielle, K. Marti, J. Jeannot, K. Nishiizumi, and M. Caffee, *Earth Plan. Sci. Lett.*, **170**, 93 (1999).

26. J. Palous, J. Ruprecht, O. B. Dluzhnevskaia, and T. Piskunov, *Astron. Astrophy.*, **61**, 27 (1977).

27. Lin C. C., Shu F. H., *Astrophys. J.***140**, 646 (1964).

28. J. Binney, S. Tremaine, *Galactic Dynamics*, (Princeton Univ. Press, Princeton, 1988)

29. L. Blitz, M. Fich, and S. Kulkarni, *Science*, **220**, 1233 (1983).

30. H. Voshage, H. Feldmann, *Earth Planet. Sci. Lett* **45**, 293 (1979).

31. V. Ramanathan, R. D. Cess, E. F. Harrison, P. Minnis, and B. R. Barkstron, *Science*, **243**, 57 (1989).

32. G. A. Bazilevskaya, *Sp. Sci. Rev.*, **94**, 25 (2000).

33. D. Rind and J. Overpeck, *Quat. Sci. Rev.*, **12**, 357 (1993).

34. J. C. Crowell, *Pre-Mesozoic Ice Ages: Their Bearing on Understanding the Climate System*, volume 192. Memoir Geological Society of America (1999).

35. L. A. Frakes, E. Francis, and J. I. Syktus, *Climate modes of the Phanerozoic; the history of the Earth's climate over the past 600 million years*. Cambridge: Cambridge University Press (1992).

36. G. E. Williams, *Earth and Plan. Sci. Lett.*, **26**, 361 (1975).

37. J. M. Scalo, In *Starbursts and Galaxy Evolution*, p. 445 (1987).

38. H. J. Rocha-Pinto, J. Scalo, W. J. Maciel, and C. Flynn, *Astron. Astrophy.*, **358**, 869 (2000).

39. van den Bergh S., *Proc. Ast.. Soc. Pac.*, **106**, 689 (1994).

40. R. E. Sloan et al., *Science*, **232**, 629 (1986)

41. R. E. Sloan, J. K. Rigby Jr., L. van Valen, and L.-D., *Science*, **232**, 633 (1986).

42. B. Galbrun, *Earth Planet. Sci. Lett.*, **148**, 569 (1997).

43. D. M. McLean, *Science*, **201**, 401 (1978).

44. B. T. Huber, *Science*, **282**, 2199 (1998).

COSMIC RAY ENERGY SPECTRA AND COMPOSITION NEAR THE "KNEE"

John P. Wefel

Department of Physics and Astronomy, Louisiana State University
Baton Rouge, LA 70803 USA

There has been renewed interest in recent years in the energy region around 10^{15} eV, the region containing a major feature in the cosmic ray spectrum called the "Knee". Discovered over forty years ago through air shower measurements, the origin of the knee and its underlying physics/astrophysics remain largely unknown. Direct particle identification experiments are being extended to ever higher energies, while the air shower data and interpretation are improving. This brief review summarizes the current results and looks at various theories and models which have been suggested recently to help unravel the astrophysics of the energy region around the knee.

1. Introduction and History

The Galactic Cosmic Radiation (GCR) is a dilute "gas" of relativistic particles that fill our galaxy and extend well beyond the disk into a galactic halo. A few percent of the cosmic rays are electrons that produce synchrotron radiation in the galactic magnetic field. Looking at other galaxies similar to our own, we observe radio synchrotron halos around them as well. The bulk of the cosmic rays, however, are nuclei, consisting of all of the naturally occurring elements from H through U. As with the general galactic abundances, H and He are the most numerous with the heavy nuclei (Z>2) accounting for about 1% of the particles. Tracing the composition of the GCR has been a principal goal in cosmic ray research over the past half century. Primary particles provide information on the source regions and on nucleosynthesis processes. Secondary species, made by nuclear fragmentation of the primaries, trace the confinement and transport of the particles within the galaxy and its halo.

The GCR span an enormous range from below 10^8 eV to beyond 10^{20} eV with an energy spectrum that is a power-law over many decades, as illustrated in Figure 1 for the "all-particle" spectrum. At the lowest energies, we cannot observe the particles directly since the out-flowing solar wind excludes these particles from entering our Heliosphere and being observed at Earth. This

128

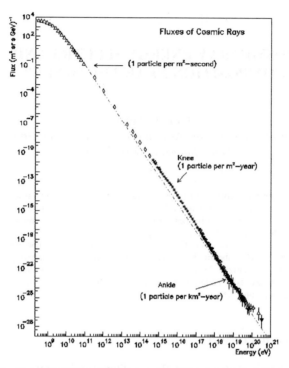

Figure 1. All particle differential spectrum of the GCR showing the knee and the ankle.

heliospheric modulation decreases with increasing energy and introduces a solar cycle variation into the particle intensity at low energy.

The flux in the range 50-500 MeV/nucleon is large enough that these particles can be studied with small balloon or satellite instruments. It is in this range that we have the best composition information, recent work having succeeded in resolving many of the isotopes of the individual elements up through the iron peak [1], giving a detailed picture of the GCR– the low energy "baseline" composition. At higher energy, isotopic separation is more difficult, and our information is limited to elemental composition through the iron peak and groups of elements in the ultra-heavy (Z>30) region of the spectrum. As Figure 1 indicates, the flux declines rapidly with increasing energy and beyond several TeV, composition has been reported for groups of elements (e.g. CNO). At sill higher energies, measurements rely on the analysis of air showers produced in the atmosphere, and only the mean composition of the overall beam (denoted as <ln A>, where A is the particle mass) is available. Beyond about 10^{10} eV the GCR follow a power law energy spectrum (differential spectrum $\propto E^{-2.75}$) up to a few times 10^{15} eV. Here there is a rather abrupt change, called the "knee", with the spectrum beyond the knee becoming softer ($\propto E^{-3.2}$). Beyond the knee, the steeper spectrum continues until a few times 10^{18} eV (the "ankle") where it hardens again and continues to beyond 10^{20} eV. It is convenient to eliminate the steeply falling spectrum in Figure 1 to better display these features. This is done by multiplying the flux by $E^{2.75}$ to

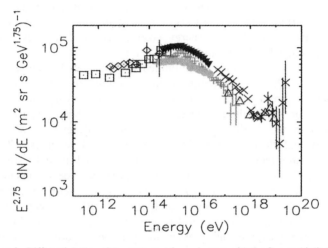

Figure 2. Differential all particle spectrum of cosmic rays, with the flux multiplied by $E^{2.75}$, showing the knee at a few 10^{15} eV and the ankle at a few 10^{18} eV. Data are selected from both direct measurements (below 10^{15} eV) and air shower measurements.

obtain a plot such as Figure 2 which shows, for selected recent data, the knee and ankle features. It should be noted that in the region around the "knee" the particles are well confined by the galactic magnetic field. Beyond the ankle, however, the particle gyro-radii become comparable to galactic dimensions, and these highest energy cosmic rays may well be extra-galactic. The ultra-high energy events are described elsewhere in this volume. Here we focus on the "knee region", roughly 10^{13} - 10^{17} eV, which holds the "key" to understanding the sources, acceleration mechanisms and modes of transport for the high energy cosmic rays in our galaxy.

In the vicinity of the knee, one expects about one particle per square meter per year (c.f. Figure 1). This necessitated detectors of many square meters located on the surface of the Earth. Cosmic rays interact in the Earth's atmosphere and develop electron-hadron cascades which spread laterally as they develop and propagate downward. The total number of particles in these "air showers" is related to the total energy of the primary cosmic ray and to its mass (charge). Thus, by deploying many detection stations over areas of square kilometers, it is possible to study the cosmic rays in the very high energy region. This was the approach followed historically in the late 1940's and 1950's which led to the discovery of both the knee and the ankle in the cosmic ray spectrum. One can site the pioneering work at Moscow State University under the late George Khristiansen, or Volcano Ranch in the US, developed by the late John Linsley or the group at Haverah Park in England. Researchers from many countries, too numerous to cite, contributed to tracing the GCR energy spectrum with air showers. Today, air shower arrays continue to provide data on the highest energy cosmic rays, and the new arrays have increased tremendously in sophistication and in scale.

Air showers, however, have limited sensitivity to the composition of the cosmic rays. Therefore, experiments that determine the charge and energy of each particle are necessary, but these require deploying the apparatus on spacecraft or high altitude research balloons. Such experiments were developed in the 1960's by N. Grigorov and colleagues at Moscow State University and flown on the "PROTON" series of satellites. These large ionization calorimeter instruments measured the all particle spectrum and separately, the H and He components. Continued development of space technology led, in the next decades, to three space experiments: the HEAO-3 satellite, SOKOL on the Cosmos satellites and CRN on Spacelab-2. Using ionization chambers, calorimeters Cherenkov counters and transition radiation detectors, these experiments pressed the particle-by-particle detection up to energies close to 1 TeV/nucleon and to elements as heavy as iron. Table 1 summarizes the properties of these space experiments.

In the late 60's large balloon experiments utilizing Cherenkov counters and ionization calorimeters were flown by US and German researchers. In the following decade, emulsion chamber technology was developed for balloon flight, and several series of such experiments took place – MUBEE, JACEE, RUNJOB – some of which are still publishing results. In the late 1980's the Long Duration Ballooning (LDB) techniques were developed allowing flights of 7-14 days, and these became the mainstay of the emulsion chamber experiments, allowing large exposures to be accumulated through yearly flights. In addition, a ring imaging Cherenkov instrument was flown on a balloon giving a new measurement of the helium spectrum to extend earlier work into the emulsion chamber energy regime.

In the mid-1990's several new balloon instruments were developed specifically for LDB applications. ATIC (Advanced Thin Ionization Calorimeter) uses a fully active Bismuth Germanate (BGO) calorimeter

Table 1. Space Experiments

Experiment		Species	Technique	Energy/ nucleus (eV)	Effective Geometry Factor (m²-sr)	Exposure Factor (m²-sr-days)
SEZ Proton 1-4	1965 -68	All, H, He	Calorimeter	10^{11} - 10^{15}	0.05 - 10	5-2000
HNE HEAO-3	1979 -80	$16 \leq Z \leq$ 28	Ionization/ Cherenkov	3×10^{10} – 10^{13}	1.2	370
French- Danish HEAO-3	1979 -80	$4 \leq Z \leq$ 28	Cherenkov	3×10^{10} – 2×10^{12}	0.14	33
CRN Spacelab 2	1985	$5 \leq Z \leq$ 26	TRD	7×10^{11} – 3×10^{13}	0.1–0.5 (low Z) 0.5-0.9 (high Z)	0.3-3
Sokol COMOS 1543;1713	1984 -86	$1 \leq Z \leq$ 26	Calorimeter	2×10^{12} – 10^{14}	0.026	0.4

Table 2. Balloon Experiments

Experiment	Species	Technique	Energy/nucleus eV	Eff. Geom. (m²-sr)	Exposure (m²-sr–days)
Ryan et al. ConUS 1969-70	$1 \leq Z \leq$ 26	Calorimeter Cherenkov	5×10^{10} - 2×10^{12}	0.036	0.01
JACEE ConUS + LDBs 1979-95	$1 \leq Z \leq$ 26	Emulsion Chambers	$10^{12} - 5 \times 10^{14}$	2-5	107 (H,He) 65 (Z>2)
MUBEE Short & long 1975-87	$1 \leq Z \leq$ 26	Emulsion Chambers	$10^{13} - 3 \times 10^{14}$	0.6	22
RUNJOB LDBs 1995-99	$1 \leq Z \leq$ 26	Emulsion Chambers	$10^{13} - 3 \times 10^{14}$	1.6	43
ATIC LDB 2000-01	$1 \leq Z \leq$ 28	Calorimeter	$10^{10} - 10^{14}$	0.23	3.5
TRACER ConUS	$8 \leq Z \leq$ 28	TRD	$10^{11} - 3 \times 10^{14}$	5	5.8

following a Carbon target to measure the total particle energy. A Silicon-matrix detector at the top and scintillator strip hodoscopes on top, bottom and within the Carbon target, measure the particle charge and trajectory. ATIC flew in Antarctica in 2000-01, and will be flown again in 2002-03. A different approach is taken by TRACER (Transition Radiation Array for Cosmic Energetic Radiation) which uses transition radiation to measure the Lorentz factor of the particles. Combined with scintillators to determine particle charge, TRACER can study the spectra of elements above nitrogen. TRACER has had a short flight and should have an LDB flight in the near future. Table 2 summarizes the characteristics of many of the balloon experiments.

Results from both the satellite and balloon experiments have been summarized [2-7], while the two newest experiments were discussed at the International Cosmic Ray Conference in Hamburg [8].

2. The "Problem" of the Knee

A steepening in a power-law spectrum denotes an "absence" of events, usually ascribed to a new loss process that begins at an energy near the steepening. An early explanation involved the loss of cosmic rays from the galaxy. Since we now know that particles in the knee energy range are well confined in the galaxy, it is doubtful that a new propagation loss from the galaxy occurs. However, cosmic rays do spend some part of their lifetime in the galactic halo. Suggestions have been offered that interactions with the dark matter in the halo may become effective in the knee energy range, and this could constitute a new loss mechanism.

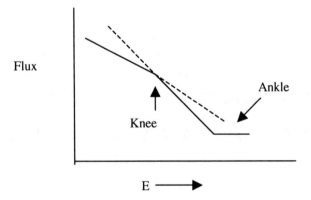

Figure 3. Schematic representation of the "knee".

Other explanations invoke "new physics" or the termination of the cosmic ray acceleration process. Figure 3 shows a sketch of the knee. Extrapolation of the low energy spectrum to above the knee, as shown by the dashed line, compared to the steeper portion of the spectrum, gives a flux difference that increases with increasing energy. Now, suppose that a new physical channel opens at these energies such that the interaction products are not observable in air shower experiments. Further, the importance of this new channel increases with energy. Then, the air shower observations would interpret the data as shown in Figure 3, i.e. a spectral steepening, while the actual spectrum in the galaxy would be the dashed continuation of the low energy portion of the spectrum. Suggestions for such a new channel include the production of new, "non-standard model" particles, or an increased production of high energy muons. There is no evidence at the highest accelerator energies of such "new physics", but these ideas cannot be ruled out until new experiments are performed, probably at the Large Hadron Collider.

Another class of explanations attributes the knee to the termination of the cosmic ray acceleration process, as is expected in models of shock acceleration in, for example, supernova remnants. In this approach the acceleration process reaches a maximum energy near 10^{15} eV resulting in a rapid turn-over of the spectrum. However, we know that the GCR continue to beyond 10^{20} eV, so an additional component or new acceleration mechanism is needed to supply the particles beyond the termination region. Moreover, if the steeper spectrum beyond in knee is extrapolated to lower energy (c.f. Figure 3) the flux would quickly dominate the spectrum. Thus, any "new component" must turn-on and become important just in the energy region of the knee where the low energy component is terminating – a difficult, "ad hoc" matching of components.

We are accustomed to thinking of cosmic rays as coming from one class of sources, currently supernova remnants (see next section). However, the actual sources may involve a variety of astronomical objects or even a plethoria of supernova remnants of various types. Each of these may show a different high energy cut-off. Thus, by assuming a mixture of astrophysical objects, it is possible to construct a model in which the entire knee region is the result of terminating "accelerators," giving rise to the spectral steepening and eventually falling below an extra-galactic component in the vicinity of the ankle. To reproduce the somewhat abrupt structure in the knee region (c.f. Fig. 2) it has been suggested that superposed upon this general terminating spectrum are particles from a nearby source such as a relatively recent supernova [9]. Other sources being considered include the explosions giving rise to gamma ray bursts, pulsars and other accretion driven objects.

In models with the effects of a number of sources/accelerators superposed, one would expect to observe some structure in the measured energy spectra, quite possible in the knee region, and such an observation would be of critical importance in understanding the knee. Moreover, the various types of models discussed each imply a cosmic ray composition which changes with energy up to and through the knee energy region. Measuring this energy dependent composition is a major challenge, but the composition may well be the crucial parameter for understanding the astrophysics of the knee.

3. The Standard Model: Supernova Remnant Acceleration

Cosmic rays and supernovae have long been associated due to the energy input that supernovae can provide. Cosmic rays interact with the ambient interstellar medium as well as escape from the galaxy. The mean confinement lifetime of the particles is about 15 million years [1,10], so cosmic rays must be replenished. This requires an energy input into cosmic rays of 10^{40}-10^{41} ergs/second. This rather large power requirement led, historically, to the presumed connection to supernova explosions which provide an overall power of about 10^{42} ergs/second. With 1-10% of this energy appearing as accelerated particles, the cosmic ray energy density can be maintained.

The acceleration mechanism that has become the "standard" for the past several decades is diffusive shock acceleration -- first order Fermi acceleration -- operating at the discontinuity where the outward moving blast wave from the supernova explosion interacts with the surrounding medium. Here the magnetic fields confine the particles, forcing many crossings of the shock boundary, with the charged particle receiving an acceleration upon each crossing [11]. The acceleration theory has been well developed and has been tested with direct observations of particles accelerated at shocks within our Heliosphere. The theory predicts power-law spectra with the same power law index (in magnetic rigidity) for all nuclear species. The expected index is in the range 2.0-2.2. In addition, there is a maximum energy for the accelerated particles due to the finite lifetime and maximum size of a supernova remnant. For a typical supernova remnant and an assumed magnetic field of 3 μG, this maximum energy is $Z \times 10^{14}$ eV where Z is the charge of the nucleus [12].

In environments with large magnetic fields, the maximum energy can be larger. For example, "bare" supernova explosions expanding into the local interstellar medium will be different than "clothed" events where the blast wave interacts with previously expelled shells of matter from the star [13]. Further, most of the massive stars are formed in groups (e.g. O-B associations) so that the explosion of one star may encounter the remains of shells from earlier supernovae. Matter densities and magnetic field strength are expected to vary in these different environments, so that acceleration beyond the $Z \times 10^{14}$ eV limit cited above is quite feasible. For a more extensive treatment of the maximum energy achievable, see [14].

4. Experimental Data

Early results from the Proton satellite experiments indicated a possible break, or steepening, in the energy spectrum of Hydrogen at energies of a few TeV. Subsequent experiments, however, were unable to confirm such a change, finding a continuous
spectrum up to near 100 TeV. However, a spectral difference between Hydrogen and Helium has been reported as shown by the top two curves in Figure 4, which summarizes measurements of H, He and groups of heavier nuclei. (Note that the flux has been multiplied by $E^{2.75}$, and the plot is in terms of kinetic energy.) The differential spectral index for Hydrogen is about -2.80 while the index for Helium appears to be closer to -2.68, based upon the JACEE data [6], giving a two sigma difference in the measured indices. However, the recent results reported by the RUNJOB experiment [7] show no difference in the Hydrogen and Helium spectral index.

For the heavier nuclei, spectral differences are observed as well. The CNO group shows a harder spectrum than Helium as does the medium heavy elements and the iron peak elements. These last two groups have power law spectral indices that appear to be between He and CNO. Extrapolation of this behavior up to the knee predicts a radically different relative composition as compared to the cosmic rays around 100 GeV/nucleon.

However, we must be careful in the interpretation of the existing data. The current results have been derived from different experiments and, often, many balloon flights of a single instrument have been added together to form an overall dataset. Moreover, the statistical limitations on the current data are evident in Figure 4. Clearly, qualitative and quantitative improvements in the experimental situation are needed.

Even with these experimental limitations, it is instructive to compare the current data to the predictions of the SNR acceleration model. Here there are several major inconsistencies. The spectra of the different nuclear species do not show the same spectral index, as the model predicts. The Hydrogen spectra shows no apparent cut-off at 10^5 GeV (10^{14} eV), and the corresponding Helium maximum energy cut-off at 2×10^{14} eV (5×10^{14} GeV/nucleon) also is not evident in Figure 4. The observed spectral index is larger than the theory predicts, but this is due to energy (rigidity) dependent diffusion/propagation in the galaxy prior to observation at Earth. An interesting exercise is to extrapolate the spectra of Figure 4 to a cut-off of energy $Z \times 10^{15}$ eV, sum the individual groups to form an all-particle spectrum, and compare the result to the measured all-particle spectrum. The "fit" is quite good. This might indicate a shock acceleration cut-off an order of magnitude higher in energy than

136

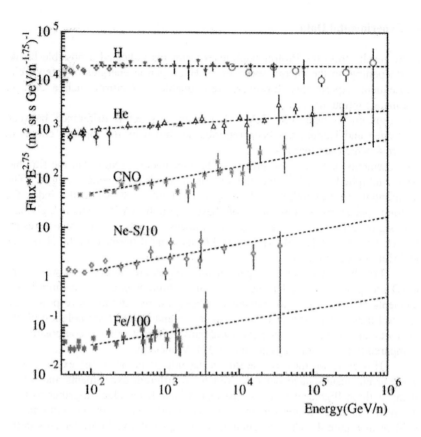

Figure 4. Selected data for the differential energy spectra of the element groups, H, He, CNO, Ne-S, and the iron group. Note that the flux has been multiplied by $E^{2.75}$ to display the results.

predicted, but it does not explain the origin of the particles well beyond 10^{15} eV.

Moving to higher energy than the direct measurements of Figure 4 requires reliance upon air shower experiments which can traverse the knee energy region. The sensitivity to composition in the air shower analyses is relatively weak, resulting in presentation in terms of <ln A>. A compilation of recent air shower results on composition is shown in Figure 5, with the results of the direct experiments converted to <ln A> at the low energy side. The data

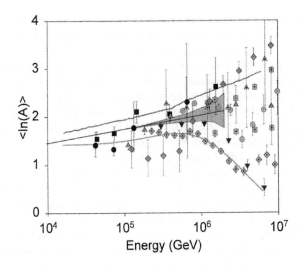

Figure 5. Relative composition <ln A> as a function of total particle energy from a number of recent direct and air shower experiments.

from the different air shower experiments display an enormous scatter [15], with the two curves in Figure 5 showing the results of different models.

The upper curve predicts a composition becoming increasingly dominated by heavy nuclei, i.e. the lighter components terminate earliest with increasing energy. The lower curve models a scenario in which the SNR acceleration process terminates and a new component, assumed to be Hydrogen, becomes dominant, leading to a decreasing <ln A> as the knee is transited. Clearly, some of the current data can support either class of models. What is required is (a) extend the direct measurements to as high an energy as possible to anchor the air shower results (the shaded region in the center of Fig. 5 being a "goal") and (b) improve the interpretation of the air shower data. This is the challenge for the coming years.

5. Summary and Prospects

After half a century, the knee remains an enigma! There are more models than there is data, and the current results are not fully consistent. Extending direct measurements to higher energy, with good charge resolution, is urgently need to both understand the knee region and provide a normalization point for the air shower experiments. The ATIC and TRACER LDB experiments will provide new results in the next few years. A year or so later the CREAM experiment [16] which combines calorimetry and transition radiation measurements will fly as the first payload in the Ultra-Long Duration Balloon (ULDB) program. A new balloon, closed at the bottom, is being developed for the ULDB effort. The plan is to complete many circumnavigations of the Earth for a total flight duration of ~100 days. All data is to be telemetered to the ground via satellite link, protecting the results from the possibility that the apparatus may not be recovered. CREAM will fly in late 2003 or 2004 depending upon the readiness of the balloon, flight systems and the instrument.

Even with the ULDB capability, balloon systems still have limited exposure. Better would be a space experiment with an exposure of ~3 or more years. Such an investigation is ACCESS which was proposed to the NASA Explorer Program, MIDEX competition, this past year. The idea is to put a ~5 ton instrument on either a free-flying satellite or on the experiment attach point on the International Space Station. Proposed was an instrument that was a combination of a calorimeter and a transition radiation detector, which could provide an order of magnitude increase in exposure over the balloon experiments. Unfortunately, ACCESS was not selected in this past round. ACCESS may be re-proposed in 2004 for the next MIDEX competition.

Larger instruments are under consideration in Russia, PROTON-S and INCA, which would be in the vicinity of 10 tons. These use the calorimeter approach but modified to include neutron calorimetry. Both would obtain reasonable statistics above 10^{16} eV, well beyond the knee. However, these instruments are still in the concept and development phases. No mission, and no launcher, has been identified for these very large payloads.

In most cases, the new experiments, either balloon or space, involve international collaborations that have formed to further this important science investigation. Each of the experiments was described in papers at the Hamburg ICRC and the reader is referred there [8] for further details.

The air shower analysis is improving rapidly with the development of new Monte Carlos and multi-parameter analysis. One of the most advanced is the KASCADE group who has performed multi-variant analysis using the shower parameters electron size, muon size, hadron size and most energetic hadron.[17] It is not always possible to get complete consistency, but the techniques being developed hold great promise for the future [15]. Further, data from other arrays has improved [8]. Moreover, there are several arrays undergoing expansion and improvement which will provide new data in years to come.

Supernova remnant shock acceleration may well account for the bulk of the cosmic rays below the knee, but these models have difficulty in fully explaining the knee region. In fact, such discrepancies have led to suggestions that the venerable supernova remnant model should be abandoned [18]. With the x-ray data from Chandra revealing details of the CRAB and VELA pulsars [19], there is renewed interest in pulsar acceleration models to complement the supernova remnant process. Also, gamma ray burst sources are being considered, as well as the galactic center region.

The data and models continue to improve, and we expect new information to become available in the next few years from the new experiments. The interest remains high, since the knee is such a challenging astrophysical problem!

Acknowledgements

The author thanks the organizers of the International School of Cosmic Ray Astrophysics for their kind hospitality in Erice. This work was supported, in part, by NASA Grant NAG5-5306.

References

[1]   *Acceleration and Transport of Energetic Particles Observed in the Heliosphere: ACE 2000 Symposium*, ed. Mewaldt, R. A. et al., AIP Conf. Proc. 528 (NY, 2000 American Institute of Physics).

[2]   Swordy, S. P. in *Proc. 23rd Int. Cosmic Ray Conf.*, eds. Leahy, D. A., Hicks, R. B. and Venkatesan, D. (Singapore, 1993, World Scientific) 243.

[3]   Zatsepin, V. I., Lazareva, T. V., Sazhina, G. P. and Sokolskaya, N. V. *Phys. Atom. Nucl., 57*, 645 (1994).

[4]   Gaisser, T. K. et al. *Opportunities in Cosmic Ray Physics and Astrophysics*, Nat. Res. Council, (Washington, DC, 1995, Nat. Acad. Press).

[5]   Wiebel-Sooth, B. and Biermann, P. L. *Astron. And Astrophys.*, **VI/3c**, 37 (1999).

[6]   Asakimori, K. et al., *Astrophy. J.*, **502**, 278 (1998).

[7]   Apanasenko, A. V. et al. *Proc. 27th Int. Cosmic Ray Conf.* (Katlenburg-Lindau, Copernicus Gesellschaft) **5**, 1626 (2001).

[8]   *Proceedings 27th Int. Cosmic Ray Conf.* (Kattenburg-Lindau, Copernicus Gesellschaft) **5** and **6** (2001).

[9]   Erlykin, A. D. and Wolfendale, A. W. *J, Phys. G, 23*, 979 (1997).

[10]   Garcia-Munoz, M., Mason, G. M. and Simpson, J. A., *Astrophys. J.*, **217**, 859 (1977).

[11]   Baring, M. G et al., *Astrophys. J.*, **513**, 311 (1999).

[12] Lagage, P. O. and Cesarsky, C. J., *Astron. and Astrophys.*, **118**, 223 (1983).
[13] Völk, H. J. and Biermann, P. L., *Astrophys. J. Lett.*, **333**, L265 (1988).
[14] Dermer, C. D., *Proc. 27th Int. Cosmic Ray Conf.* (Katlenburg-Lindau, Copernicus Gesellschaft) **6**, 2039 (2001).
[15] Swordy, S. P. et al., *Astroparticle Phys.*, in press (2002). See also http://lanl.arxiv.org/abs/astro-ph/?0202159.
[16] Ganel, O., et al., *Proc. 27th Int. Cosmic Ray Conf.* (Katlenburg-Lindau, Copernicus Gesellschaft) **6**, 2163 (2001).
[17] Roth, M., et al., *Proc. 27th Int. Cosmic Ray Conf.* (Katlenburg-Lindau, Copernicus Gesellschaft) **1**, 88 (2001).
[18] Parizot, E., Paul, J. and Bykov, A., *Proc. 27th Int. Cosmic Ray Conf.* (Katlenburg-Lindau, Copernicus Gesellschaft) **6**, 2070 (2001).
[19] See http://chandra.harvard.edu/photo/category/snr.html

ENERGETIC PARTICLE POPULATIONS INSIDE AND AROUND THE SOLAR SYSTEM

PÉTER KIRÁLY

KFKI Research Institute for Particle and Nuclear Physics,
H-1525 Budapest, P.O.Box 49, Hungary
E-mail: pkiraly@sunserv.kfki.hu

The heliosphere is a very special region for us. It is mostly controlled by the Sun, and is shielded from the plasma and the moderately energetic particles of the interstellar environment. Compared to the main sources of cosmic rays the Sun and the heliosphere are very inefficient energetic particles producers. Because we are so much closer to the Sun than to any other star, those locally produced energetic particles are still competitive in their influence on Earth, and are also much more accessible for detailed study than the much more energetic emissaries of far-away violent events. Several clues on fundamental energetic particle acceleration and propagation processes are due to heliospheric studies, and there is ample scope for progress. Recent results and further prospects will be reviewed.

1. Introduction

The prime purpose of high energy astrophysics is the study of very energetic processes in the Universe. Cosmic ray (CR) physics, as reflected in the topics discussed at the biennial International CR Conferences, is not simply a subfield of that discipline, but it also deals with particles and processes of local origin, even if the energies involved are much lower. In particular, acceleration and propagation of energetic particles of solar and heliospheric, and even magnetospheric origin are considered as important topics. While the inclusion of those topics is partly a result of the history of CR physics, there is also some sound logics behind it. CR sources are associated with distant astronomical objects, only observable by telescopes sensitive to certain frequency bands of electromagnetic radiation. Suppose someone in Erice, on the top of the hill, was interested in the biological details of creatures living in the sea surrounding Sicily. A powerful telescope certainly would give him or her a few glimpses of big fish well illuminated near the surface, but most of those fish would soon submerge again to invisibility. The alternative strategy of getting hold of a bucket (or even a drop) of sea water and observe it in great detail might lead to a deeper knowledge on certain aspects of sea life, even if no big fish is involved. The tiny creatures that live in that water could then be studied by some kind of microscope, and their behaviour and interactions could also be followed for an extended period. Similarly, an intensive study of heliospheric acceleration and propagation processes, partly by *in situ* observation, and partly by "remote sensing" at distance scales that are minute compared to astronomical ones, can provide clues that may not be accessed by observation through telscopes. That does not mean, of course, that "big fish" should not be observed.

2. The heliosphere and its environment

2.1. *The solar sytem and the heliosphere*

The term "solar system" is poorly defined. It usually refers to the Sun and planets, also including their moons and some minor constituents like asteroids, meteorids, comets and protocomets, dust. In spite of the recent progress in observing extrasolar planets or exoplanets around nearby stars, our knowledge on planetary systems around stars mainly comes from a single example, our own solar system.

It is often stated that the two Voyager probes, celebrating their silver jubilee in 2002, left the solar system when their radial distances from the Sun exceded those of all major planets. But what about the comets that return periodically, even if that period is very long? According to Newtonian mechanics, they should still be considered components of the solar system, held together by the gravitational attraction of the Sun. The aperiodic comets, on the other hand, are just visitors in the solar system and do not belong to it, although the two populations cannot be separated purely by their spatial position.

For the plasma regions surrounding the Sun, the criterion of gravitational binding is still less applicable. Therefore the new term "heliosphere" was introduced to include all plasma structures that are under the influence of the Sun and of the solar wind (SW) emanating from it, as distinct from regions influenced primarily by the interstellar (IS) wind, in which the elongated bubble of the heliosphere is embedded. The concept of the heliosphere is similar to that of the terrestrial magnetosphere, to which magnetic fields of terrestrial origin are confined. Telescopes are usually of little use for mapping the boundaries of either the magnetosphere or the heliosphere. Those fairly sharp but dilute magnetic and plasma boundaries emit namely very little electromagnetic radiation, and their identification is much easier by *in situ* observations of magnetic fields, plasma, and energetic particles. This does not exclude a limited use of electromagnetic signatures in the identification of some heliospheric boundaries, and some tentative identifications have already been done (Gurnett *et al* 1993, Kurth and Gurnett 1993, Ben-Jaffel *et al* 2000, Ratkiewicz and Ben-Jaffel 2002).

2.2. Dimensions and basic structure of the heliosphere

The size of the heliosphere, even if not precisely known, is certainly very small compared to the distances to nearby stars — of the order of a light day, or, in the downwind direction, perhaps several light days, while the distances to nearby stars amount to several light years. Heliospheric structure is determined jointly by various flows entering through its inner (solar) and outer (IS) boundary regions, and by processes taking place in between.

The heliosphere is separated from the Sun and from the local IS medium (LISM) by complex regions. On the solar side, the convective zone, photosphere, chromosphere, transition region, and corona provide a very dynamic and structured inner environment that gives rise to the outward flowing supersonic, magnetized solar wind (SW), the dominant component of the heliosphere beyond a heliocentric distance of a few solar radii. The source region of the SW can be studied only by remote sensing (for a good early review, see Bird and Edenhofer, 1990; for more recent reviews, based on SOHO and partly on Ulysses results, see Brekke and Fleck, 2002, and Fisk, 2001). It can only be hoped that the Solar Probe of NASA, planned to approach the solar surface to a few solar radii, will be sooner or later realized, opening up at least the outer part of the SW source region for *in situ* study. ESA's Solar Orbiter mission will monitor the Sun from a distance of some tens of solar radii. Up to now, the Helios probes in the 1970's approached the Sun most (0.29 AU or 62 solar radii), while Voyager–1, the outermost member of our small outer heliospheric flotilla, was at a heliocentric distance of 85 AU at the time of the Erice Summer School. It overtook Pioneer–10 in 1998 (Pioneer-10 was launched in 1972 and is still alive though weak; it was last contacted in March 2002, just after its 30-eth birthday).

Effects propagating inward from the LISM through the bow shock (BS, separating supersonic ionized IS wind from the subsonic one, diverted and slowed by the heliosphere), the heliopause (HP, the surface separating SW matter and magnetic field from IS fields and the charged component of the

IS wind), and the termination shock (TS, separating supersonic and subsonic SW) are less obvious than those propagating outward (see *e.g.* Király 1998, 2001). Effects of solar activity, mostly mediated by the SW and by its frozen-in interplanetary magnetic field (IMF), contribute much more to heliospheric variability than inward effects of the presumably slowly changing LISM (although, in fact, the variability characteristics of the LISM are still poorly known). The speed of the partially ionized LISM relative to the solar rest frame is only about 26 km/s, while the SW speed at solar minimum is 400 to 800 km/s in the solar equatorial and polar regions, respectively. The relative importance of inward and outward propagating heliospheric effects differs from the case of the otherwise analogous terrestrial magnetosphere, where external (solar–interplanetary) influences are more variable than the slowly changing terrestrial magnetic fields of internal origin.

The most important parameters that set the scale size of heliospheric structures are the SW dynamic pressure or ram pressure (the product of SW density and squared speed), LISM speed in the solar frame and its thermal pressure, interstellar ion density and temperature, and the magnitude and direction of the local IS magnetic field. While SW parameters as well as LISM speed are directly measured, the degree of ionization of the LISM and in particular the external magnetic field are only poorly known. Thus pressure equilibrium calculations (e.g. Axford and Suess 1994) give only tentative results. Some recent estimates for parameters of the local interstellar cloud (Frisch 2002) are as follows: relative Sun–cloud speed 26 km s^{-1}, temperature 7000 K, neutral H density 0.24 cm^{-3}, electron density 0.13 cm^{-3}. Typical estimates for the heliocentric distances of characteristic surfaces in the upwind direction of the LISM: 85–110 AU for the TS, 120-160 AU for the HP; for the BS, even its existence is unclear, and it may also include separate subshocks in the distance range of 150 to 300 AU. Every 10^4 to 10^5 years the heliosphere is expected to meet a cloud of about 10 cm^{-3} density. Immersion in such a cloud would reduce its scale size by about a factor of 10, and the boundary regions might also become unstable (Zank and Frisch 1999). Even more extreme densities like those of Fred Hoyle's Black Cloud would probably completely disrupt the heliosphere, but such encounters should be extremely rare.

Several indirect clues (Stone and Cummings 2001) indicate that Voyager–1 will reach the TS between 2003 and 2008, and might even cross it several times during that period, as the speed of inward and outward motions of the TS may excede that of our most distant probe (Whang and Burlaga 2000, Stone and Cummings 2001). There is hope that Voyager–1 will survive to cross the HP as well. For crossing the BS before its power supply runs out around 2020, the chances are less, and they are diminishing as the TS crossing is delayed.

2.3. *Some complications*

There are several complications omitted in the above schematic description. Energetic particles streaming through plasma often give rise to instabilities, thus the test particle description is not strictly valid. Bundaries, even if thin, are not 2-dimensional structures, and may be strongly modified by instabilities and wave generation. Inner and outer heliospheric boundary regions are accessible only by remote sensing; *in situ* observations by spacecraft instrumentation are at present restricted to part of the supersonic SW region, between solar distances of 0.3 AU (Helios) and 85 AU (Voyager–1). One might expect that the inner boundary region should be well known and understood, due to long-term observation of the Sun over all wavelengths of the electromagnetic (EM) spectrum, both from ground-based and space instruments. Most of the observations, however, refer to much lower levels of the solar atmosphere than where the SW is accelerated to supersonic speeds (at 2-10 solar radii, sometimes even further out). Magnetic fields, as well as velocity and density variations are seen only in projection, with limited spatial resolution, and the 3D structure can be only tentatively inferred. Most of the electromagnetic (EM) radiation is related to electrons, while the dynamically dominant part of the SW consists of ions. Gamma-ray signatures of the interaction of accelerated ions with ambient gas are actually seen in flares, but no detailed information on the behaviour of ions in quiet regions and in coronal holes exists. The scarcity of EM information is even more acute for the outer boundary region. Although solar UV light backscattered from interstellar neutral atoms entering the heliosphere is well observed (mainly Lyman-α and He-I lines), those observations provide little informationon on the position and structure of boundaries.

The solar activity cycle (SAC) reverses large-scale solar magnetic dipole fields with a periodicity of about 11 years (a complete magnetic cycle or Hale-cycle lasts 22 years). On the solar surface the SAC is best observed through variations of the number and latitudinal distribution of sunspots. The SW drawing out coronal fields from the rotating Sun results in an Archimedian spiral pattern of the interplanetary magnetic field, and also causes an approximately 26-day recurrence tendency (in the inertial frame) of solar wind structures. Fast solar wind streams overtake slow ones, creating stable corotating interaction regions (CIRs) at relatively quiet periods, bounded by forward and reverse shocks in the low-latitude SW at radial distances of about 2 to 10 AU. Beyond 10 AU, CIRs still persist but widen and tend to form merged interaction regions (MIRs). As the SAC progresses toward solar maxima, the dipole becomes more tilted and the separating line between inward and outward oriented fields is also more complex. Magnetically unipolar fast SW streams emanating from polar coronal holes at solar minima become fragmented at maxima, and the time scale of intrinsic variation becomes shorter than the solar rotation time. Coronal mass ejections (CMEs), i.e. massive plasma clouds catapulted into interplanetary space by coronal processes also become more frequent. The largest CMEs give rise to outward moving magnetic barriers (global merged interaction regions or GMIRs), and may sweep through the entire heliosphere. Those barriers scatter and reflect both inward and outward propagating energetic particles, causing step-like decreases in the galactic CR flux, and forming expanding reservoirs for particles of internal origin.

3. Energetic particle populations in the heliosphere

3.1. *Is there a baseline population?*

There is a wide variety of suprathermal and energetic particle populations in the heliosphere, from SW up to CR energies. The bulk of SW ions have keV amu^{-1} energies, essentially determined by SW speed; the component of freshly ionized atoms "picked up" by the SW extends (without further acceleration) to about twice the SW speed, i.e. to about 4 times the SW energy. Any particle with energies from SW to a few hundred keV is called *suprathermal*, above that *energetic* (although the terminology is rather vague).

Intensity levels of suprathermal and energetic particles of solar and heliospheric origin in the near-Earth interplanetary space never seem to drop below some energy-dependent threshold, even when solar–heliospheric activity is at its lowest level. Those minimum intensities are often attributed to a separate component called quiet-time or baseline population. It is not clear, however, to what extent instrumental background effects contribute to it even in the best data sets. For some discussion, see e.g. Király and Kecskeméty 1998, Logachev *et al* 1998, and Zeldovich *et al* 1998. The spectrum of the baseline component seems to decrease from 10 to 20 keV amu^{-1} (*i.e.* from a few times the SW energy) to several MeV as a power law. Since modulated CR energy rises almost proportionally to energy up to a few hundred MeV, and it also forms a fairly stable population, the sum of the two differential spectra has to have a minimum somewhere. At solar minima, it happens to be slightly below 10 MeV for protons at a level of about 10^{-1} m^{-2} s^{-1} st^{-1} MeV^{-1}. Flux levels in major solar events excede the minimum by 5 to 7 orders of magnitude, and even minor solar or heliospheric events may cause order of magnitude increases. In the total fluence (integrated flux) over extended periods the baseline component is quite negligible. A better understanding of its origin, however, is important from the point of view of basic acceleration processes going on even when other signs of solar–heliospheric activity are absent. There have been recent advances on the dependence of the baseline component on heliospheric radius and latitude (see Kecskeméty *et al* 2001, Mason 2001). It appears that in the ecliptic the baseline component is lowest at about 0.6 AU, while in the unipolar field regions far from the ecliptic the baseline is even lower.

3.2. Solar energetic particles

Solar processes are the direct or indirect drivers of most acceleration phenomena in the heliosphere. The traditional name "solar energetic particles" (or SEPs) refers to particles accelerated in the vicinity of the Sun, and then escaping to interplanetary space (those unable to escape can also be detected by their EM signatures). The prime site of their acceleration is either in chromospheric or coronal flares, or in CME shocks propagating outward. SEPs were first observed in ground-level events (GLEs) in 1942, and associated with solar flares by Forbush (1946). Before the space age, SEP events were only detected when energy spectra extended beyond about half GeV, and some events containing particles of several tens of GeV were occasionally also observed. It was realized and emphasized only much later that the largest events were not directly related to flares, but to shocks generated by CMEs (Kahler 1978, Gosling 1993).

Solar energetic particle events are thus subdivided into flare (or impulsive) and CME (or gradual) events, where the traditional names put in parentheses refer to the duration of EM and particle emissions. The two classes of events differ in many of their characteristics. Flare events are shorter, smaller, their composition is both variable and quite distinct from that of the corona or the SW, their charge states reflect a flare-heated source region of 3 to 5 MK temperature (typically full ionization up to Si, while Fe ions have a charge of about 20). Only fast CMEs (about 1 to 2 % of all CMEs) accelerate particles efficiently. CME particle events are more extended in time than flare-originated ones, usually give rise to larger events, the composition is closer to that of the SW, and the charge states correspond to lower temperatures (1.5 to 2 MK, typical ionization state of Fe is about 12). The time profiles of CME SEP events often display a peak around the time of shock passage through the observation point. That peak often indicates local shock acceleration, but also shock compression of ions accelerated closer to the Sun may play a role.

Some CME shocks travelling outward continue to accelerate particles beyond 1 AU, thus not all CME particles arrive from the vicinity of the Sun, although the most energetic ones do. The sharp distinction between flare and CME events has been questioned recently by several workers (see e.g. Cane 2002), and a more complex paradigm is likely to emerge. In addition to the shock accelerated component giving the major contribution, a flare-related component appears to be always present in major CME events, probably due to reconnection giving rise to the CME. It is also found that the largest events are not due to single shocks, but to the interactions of major shock events with previous weaker ones (Gopalswamy *et al* 2002, Kallenrode and Cliver 2001). Such large events (also called rogue events) often dominate the total particle fluence of entire solar cycles. One recent example of such events is the much discussed "Bastille-day event" on 14 July 2000.

Magnetic energy of coronal loops tapped by reconnection provides most of the heating and nonthermal EM radiation observed in flares, but the detailed processes behind the very fast energy release are still unclear. At the reconnection site strong electric fields accelerate prompt electron and possibly also proton beams, later transforming their energy partly into a variety of waves (or turbulence), partly into heating the plasma. The recently launched RHESSI and TRACE X and gamma telescopes have identified some fine details of that process, showing that electron acceleration really precedes heating. A second generation of particles is later accelerated by waves through the second-order Fermi process. A substantial fraction of accelerated particles is guided downward into denser regions, where they loose their energy by bremsstrahlung and are absorbed, emitting EM radiation in various wave bands. Some particles escape along open field lines, giving rise to impulsive SEP events. The ratio of absorbed and escaping fractions varies widely. An important difference between flare and CME-generated events is that the energy input in flares occurs in active regions of the lower corona dominated by magnetic fields and waves, thus shocks are expected to play a minor role relative to direct induction electric fields, resonant wave–particle interactions, and second-order Fermi acceleration. One important peculiarity of impulsive events is the huge enhancement of the ^{3}He/^{4}He isotope ratio up to several MeV, first observed by Hsieh and Simpson (1970). While in the photosphere, corona and SW ^{3}He/^{4}He $\sim$ 0.0004, the ratio is above 1 in some events. ^{3}He-rich events were found to be associated with 10 to 100 keV electron beams and type III radio bursts, and also with heavy ion enhancements. The now favoured explanation of the effect is in terms of electromagnetic ion cyclotron (EMIC) waves, proposed by Temerin and Roth (1992). The special property of ^{3}He that selects it for preferential acceleration is its unusual charge to mass (Q/M) ratio of 2/3. Waves generated in an extended frequency and wave number range are strongly damped for ratios near 1 (H^+) and 0.5 (He^{++} and other fully ionized species), as those abundant species use up the wave energy very fast (Reames 1999).

In the field of acceleration in CME shocks it is a very important new development that the particles accelerated there should often not be considered as test particles, but they also modify shock structure (Reames and Ng 1998, Reames 1990 and 2000). Of course such "cosmic ray modified shocks" have been theoretically discussed earlier, but this is the first time that their effects have been actually seen. Particles streaming out of the shock generate waves that throttle the streaming below some characteristic energy. Spectra are flattened below this "knee". As different particle species having the same velocity resonate with different waves, the process affects composition as well. Understanding CME-generated energetic particle transport in terms of self-generated waves near the shock apparently solves some earlier discrepancies of interplanetary propagation. Obviously, a close examination of self-generated wave effects is very important for understanding CR generation as well.

3.3. Acceleration and propagation in the inner heliosphere

As already mentioned in the previous section, the acceleration by outward travelling shocks of CME origin is not restricted to the solar neighbourhood, but often continues beyond 1 AU. Another class of inner heliospheric energetic particles is accelerated in CIR shocks, usually forming beyond 1 AU. The shock-accelerated particles then stream inward at 1 AU, and are associated with fast, recurrent SW streams. CIR acceleration extends at least to a heliocentric distance of 10–15 AU, with a maximum between 3 and 5 AU. Under slowly changing heliospheric conditions CIRs are nearly stationary in the frame corotating with the Sun. The CIR peaks seen by a near-Earth spacecraft are thus not proper 'events', but represent the changing magnetic connection between the source region and the spacecraft. For suprathermal particles less energetic than 100 keV amu^{-1}, however, inward propagation is strongly inhibited by scattering on magnetic irregularities (Mason 2001). Thus the suprathermal component has to be produced fairly close to the observation point, and not at several AU where CIR shocks accelerate efficiently. The large contribution of singly ionized He at several AU found by Glockler et al (1994) compared to low H$^+$ content at 1 AU also supports that claim. The energy spectra of CIR particles steepen above 1 MeV/amu. Ulysses recorded a unique series of 36 consecutive CIR enhancements between June 1992 and December 1994, at southerly heliospheric latitudes between 0 and 80 degrees. In those in $situ$ observations acceleration was found to be more efficient in the reverse shock of the forward–reverse shock pair of the CIRs. For recent reviews on CIRs and their successors in both the inner and outer heliosphere, see Gazis et al 2000.

At 1 AU corotating interaction regions contain steep field gradients, although shocks are not yet present. Jokipii (2001) pointed out that those gradients are sufficient for acceleration when some conditions on scattering and field configuration are fulfilled. Due to their different charge, speed, and phase space distribution, pick-up ions were shown to be much more efficiently injected into the CIR accelerator than SW ions (Gloeckler et al 1994). It was thus demonstrated that some of the pick-up ions are accelerated prior to drifting out to the TS. A new, inner heliospheric source of pick-up carbon was discovered by Geiss et al in 1995, followed by other elements. As C is supposed to be bound up in grains in the LISM, practically no neutral C atoms are likely to enter the heliosphere, thus a pick-up component was not expected. Recently Gloeckler et al (2000) showed that inner source pick-up abundances can be explained in terms of SW composition (exept for H). The result was interpreted in terms of absorption of SW ions by dust, and their re-emission as neutral atoms. Those atoms are then efficiently ionized in the inner heliosphere. The importance of dust in heliospheric acceleration reminds one of acceleration from evaporating dust particles in SN shocks, suggested as an explanation of the FIP fractionation in CRs (see $e.g.$ Ellison et al 1997).

As already mentioned, one of the most surprising results of Ulysses was that periodic CIR energetic particle enhancements under solar minimum conditions continued to high heliospheric latitudes, far beyond the latitudes where CIR shocks ceased to exist. During the 2000 to 2001 fast latitude scan of Ulysses solar maximum conditions prevailed, and CMEs predominated. A close association of the energetic particle events seen by Ulysses at high latitudes and by the Advanced Composition Explorer (ACE) near the ecliptic was, however, not less surprising.

3.4. *Messengers from the outer heliosphere and beyond*

Most neutral atoms, as well as ACRs and CRs arrive to the inner heliosphere from the boundary regions or from IS space. ACRs were first observed in 1973 and interpreted as derivatives of interstellar neutral atoms by Fisk *et al* in 1974. That was the start of a success story, revealing increasing detail on both ACRs and their progenitors ever since. The ionization of neutrals due to solar UV radiation and charge exchange reactions with SW ions create singly-charged ions picked up by the SW and carried outward. The SW is measurably slowed by the mass-loading effect of those ions (by about 60 km s^{-1}, see Wang and Richardson 2002). Those PU ions of mostly interstellar ancestry (but containing some inner and perhaps outer heliospheric source ions as well) are then preferentially accelerated at the TS and partly return into the inner heliosphere as ACRs, still only singly ionized below some energy threshold. An interesting trap for ACRs is the terrestrial magnetosphere, where some of those singly charged ions loose additional electrons due to interactions with the upper atmosphere, and the higher charge tightens their trapping. Some persistent ACR radiation belts are thus formed. ACRs also charge exchange in the heliosheath with interstellar atoms, and return to the inner heliosphere as energetic neutral atoms (ENAs), providing a chance to remotely sense the regions with weak EM signatures but intense charge exchange. ENAs arriving from the tail region of the heliosheath have been recently discovered by a Ulysses team (Hilchenbach *et al* 1998).

The outward flowing magnetic disturbances modulate (i.e. reduce the intensity) of both CRs and ACRs. As the ACR energy spectrum is much softer, the shielding effect of heliospheric magnetic fields is much more efficient for them. In fact, at solar maxima ACRs practically disappear from the inner heliosphere (the flux ratio between solar minima and maxima is at least 100). Thus ACRs provide a more sensitive probe for the state of disturbence of the heliosphere than CRs do, but CRs of GeV energy give better information about the domains of the boundary regions that are beyond the site of acceleration of ACRs. A recent summary of modulation effects was given by Heber (2002). One should also mention that modulation also depends on the dominant polarity of the Sun, because the predominantly positive particles arrive along different routes for the two alternating polarities. Electrons of several MeV energies observed near Earth have also a large contribution from Jupiter, thus the analysis of their intensity variation is more complicated.

For CR researchers looking for the origin of CRs it would be important to know how to demodulate the locally observed CR spectrum, *i.e.* how to calculate fluxes in local IS space from those measured inside the heliosphere. Unfortunately this cannot be reliably done for local interstellar energies much lower than one GeV, because the energy loss processes involved in modulation are of statistical nature, and even the lowest energy CRs observed near Earth come from a population of several hundred MeV energy. It is hoped that as the Voyagers continue their route out of the heliosphere a better understanding of low-energy CR spectra may result.

4. Discussion and conclusions

Our heliosphere is a fairly well-equipped laboratory for the study of universal injection, acceleration, and propagation processes. Several results discussed above seem relevant to CR astrophysics. Pre-acceleration in the SW shows how ion species are selected for efficient further acceleration. Flares are archetypes of reconnection processes in magnetic field dominated plasma, even for those occuring on much larger scales. Generation of intense particle beams, wave generation by the beams, particle scattering on the waves and the resulting second-order Fermi acceleration, nonlinear cascading of the waves and generation of turbulence, are all generic processes occurring in many active sites of the Universe. Blast waves and shock formation in plasma, diffusive and drift acceleration, and in particular confinement of upstream particles by self-generated waves are undoubtably important processes in cosmic ray sources or in the interstellar medium processed by powerful supernova shocks. Magnetic bottles created by previous mass ejections should also be ubiquitous features, found *e.g.* in and around galactic arms or in molecular clouds giving rise to a series of supernovae. Current sheets, termination shocks, and corotating shocks reminiscent of heliospheric CIRs should also occur in many rotating systems spewing out some sort of magnetized wind. Magnetospheric boundaries and bow shocks are probably even more common.

Several space probes have contributed or are going to contribute to the understanding of the energetic component of the heliosphere. The Voyagers are sampling the effects of high solar activity in the outer heliosphere, and are approaching the most extensive heliospheric shock. Ulysses has seen a very different latitudinal variation recently than earlier, during its solar minimum fast latitude scan. SOHO continues its complex analysis of the Sun and of its immediate environment. Very efficient new instruments aboard TRACE and RHESSI provide unprecedented fine detail on energetic solar processes. IMAGE is just opening a new window — neutral atom spectroscopy. IMP-8 has finished its 28-year monitoring at 1 AU last October, but data taking was resumed in February 2002 with more restricted coverage. New missions are also being prepared or planned for most of the fields discussed in this review. The two STEREO spacecraft will study CMEs and their energetic particles from two different directions. Solar-B, the successor of Yohkoh is to continue its work with improved resolution and spectral coverage. The heliospheric radial distance coverage of *in situ* observations will be extended both inward and outward, beyond the records set by the Helios and Voyager probes. NASA's Solar Probe is planned to sample the outer solar corona at about 4 solar radii. ESA's Solar Orbiter will study the Sun and its plasma environment from a distance of 45 solar radii, both from the solar equatorial plane and from higher latitudes. To study the distant heliosphere and our IS environment an Interstellar Probe is being planned.

All those efforts are being spent on a single one in at least 10^{11} stellar environments in our Galaxy. Will it be worth it? Shall we better understand the ocean by knowing a single droplet? We very much hope so. But only future will tell.

Acknowledgments

It is a pleasure to express my thanks to the Organizers of the 13th International Course of Cosmic Ray Astrophysics for invitation and support. Hungarian National Grant OTKA-T-030078 is acknowledged for financial backing. The International Space Science Institute (ISSI, Bern) is also thanked for supporting an earlier project relevant to the present review.

References

Axford W I and Suess T 1994 *EOS* **75** 587
Ben-Jaffel L Puyoo O and Ratkiewicz O 2000 *Astrophys. J.* **533** 924
Bird K and Edenhofer P 1990 in *Physics of the Inner Heliosphere I* ed R Schwenn and E Marsch (Berlin, Heidelberg: Springer) p 13

Brekke P and Fleck B 2002 *27th Int. Cosmic Ray Conf. (Hamburg) Invited, Rapporteur and Highlight Papers* ed R Schlickeiser p 21

Cane H V 2002 *27th Int. Cosmic Ray Conf. (Hamburg) Invited, Rapporteur and Highlight Papers* ed R Schlickeiser p 311

Ellison D C, Drury L O'C and Meyer J-P 1997 *Astrophys. J.* **487** 197

Fisk L A, Kozlovsky B and Ramaty R 1974 *Astrophys. J.* **190** L35

Fisk L A 2001 *J. Geophys. Res.* **106** 15849

Forbush S E 1946 *Phys. Rev.* **70** 771

Frisch P C 2002 ArXiv:astro-ph/0207002 v1

Gazis P R 2000 *J. Geophys. Res.* **105** 19

Geiss J, Gloeckler G, Fisk L A and von Steiger R 1995 *J. Geophys. Res.* **100** 23373

Gloeckler G, Geiss J, Roelof L A, Fisk L A, Ipavich F M, Ogilvie K W, Lanzerotti L J, von Steiger R and Wilken B 1994 *J. Geophys. Res.* **99** 17637

Gloeckler G, Fisk L A, Geiss J, Schwadron N A and Zurbruchen, T A 2000 *J. Geophys. Res.* **105** 7459

Gopalswamy N Yashiro S Michalek G Kaiser M L Howard R A Reames D V Leske R and Rosenvinge T von 2002 *Astrophys. J.* **572** L103

Gosling J T 1993 *J. Geophys. Res.* **98** 18937

Gurnett D A Kurth W S Allendorf S C Poynter R L 1993 *Science* **262** 199

Heber B 2002 *27th Int. Cosmic Ray Conf. (Hamburg) Invited, Rapporteur and Highlight Papers* ed R Schlickeiser p 118

Hilchenbach M *et al* 1998 *Astrophys. J.* **503** 916

Hsieh K C and Simpson J A 1970 *Astrophys. J. (Lett.)* **162** L191

Jokipii J R 2001 *27th Int. Cosmic Ray Conf. (Hamburg)* ed R Schlickeiser p 3581

Kahler S W, Hildner E and Van Hollebeke M A I 1978 *Solar Phys.* **57** 429

Kallenrode M-B and Cliver E W 2001 *27th Int. Cosmic Ray Conf. (Hamburg)* ed R Schlickeiser p 3314

Kecskeméty K Müller-Mellin R and Kunow H 2001 *27th Int. Cosmic Ray Conf. (Hamburg)* ed R Schlickeiser p 3108

Király P 1998 Nucl. Phys. B (Proc. Suppl.) **60B** 12

Király P Kecskeméty K 1998 Proc. 16th European Cosmic Ray Symp. Alcalá, Spain Ed. Medina J p.173

Király P 2001 *J. Phys. G: Nucl. Part. Phys.* **106** 1579

Kurth W S and Gurnett D A 1993 *J. Geophys. Res* **98** 15129

Logachev Yu Zeldovich M Kecskeméty K Király P 1998 Proc. 16th European Cosmic Ray Symp. Alcalá, Spain Ed. Medina J p.181

Mason G M 2001 *Space Sci. Rev.* **99** 119

Ratkiewicz R and Ben-Jaffel L 2002 *J. Geophys. Res.* **107**

Reames D V, Ng C K 1998 *Astrophys. J.* **504** 1002

Reames D V 1999 *Space Sci. Rev.* **90** 413

Reames V R 2000 *Astrophys. J.* **540** L111

Steinacker J, Meyer J-P, Steinacker A, Reames D V 1997 *Astrophys. J.* **476** 403

Stone E C and Cummings A C 2001 *27th Int. Cosmic Ray Conf. (Hamburg)* ed R Schlickeiser p 4263

Temerin M and Roth I 1992 *Astrophys. J.* **391** L105

Wang C and Richardson J D 2002 *J. Geophys. Res.* in press

Whang Y C and Burlaga L F 2000 *Geophys. Res. Letts* **27** 1607

Zank G P and Frisch P C 1999 *Astrophys. J.* **518** 965

Zeldovich M Logachev Yu Kecskeméty K Király P 1998 Proc. 16th European Cosmic Ray Symp. Alcalá, Spain Ed. Medina J p.177

ON THE ORIGIN AND PROPAGATION OF THE ULTRAHIGH ENERGY COSMIC RAYS

MARIA GILLER

Department of Physics and Chemistry, University of Lodz, Poland
E-mail: mgiller@kfd2.fic.uni.lodz.pl

The ultrahigh energy (UHE) cosmic rays ($E > 10^{19}$ eV) are of particular interest for astrophysicists. Not only because of the highest parlicle energies involved but also because quite a lot can be predicted about their propagation in the extragalactic and Galactic space .the best example is the prediction of the cutoff in the energy spectrum. In these lectures we review in some detail the UHE cosmic ray propagation, their deflections and delays in the extragalactic chaotic fields together with energy losses on the universal radiations. The latest experimental results on energy spectrum and distribution of the arrival directions are reviewed and confronted with various assumptions about the source spatial distribution. The probabilities of multiple events observed by AGASA are calculated and seem to indicate an existence of point sources, although their nature is not yet clear.

1. Introduction

Cosmic ray (CR) energy spectrum extends from below 1 GeV up to values estimated as over 10^{20} eV. These lectures are devoted to the high-energy end of the spectrum, i.e. to cosmic rays with $E > 10^{19}$ eV. This energy region is a subject of a particular interest as (contrary to CRs with $10^{16} - 10^{19}$ eV) quite a lot can be predicted about particle propagation in the Galactic and extragalactic space. The most important prediction was made by Zatsepin and Kuzmin (1) and Greisen (2) just after the discovery by Wilson and Penzias in 1964 (Nobel Prize in 1978) that the extragalactic space is pervaded by microwaves interpreted quickly as the black body radiation, a relict of the hot early Universe. The prediction was (known now as the GZK effect) that the Universe becomes nontransparent for CRs with energies above $\sim 5 \cdot 10^{19}$ eV because of particle interactions with the microwave background. If the observed CRs below that energy were to be produced by extragalactic sources (together with CRs above $\sim 5 \cdot 10^{19}$ eV) then the observed spectrum should be dramatically cutoff at high energies. The observation of such a cutoff would be almost a proof of an extragalactic CR origin in this energy region. As we shall see later, however, the experimental situation is not yet conclusive.

Another prediction which can be made is that about the angular distribution of the particle arrival directions. The Galactic magnetic field seems to be too weak to confine and isotropize CR protons above $\sim 10^{18}$ eV, so that if CRs of those energies were produced in the Galaxy, their arrival directions should be rather strongly correlated with the Galactic disk. This prediction is not as absolute as the previous one (the GZK cutoff) because the extent of the Galactic magnetic halo is not well known. Moreover, if a CR source was not producing particles continuously and the magnetic halo was very large (~ 20 kpc) there could be an isotropic flux of $> 10^{19}$ eV iron nuclei observed at the Earth [3]. The observations do not indicate any particle excess from the Galactic disk, this being after all a point for the extragalactic origin.

The third prediction concerns a possibility of detecting relatively nearby extragalactic point sources. Charged particles with such high energies should travel in the extragalactic space along almost straight lines, being only slightly deflected by the magnetic fields, indicating directions towards their origin sites. Indeed, the observations of the AGASA experiment of multiple events [4] (two or three showers arriving from almost the same direction) is a rather strong indication that, despite the small statistics, we are beginning to see some point sources.

In these lectures all the above aspects of the UHE CRs will be discussed in more detail.

However, I owe to the Erice School participants one explanation why this text is a bit different from what I presented in Erice in June 2002. At that time the only published CR energy spectrum at the highest energies, was that from the AGASA experiment [5], extending up to $\sim 3 \cdot 10^{20}$ eV without any cutoff. Together with the observed particle isotropy that really was a conundrum. Thus the main stress of my talk was put on finding a way out of it, by mainly reviewing papers which were trying to do this.

In August 2002, however, new results from the HiRes experiment were published [6]. According to the authors the CR energy spectrum, measured by the fluorescence light technique, does steepen above $\sim 6 \cdot 10^{19}$ eV, agreeing well with the prediction of the universal origin. So now we are facing two different experimental results, leading to different conclusions about the CR origin! Thus, the discrepancy must be due to the different experimental methods and ways of determining the shower primary energy. More attention will be given here to this problem than it was done in Erice (although, as I must admit, without much of a conclusion).

2. The energy spectrum

In this chapter we shall describe in more detail what sort of energy spectrum one should expect if CR particles were of the extragalactic origin suggested by their arrival direction distribution. We shall show that CR propagation in the extragalactic space changes dramatically above a few times 10^{19} eV, modifying the particle energy spectrum at production. It is because CRs (both protons and heavier nuclei) interact with the extragalactic radiations (mainly with the cosmic microwave background, CMB, but also with the infrared radiation) and lose energy. We do not know whether the highest energy CRs arriving to us are protons or heavy nuclei, so we shall discuss both possibilities. This subject has been already described in many papers and reviews, see e.g. [7].

2.1. Energy losses of protons

If a high energy proton interacts with a CMB photon if may produce a new particle X. If the particle has mass M then the threshold energy for the process

$$p + X \to N + X \tag{1}$$

where N is a nucleon, can be derived from kinematics only:

$$E_{th} = \frac{M(m_p + M/2)}{\varepsilon(1 - \cos\theta)} \tag{2}$$

where ε is the photon energy, m_p is the proton mass and θ is the angle between the momenta of the proton and the photon in the laboratory system. The "new particle" X can be an electron-positron pair ($M \cong 1$ MeV). Adopting for ε the average CMB photon energy ($6 \cdot 10^{-4}$ eV) we get that the pair production can occur if the proton energy $E > 10^{18}$ eV. The produced particle can be a pion (π^0 or π^+) but only if $E > 10^{20}$ eV. (In reality the processes start at lower energies, because there are a lot of photons with $\varepsilon < \bar{\varepsilon}$ and/or with $\theta > \pi/2$).

For the CR propagation problem it is important to know how these interactions affect the proton energy. The relevant parameter is the particle ebergy attenuation length λ_{att}, defined as

$$\frac{1}{\lambda_{att}(E)} = \frac{1}{E} \left\langle \left| \frac{dE}{dl} \right| \right\rangle = \left\langle \frac{k}{\lambda_{int}} \right\rangle \tag{3}$$

where λ_{int} is the mean interaction length with photons of a given energy, incident at a given angle and k is the inelasticity coefficient in those interactions. The averaging is over all photon energies and angles.

The inelasticity k at the threshold can be derived from kinematics only:

$$k_{th} = 1 - \left[1 + 2x \left(1 - \frac{x}{2} \right) \right]^{-1/2} \qquad (4)$$

where $x = M/m_p$. If $x \ll 1$ (well fulfilled for pair production) then $k_{th} \approx x$ and we have that $k_{th} \approx 10^{-3}$ and 0.11 for pair and pion production respectively. Above the threshold k does not change much and if we assume that it is not correlated with λ_{int} then

$$\lambda_{att}^{-1}(E) \approx \langle k(E) \rangle \cdot \langle \lambda_{int}^{-1}(E) \rangle \qquad (5)$$

The integration over photon energies ε and angles θ is represented by the following expression

$$\langle \lambda_{int}^{-1}(E) \rangle = \int_{\varepsilon_{min}}^{\infty} d\varepsilon \int_{-1}^{1} [\nu + c \cdot \cos(\pi - \theta)] \frac{\rho(\varepsilon)}{4\pi} \sigma(s) d(\cos \theta) \qquad (6)$$

where $\rho(\varepsilon)d\varepsilon$ is the number density of photons with energies $(\varepsilon, \varepsilon + d\varepsilon)$ per unit solid angle and ν is the proton velocity. The velocity in brackets is the relative velocity of the proton and photon as seen in the laboratory system. The cross section for a given process depends only on the total energy $\sqrt{s}$ in the center of mass system. Actually the integral over $\cos \theta$ in (6) can be simplified a bit when s is the new variable

$$\langle \lambda_{int}^{-1}(E) \rangle = \frac{1}{8E^2} \int_{\varepsilon_{min}}^{\infty} \frac{d\varepsilon \rho(\varepsilon)}{\varepsilon^2} \int_{s_{min}}^{s_{max}} \sigma(s)(s - m_p^2) ds \qquad (7)$$

where s depends on ε.

The cross section for pion production is shown in Fig.1 [8]. It can be seen that it has a resonant character with a maximum for s close to the threshold (1.16 GeV^2), so that the inelasticity is not much different from that calculated above. λ_{att} for protons is shown as a solid curve in Fig.2. The drastic decrease of λ_{att} above $\sim 3 \cdot 10^{19}$ eV is due to a strong increase of the number of photons above the threshold for pion production. At 10^{20} eV $\lambda_{att} \sim 100$ Mpc, what is small on the Universe scale. At $\sim 10^{21}$ eV practically all photons are above this threshold and λ_{att} flattens out on the level of only (3–4) Mpc!

At lower energies, where even pair production becomes uneffective the Universe expansion comes to the scenes. The present rate of energy de-

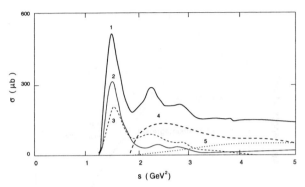

Figure 1. Cross section of photopion production. (1) – the summation of all channels, (2) – $\gamma p \to p\pi^\circ$, (3) – $\gamma p \to n\pi^+$, (4) – $\gamma p \to n$ or $p +$ doublepion, (5) – $\gamma p \to p +$ triplepion.

crease, λ_H, equals

$$\lambda_H = \frac{Ec}{|dE/dt|} = \frac{c}{H_o} \qquad (8)$$

where H_o is the present Hubble constant. For $H_o = 60$ km·s^{-1}·Mpc^{-1} we obtain that $\lambda_H = 5$ Gpc as drawn on Fig.2.

2.2. Energy losses of heavy nuclei

First important thing to notice is that a nucleus of the same energy as proton has the Lorentz factor A times smaller (A is its mass number). As the threshold Lorentz factor for producing a pair or a pion on a CMB photon does not depend on the nucleus mass (i.e. it is the same as that of the proton) then the nucleus energy threshold is $\sim A$ times higher than that of a proton. For iron we have

$$E_{th} \cong \begin{cases} 6 \cdot 10^{19} \text{ eV} & \text{for } e^+e^- \text{ production} \\ 6 \cdot 10^{20} \text{ eV} & \text{for } \pi \text{ production} \end{cases} \qquad (9)$$

if the photon energy equals to its mean value $6 \cdot 10^{-4}$ eV. We can estimate the nucleus attenuation length for this process from what we already know for protons:

$$\lambda_{att}^A(E) \cong \frac{1}{k_A \sigma_A n_\gamma} \qquad (10)$$

At the threshold energy the inelasticity is $\sim A$ times smaller than that for a proton. As the cross section fulfils the relation

$$\sigma_A(E) = Z^2 \sigma_p(E/A) \qquad (11)$$

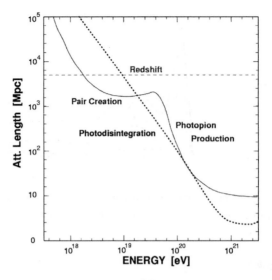

Figure 2. The attenuation lenght of cosmic rays as a function of energy. The solid curve shows the case for nucleons calculated by Yoshida and Teshima [12]. The dashed curve shows the case for iron calculated by Puget *et al* [10], but later calculations [9] show that it may be shifted up. The bound given by redshift (adiabatic energy loss) are applicable to all primares.

we obtain

$$\lambda_{att}^A(E) = \frac{A}{Z^2} \lambda_{att}^p(E/A) \tag{12}$$

$A/Z^2 \sim 0.082$ for iron meaning that these nuclei would attenuate much quicker than protons with energies 56 times smaller. At 10^{20} eV $\lambda_{att}^{Fe} \approx 3 \cdot 10^2$ Mpc for pair production .

There is, however, another process, which turns out to be more important – it is the photodisintegration of the nuclei, consisting mainly in emitting one nucleon after a collision with a background photon. The process has a resonant character and the experimental data on the cross section can be approximated as follows

$$\sigma(\varepsilon^*) = 1.45 \text{ mb} \cdot A \cdot \frac{(\varepsilon^* T)^2}{(\varepsilon^{*2} - \varepsilon_0^2)^2 + (\varepsilon^* T)^2} \tag{13}$$

for $\varepsilon^* \leq 30$ MeV, where ε^* is the photon energy in the rest frame of the nucleus, $T = 8$ MeV and $\varepsilon_0 \simeq 18$ MeV for iron nucleus. For $\varepsilon^* > 30$ MeV $\sigma_A(\varepsilon^*) \approx A/8$ mb. For a CMB photon ($\varepsilon = 6 \cdot 10^{-4}$ eV) to have

$\varepsilon^* = 18$ MeV one needs $\gamma_A = \varepsilon^*/2\varepsilon = 1.5 \cdot 10^{10}$ for a head on collision, corresponding to $E_{Fe} = 8 \cdot 10^{20}$ eV.

Another radiation background, with shorter wavelengths, makes this energy lower. The infrared photon has an energy $\varepsilon_{IR} \approx (10^{-2}$–$10^{-1})$ eV and the iron nucleus energy needed now for the resonant interaction is about $1.6 \cdot 10^{19}$ eV, i.e. typically about 50 times smaller than that with a CMB photon. The problem with the IR is that its density is still uncertain, in contrast to the CMB density following from its well measured black body temperature. The attenuation length for a nucleus needs to be defined again as it depends also on processes when the nucleus loses its identity. It is defined as [9]

$$\frac{1}{\lambda_{att}} = \frac{1}{\gamma}\frac{d\gamma}{dx} + \frac{R_1}{A} + \frac{2R_2}{A} + \frac{\langle \Delta A \rangle R_k}{A} \qquad (14)$$

where R_1, R_2 are the reaction rates (per unit path) for one-and two-nucleon emission, and R_k is the reaction rate for more than 2 nucleon loss. Its dependence on energy calculated by Puget [10] for iron is shown in Fig.2. It can be seen that at $\sim 10^{20}$ eV proton and iron have almost the same λ_{att}. However, as we shall see later it may be that the IR background is lower than that adopted here, so that heavier nuclei could reach us from further distances than protons can do.

2.3. Maximum path length

Here we shall show that if a particle loses energy quickly enough then is path length (life time) before it reaches a given energy E, is finite.

If the particle relative energy loss rate equals

$$\frac{1}{E}\frac{dE}{dt} = -h(E), \quad \text{where} \quad h(E) > 0 \qquad (15)$$

then the time needed to decrease its energy from the initial E_o to E equals

$$t(E_o \to E) = \int_E^{E_o} \frac{dE'}{h(E')E'} \qquad (16)$$

Let us assume, just for illustration, that

$$h(E) = bE^\alpha \qquad (17)$$

Then, solving (16) we obtain for $\alpha \neq 1$ that

$$t(E_o \to E) = \frac{1}{\alpha b}(E^{-\alpha} - E_o^{-\alpha}) \qquad (18)$$

If $\alpha > 0$ then $t \to t_{max} = (\alpha b E^\alpha)^{-1}$ if $E_0 \to \infty$. It means that if the relative energy loss rate grows with energy ($\alpha > 0$) then the particle, born even with an infinite energy, reaches a given energy E after a finite time, meaning that there is a maximum distance $l_{max} = c t_{max}$ from where it can arrive.

We have assumed above that there are no fluctuations in the energy loss processes so that $E(E_0, t)$ is a unique function of time for a given E_0. However, the effect of fluctuations is not negligible, so that t_{max} can be exceeded to some extent.

2.4. Predictions of the ambient CR energy spectrum

Firstly, we shall show how to calculate the expected energy spectrum for the most simple model of the Universe, i.e. the homogeneous, closed, nonevolving box. Assuming moreover that it is an equilibrium spectrum (and neglecting the fluctuations in dE/dt) we have

$$Q(E) = \frac{\partial}{\partial E}\left(\frac{dE}{dt}N(E)\right) \tag{19}$$

where $N(E)$ is the differential ambient spectrum and $Q(E)$ is the CR production rate (both e.g. per unit volume). The solution to (19) is

$$N(E) = \frac{1}{|dE/dt|}\int_E^\infty Q(E')dE' \tag{20}$$

Assuming that the production rate has a power law shape – $Q(E) \sim E^{-\gamma}$ we get

$$N(E) = \frac{\lambda_{att}(E)}{(\gamma-1)c}Q(E) \qquad (\gamma \neq 1) \tag{21}$$

Now we see that, in this model, $\lambda_{att}(E)$ gives in a straightforward way the expected spectrum, once $Q(E)$ is assumed.

The Universe, however, does evolve and fortunately we know that the CMB temperature and photon density at an epoch z were larger by $(1+z)$ and $(1+z)^3$ correspondingly. Then, for the energy losses on CMB we have that

$$h(E, z) = (1+z)^3 h\left[(1+z)E, z = 0)\right] \tag{22}$$

so that

$$\lambda_{att}(E, z) = (1+z)^{-3}\lambda_{att}\left[(1+z)E, z = 0\right] \tag{23}$$

The evolution of the IR background, important for the propagation of nuclei, is much more uncertain as it is a results of the evolution of galaxies, not known very well. Thus, our prediction of the effect of intergalactic propagation of nuclei on their energy spectrum is much less robust than in the case of protons.

Many authors have calculated in the past the expected proton (and nuclei) energy spectrum, assuming a power-law spectrum at production, $Q(E)$, and an homogeneous spatial density of the sources (for references see e.g. [13]). In figure 3a we present the result of the calculations by Yoshida and Teshima [12] (dotted line) for $Q(E) \sim E^{-3}$. The dramactic decrease of the predicted flux for $E \leq 8 \cdot 10^{19}$ eV is the result of such a behavior of $\lambda_{att}(E)$.

2.5. The observations

We will not describe here in detail the experimental methods for observing the extensive air showers produced by the UHECRs in the atmosphere, as it has recently been done very clearly and thoroughly in the review paper by Nagano and Watson [13]. We will present the measurement results, comment on their possible accuracies and compare with predictions.

Fig.3a (taken from [14]) represents the energy spectrum determined by the AGASA experiment and its exposure (for zenith angles $< 45°$). Despite large errors in the flux the highest energy AGASA data do not agree with the predictions of the universal origin of protons (see §2.4).

The conclusion about the non-existence of the GZK cutoff would be extremely important for models of the particle origin. Thus, of the same importance is a proper and critical determination of the primary energy of a shower.

In the AGASA experiment located at 667 m a.s.l. the local density of charged particles at 600 m from the shower core, $S(600)$, is used as the primary energy E indicator. The detectors are plastic, 5 cm thick scintillators of 2.2 m^2 area each, with 1 km spacing [15]. Shower simulations have shown [16] that $S(600)$ depends only weakly on the primary mass, interaction model and shower fluctuations. Its experimental determination, however, involves (among other factors) a good shower core localization and a true lateral distribution function (LDF) to be applied for different zenith angles. The AGASA collaboration has recently devoted a whole paper [14] to the energy determination problems in their experiment and concluded that $\Delta E/E = \pm 25\%$ in event reconstruction and $\pm 18\%$ as a systematic

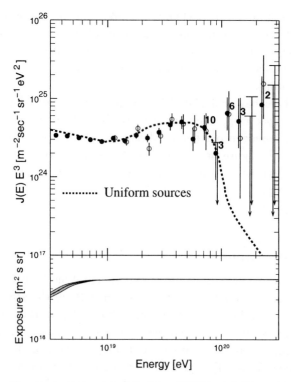

Figure 3a. Energy spectrum determined by AGASA and the exposure with zenith angles smaller than 45° (until July 2001). Open circles: well contained events; closed circles: all events. The vertical axis is multiplied by E^3. Error bars represent the Poisson upper and lower limits at 68% confidence level and arrows are 90% C.L. upper limits. Numbers attached to the points show number of events in each energy bin. The dashed curve represents the spectrum expected for extragalactic sources distributed uniformly in the Universe, taking into account the energy determination errors. The uncertainty in the exposure is shown by the shaded region. (Figure taken from [14]).

error around 10^{20} eV, confirming the observation of events above 10^{20} eV (no cutoff!). The systematic error of 18% contains all possible detector uncertainties, estimated by the authors as $\sim 9\%$, air shower phenomenology as $\sim 11\%$ and $S(600)$ theoretical uncertainty as $\sim 12\%$.

An analysis of $S(600)$ uncertainty has also been done by Capdevielle et al [17]. These authors obtain, however, that depending on the LDF used, $\Delta S/S(600)$ for some actually registered showers is $\sim 30\%$, whereas the intrinsic $S(600)$ fluctuation are $20 - 30\%$ – adding up to as much as $\sim 40\%$, being more than the uncertainty given by AGASA. One should also keep in mind that the energies studied here are $4 \div 5$ orders of magnitude higher than those achieved by accelerators, so that our bunch of models used for simulations may predict systematically shifted shower parameters. Thus, we think that 40% should be considered as a lower limit to the energy uncertainly (E is almost proportional to $S(600)$).

The High Resolution Fly's Eye experiment [18] (HiRes) uses a different technique for studying the UHECRs. By measuring the shower image in the fluorescence light excited in the atmosphere by charged particles of the shower, it is possible to reconstruct the electromagnetic cascade curve, $N_e(X)$, where X is the atmospheric slant depth in $g \cdot cm^{-2}$. Determining the primary energy from the whole cascade curve, when N_e is known on many levels in the atmosphere rather than on the ground level only, is believed to be much more accurate and much less model dependent. To reconstruct $N_e(X)$ is not an easy task, however, since there are many factors contributing to its uncertainty (as atmospheric transmission, scattered Cherenkov light contribution to the fluorescence flux or absolute calibration of the telescopes.)

The results obtained by HiRes [6] (published after the Erice School) are shown in Fig.3b, together with the AGASA [14] and the Haverah Park spectrum [11]. The HiRes data extending to the highest energies and complying with the predicted cutoff curve, comes from the first of their two detectors, HiRes I, operating since 1997. The data from HiRes II having $\sim$ twice as large angular field of view and accurate time distributions of the light signals due to FADC, contain fewer events but are in very good agreement with HiRes I.

At first sight there is a large discrepancy between AGASA and HiRes, particularly for $E \geq 6 \cdot 10^{19}$ eV. This figure is, however, misleading to some extent as the vertical error bars (as presented by the authors) represent (apparently) the flux uncertainties only, or rather statistical fluctuations of the number of events in a given energy bin. The uncertainties in the energy determination have not been shown by the authors. Thus at one of the AGASA points (at $4.5 \cdot 10^{19}$ eV), we have put the hypothetical error bars of the quantity $E^3 J(E)$, if the energy uncertainty was 30%. However, if the energy uncertainty had a statistical character only then the disagreement between the results of the two experiments would not disappear. The comparison of the two results strongly suggest that, at least for $E < 5 \cdot 10^{19}$ eV, there is a systematic error in either energy or

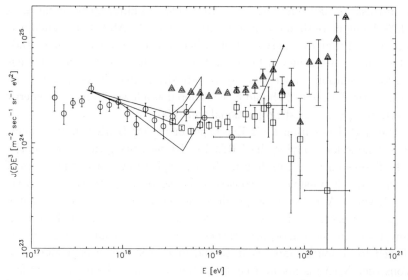

Figure 3b.   Differential energy spectrum multiplied by E^3, observed by: HiRes I (squares) and HiRes II (circles), AGASA (triangles) and Haverah Park (region marked by lines).

flux determination in at least one of the experiments. In figure 3c we assumed that there was a 30% overestimating systematic error in energy ($E_{true} = 0.7 \cdot E$) in the AGASA experiment. Then $E^3 J(E)$ goes down by factor of 2 ($0.7^2 \simeq 0.5$). Five of the 10 AGASA showers with $E > 10^{20}$ eV are shifted now below this value, and the agreement between the two experiments is very good up to $\sim 6 \cdot 10^{19}$ eV and almost satisfactory above it. HiRes observed 1 event above 10^{20} eV whereas AGASA would be left with 5 (the exposures of the two experiments are similar). Decreasing the energy by 40% would leave only 2 events with $E > 10^{20}$ eV. Assuming that the AGASA energy overestimation is much greater (say factor of 2) would shift its highest energy point down close to the prediction curve but then the points with $E < 10^{19}$ eV would go too much below the HiRes data, so that this assumption is rather unlikely. Anyway the problem whether there is the cutoff in the spectrum would not be resolved.

Certainly, more data is being collected by HiRes II but, even if the HiRes I results were confirmed by them in the near future, the cutoff problem would still be there.

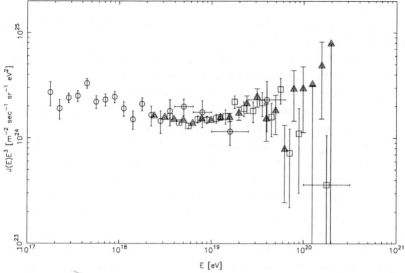

Figure 3c.   As in Fig.3b, but with the AGASA data points shifted in energy, down by factor 0.7. The HP data have been omitted for a clearer comparison with HiRes.

2.6. Attempts to explain the AGASA spectrum (no cutoff)

Before the HiRes spectrum was published there appeared many papers trying to reconcile the flat AGASA spectrum with a UHECR origin model. Here we shall discuss some of the recent ones to illustrate various possibilities.

Berezinsky et al.[19] consider a uniform distribution of the proton sources in the Universe, with and without evolution of the source emissivity. They describe in great detail the adopted formulas for the proton energy losses, both for e^+e^- and pion production. A very good agreement with the AGASA data is obtained in the energy region from 10^{17} eV up to 10^{20} eV for a production spectrum $E^{-2.45}dE$ and source evolution $\sim (1+z)^4$. However, the three highest energy data points stick out above the cutting-off predictions, having (according to the authors) another origin.

Stecker and Salamon [20] on the other hand, assume a universal production of iron nuclei. They calculate the intergalactic propagation of the nuclei using their new determination of the infrared background (from the IRAS data) and a more accurate treatment of the photodisintegration of individual nuclides. The cutoff energy in the obtained spectrum for all produced nuclei depends on the propagation time and exceeds 10^{20} eV for source distances ≤ 200 Mpc, much larger than the author's previous estimations. They do not produce the final spectrum expected from all sources in the Universe but conclude that heavy nuclei origin could be important for understanding the highest energy air shower events.

To a similar conclusion arrive Szabelski et al. [21]. Assuming a flat ($\gamma = 2.1$) production spectrum of iron nuclei they can explain the world's data with $E > 10^{19}$ eV, normalizing at 10^{19} eV (the data from various experiments have also been "normalized" to each other). This scenario implies, however, that the CR origin below $\sim 3 \cdot 10^{18}$ eV must be different.

Another approach is undertaken by Medina-Tanco [22] where the reason for discrepancy lies neither in a wrong primary mass assumed, nor in the slope of the production spectrum but in the spatial distribution of the sources. The author assumes that CRs are produced by luminous matter, i.e. by normal galaxies. This matter is not distributed isotropically, at least for $z < 0.001$, but there is an excess of it in that region, as compared with the mean density for higher z. If CR source intensity was proportional to the density of galaxies and the production spectrum was $\sim E^{-3}dE$ then the observed spectrum would cutoff at a higher energy, say $\sim 10^{20}$ eV (Fig.4) and the AGASA points would just be acceptable.
This approach, although reminding us about the importance of the local distribution of CR sources, has a drawback: once normal galaxies are assumed to produce high energy particles, one should *not exclude our Galaxy*, as it has been done in that paper.

Let us estimate, assuming a most simple model, how big a contribution of the Galaxy to the total flux would be. The number N of cosmic ray particles in the Galaxy equals

$$N = N_G + N_{ex} \tag{24}$$

where N_G and N_{ex} are particles of galactic and extragalactic origin respectively. Assuming that extragalactic particles pervade freely the insides of galaxies (in particular, that of ours) we can write

$$\frac{N_G}{N_{ex}} = \frac{QT_G}{V_G n_g Q T_{ex}} \tag{25}$$

where Q is the CR production rate per galaxy (canceling out), V_G – CR confinement volume in the Galaxy, n_g – number density of galaxies, T_G and T_{ex} – CR lifetime in the Galaxy and in the Universe correspondingly. Particular values for V_G, T_G and T_{ex} depend strongly on energy, so let us

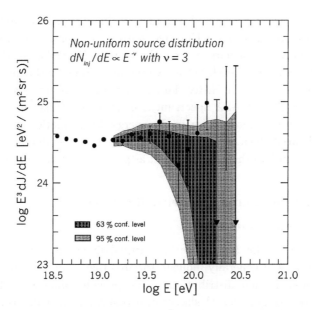

Figure 4. Differential energy spectrum ($\times E^3$). Points – the AGASA data, as in [5]. Hatched areas – predictions by Medina-Tanco [22] for the luminous matter model of CR origin. (Figure taken from [22])

fix $E = 10^{20}$ eV. V_G and T_G can be estimated only roughly from our poor knowledge of the large scale Galactic magnetic field and the extension of its halo. Here we assume that

$$V_G = \pi(15 \text{ kpc})^2 \cdot 10 \text{ kpc} \simeq 7 \cdot 10^{-6} \text{ Mpc}^3 ,$$
$$T_G \leq 10^5 \text{ yr} , \tag{26}$$
$$T_{ex} = \frac{\lambda_{att}}{(\gamma - 1)c} = \frac{100 \text{ Mpc}}{2c} \simeq 1.6 \cdot 10^8 \text{ yr} ,$$

where γ is the power index of the differential energy spectrum, $n_g = 2 \cdot 10^{-2} \text{Mpc}^{-3}$, and obtain $N_G/N_{ex} \simeq 5 \cdot 10^3$!

Even if one increases the (local) n_g by factor of 10 and decreases T_G by 10 (T_G can hardly be smaller than 10^4 yr) it will still be that $N_G/N_{ex} \gg 1$. This ratio should, however, be smaller for lower energies (T_{ex} increases strongly). Nevertheless, above the cutoff the contribution from the Galaxy should be dominant. Thus, some correlation with the Galactic disk should be expected, contrary to observations (see later).

A way to save the Medina-Tanco idea would be to assume that UHECR sources are distributed in the Universe *in a similar way as luminous matter*, but *normal galaxies* as ours, are *not* the sources. Nevertheless, the highest energy AGASA points stick out above the predictions of this model.

In view of the latest HiRes data it may turn out, however, that the reason of the flat AGASA spectrum is much more trivial and connected with a hidden systematic error in the energy determination of the largest showers. It would be a malicious joke of nature if some systematics in HiRes produced a cutoff in the spectrum, just as predicted by the universal origin.

3. Angular distribution of arrival directions

3.1. *Large scale anisotropy*

It is expected that for $E > 10^{19}$ eV the arrival directions, particularly those of protons, should generally indicate towards their sources. Thus, a study of the shower angular distribution should provide important information about particle origin sites. A shower direction can be now determined with an accuracy of a few degrees, so that anisotropy in a large angular scale as well as in a small scale can be studied.

Any extensive air shower detector, working contantly with its characteristics unchanged with time, should observe a homogeneous distribution of the right-ascension (RA) of the registered events. However, their declination distribution depends on the detector latitude φ, on the maximum zenith angle z_{max} of the shower accepted for analysis and on the dependence of the detector response (probability of detection) on the zenith (or azimuth) angle. Assuming that above some threshold energy this response is independent of the shower angle (detection probability $= 1$ if e.g. the shower core is within a fixed distance from the center of the array) one can derive that the declination distribution $\varepsilon(\delta)$ of the registered showers, if the *incident* directions of the CR particles are isotropic, equals

$$\varepsilon(\delta) = \sin(2\delta) \cdot \sin \varphi \cdot \alpha_z + 2 \cdot \cos^2 \delta \cdot \cos \varphi \cdot \sin \alpha_z \tag{27}$$

where α_z is defined by

$$\cos \alpha_z = \frac{\cos z_{max} - \sin \delta \cdot \sin \varphi}{\cos \delta \cdot \cos \varphi} \tag{28}$$

The function $\varepsilon(\delta)d\delta$ is the effective solid angle of the strip $(\delta, \delta + d\delta)$ (meaning that each direction is multiplied by $\cos \theta_z$, to allow for the decrease of the array surface as "seen" by the shower) and it is called the exposure of

the array. This formula applied e.g. to the AGASA array describes quite
well the exposure quoted by the collaboration [4].

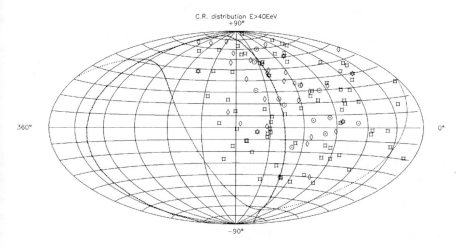

Figure 5.   Arrival directions of CR with $E > 4 \cdot 10^{19}$ eV (Galactic coordinates). Squares
– AGASA, diamonds – Haverah Park, circles – Yakutsk, stars – Volano Ranch. Dotted
line is the terrestrial equator. Thick dotted line is the Supergalactic plane.

Any large scale anisotropy should manifest itself as a deviation of the
shower declination distribution from the array exposure $\varepsilon(\delta)$. Of particular
interest are showers with the determined energies above the expected cutoff.
Fig.5 shows CR arrival directions for $E > 4 \cdot 10^{19}$ eV as measured by four
experiments AGASA [23], Haverah Park, Yakutsk and Volcano Ranch [24].
No obvious anisotropy can be seen. There is no correlation with the Galactic
disk, although there might be some small excess close to the Supergalactic
plane (SP)(Fig.6). There are three events lying on this plane, very close
to each other (a triplet) and another two events forming a doublet. This
coincidence has been noticed by Stanev et al. [25], and the evidence for
some directional correlation with SP is even stronger when analyzing 92

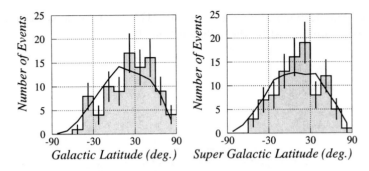

Figure 6. Distribution of 92 events with $E > 4 \cdot 10^{19}$ eV (up to 2000 yr) in Galactic and Supergalactic latitude (histograms). Lines show predictions for isotropic distribution of arriving showers. (Figure taken from Uchihori et al. [4])

world events [4]: for $E > 4 \cdot 10^{19}$ eV there are two triplets lying within $\sim 2^\circ$ on the SP and 8 doublets, two of them on SP, another one $\sim 5^\circ$ away from it. We think that these coincidences should not be disregarded, although there are another interesting hypotheses as well.

One of them is a supposition that luminous infrared galaxies (LIRGs) may be the origin sites of the UHE particles [26]. These are mainly interacting systems (colliding or merging galaxies), thus having probably favourable conditions for high energy particle acceleration [27,28]. Another strong point for this hypothesis is that the AGASA triplet coincides with such an object (Arp 299) consisting of two merging starburst galaxies. It is the brightest infrared source within 70 Mpc, at the distance 42 Mpc (see 3.2.2). Śmiałkowski et al. [26] considered in some detail this hypothesis. They have chosen ~ 2800 LIRCs (with $L_{FIR} > 10^{11} \cdot L_\odot$) from the PSCz catalogue , as the proton sources with a source spectrum $\sim E^{-2} dE$. The absolute CR intensity of a source was assumed to be proportional to its infrared emissivity. Including energy losses on microwave background and scattering in the intergalactic irregular magnetic field (with $(\overline{B^2 l})^{1/2} = 1$ nG$\cdot$ Mpc$^{1/2}$) they calculated the expected directional distribution of the arriving particles. The obtained maps, together with the AGASA point, are shown in figure 7a,b. It can be seen by eye that the coincidence of the observed showers with the map is not bad for lower energies, but rather poor for those with $E > 8 \cdot 10^{19}$ eV.

One of the statistical tests, used to check the LIRG hypothesis against isotropy, was based on the orientation matrix produced from the unit vectors indicating shower arrival directions [29]. This test was first used for CR anisotropy analysis by Medina-Tanco [30]. In principle, the test assumes that there are directional data from the whole solid angle 4π, which is not the case for the UHECR data. Thus, by using it in that case one has to be cautious in interpretation of the obtained results. In particular the two parameters γ and ξ , which are to be determined from the data, should correspond to the shape of an anisotropy pattern and its strength correspondingly, for the full sky data obtained with an homogeneous exposure. In our case, however, they lose their simple meaning, nevertheless they can still serve as parameters differentiating between various anisotropy hypotheses.

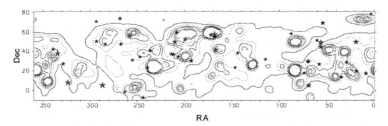

Figure 7a. Expected map of proton intensites without influence of the regular Galactic magnetic field for protons 40–80 EeV originating in LIRGs to be seen by AGASA, with superimposed 47 AGASA shower directions in this energy range (stars scaled with energy). Contours of constant flux per unit solid angle are spaced linearly, with the second lighest line describing the intensity lower by factor 0.85 than that for the highest one.

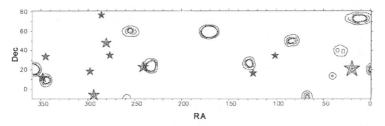

Figure 7b. As above but for $E > 80$ EeV, with superimposed 11 AGASA showers.

The dots in Fig.8 show the expected distribution of (γ,ξ) values if the arrival directions were isotropically distributed. Crosses correspond to the

LIRG scenario. On the left graph each point corresponds to 47 simulated events, on the right one – to 500 events. The two AGASA points (from two different samples, see figure caption) are marked by the cross and the open square. The error of any point due to uncertainties of the shower arrival directions is small as compared to their spread. It is seen that the predictions for isotropy and the LIRG scenario overlap significantly if the data sample is small (47) so that the data points can not distinguish between them (although the open square does lie in the region where LIRG scenario is more probable). In the near future (the Auger experiment!) the statistics will greatly improve. With 500 events the predictions for the two assumptions separate very well on the (γ, ξ) plane, making this test decisive then.

On figure 8 (left) we have also marked the point (thick cross) calculated by Medina-Tanco [30] for the same AGASA data as were used by Śmiałkowski et al. – big star. Medina-Tanco used this test to check shower isotropic distribution against his (mentioned earlier) hypothesis of UHE-CRs originating in the normal galaxies (luminous matter). The calculated by him point from AGASA data lies between the well separated regions corresponding to isotropy and the luminous matter hypothesis, leaving the question unresolved (according to the author). According to Śmiałkowski et al., however, his data point is erroneous, and should lie in the region being consistent with isotropy (big star) and, as such, excluding the luminous matter origin.

3.2. Multiple events

3.2.1. Observations

As it was mentioned earlier an intriguing feature of the shower directions distribution are the multiplets: two or three showers arriving practically from the same direction. An extensive study of this subject using the world data has been done by the AGASA group [4].

AGASA itself has observed one triple event ($\Delta\theta < 2.5°$) in their data set with $E > 4 \cdot 10^{20}$ eV. All three estimated energies are quite close: 5.4, 5.5 and $7.8 \cdot 10^{19}$ eV, being almost within energy uncertainty. This triplet coincides with the supergalactic plane. As it has been already mentioned it also comes from the direction of the most luminous infrared colliding galaxy system Arp 299. AGASA sees also 3 doublets, one on the SP. When adding Haverah Park (HP) data the doublet on the SP becomes a triplet. Away by 9° from this triplet position lies again a pair of colliding galaxies, VV338, in

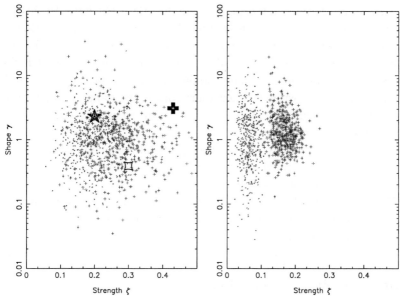

Figure 8.   Orientation matrix analysis of the large scale anisotropy. Each dot corresponds to (γ,ξ) point calculated for 47 (left) and 500 (right) "events" simulated from isotropic distribution (with AGASA exposure). Small crosses describe the same but for the LIRG hypothesis. The AGASA 47 events with $E > 4 \cdot 10^{19}$ eV are represented by the star [26], but according to [30] by the thick cross (!?). The open square corresponds to new AGASA events with $(4 \div 8) \cdot 10^{19}$ eV (by chance also 47 events). The two hypotheses separate very well for large samples (500, right)

the distance of only 5.7 Mpc [28]. At a smaller angular distance of 2° there is Mrk 359 but this is an object much more distant, lying $\sim$ 70 Mpc away. Together with the Volcano Ranch showers it adds up to 2 more doublets on the SP. One of the doublets has also arrived from a direction of colliding galaxies. It is the object VV89 at 15.5 Mpc (another one, VV101, lies far behind it, at 100 Mpc). All this suggest that colliding galaxies may play a role in the CR acceleration to ultra high energies, as it has been proposed [27] and studied [28] earlier. Anyway, the multiplet distribution on the sky does not correlate with that of the galaxies within 100 Mpc [4].

3.2.2. Multiplet probabilities

Here we want to study the obvious problem arising: are the observed multiplets a manifestation of point sources or just fluctuations of an isotropic (or a more or less smoothly depending on direction) CR flux. To this aim one has to know what is the probability that a given number of multiplets occurs. This depends, of course, on the total number N of the observed showers and on the assumed angular distribution of the arriving CR flux $\lambda(\alpha, \delta)$. Monte-Carlo methods and computers allow to calculate these probabilities quickly and without much effort (but if one wants to improve the accuracy of the result, say, by factor 2, one needs to use altogether 4 times more of the computer time).

Below we show how to derive these probabilities analytically [26] (we hope not to have to convince the reader about advantages of analytical solutions over those obtained numerically).

First, we notice that the angular distribution of the *registered* showers $\rho(\alpha, \delta)$ differs from that of the *arriving* showers because of the experiment exposure $\eta(\alpha, \delta)$, so that

$$\rho(\alpha, \delta) = \eta(\alpha, \delta) \cdot \lambda(\alpha, \delta) \tag{29}$$

where $\int_0^{2\pi} \eta(\alpha, \delta) d\alpha = \varepsilon(\delta)$ (see §3.1).

The total number of the observed showers equals

$$N = \int \rho(\alpha, \delta) d\Omega \tag{30}$$

where the integration region covers the part of the sky seen by the experiment.

The occurrence of a doublet takes place if within a small angle R around a shower direction (α, δ) there is another shower. The expected number of showers to be registered within this small solid angle equals

$$\mu(\alpha, \delta) = \pi R^2 \rho(\alpha, \delta) \tag{31}$$

As we assume that showers arrive independently the probability P_2 of detecting two of them is given by the Poisson distribution, with the mean $\mu(\alpha, \delta)$

$$P_2 = e^{-\mu} \frac{\mu^2}{2!} \tag{32}$$

so that the local density of doublets equals

$$\rho_2 = e^{-\mu} \frac{\mu \rho}{2} \tag{33}$$

Thus, the mean number of doublets from the whole sky observed can be obtained by integrating ρ_2 over the solid angle

$$N_2 = \int \rho_2(\alpha, \delta) d\Omega = \frac{1}{2} \pi R^2 \int (\eta \lambda)^2 e^{-\mu} d\Omega \tag{34}$$

The actual number of the observed doublets has also the Poisson distribution, so that the probability of observing k_d doublets is

$$p(k_d; N_2) = e^{-N_2} \frac{N_2^{k_d}}{k_d!} \tag{35}$$

The idea in the derivation of the analoguous probabilities for a given number of triplets is exactly the same, although in this case the definition of a triplet is not as obvious as that of a doublet. We refer the reader to the paper by Śmiałkowski et al. [26] for the details of this derivation. The result for the mean number N_3 of the observed triplets is

$$N_3 = \frac{1}{6} \left(\pi R^2 \right)^2 \left(1 + \frac{3\sqrt{3}}{2\pi} \right) \int (\eta \lambda)^3 e^{-\mu} d\Omega \tag{36}$$

and, of course, the actual number of triplets observed undergoes the Poissonian fluctuations.

Let us note, however, that in the above derivations the total number of showers N is not strictly fixed, being the total expected (mean) number of showers in a given period of time. As it usually is large, the normalization of $\lambda(\alpha, \delta)$ to the *actual* number of events N rather than to the *expected* one (unknown) is quite justified.

For the AGASA experiment ($\varphi = 35°47'$, $z_{max} = 45°$, $R = 2.5°$) the expected number of doublets and triplets for isotropic CRs, calculated from the above formulae, equals: $N_2 = 1.02$, $N_3 = 2.7 \cdot 10^{-2}$. The observed numbers are 2 doublets and 1 triplet (for $E = (4 \div 8) \cdot 10^{19}$ eV).

Whereas the probability of the result being 2, when the random variable has the Poisson distribution with the mean 1.02, is quite large (0.19), that for observing 1 triplet with the expected value $2.7 \cdot 10^{-2}$ equals $2.6 \cdot 10^{-2}$ (about 7 times smaller). Thus, the combined probability (for the observed set of multiplets to occur) is $\sim 4.9 \cdot 10^{-3}$.

This number is small but its meaning must be treated with caution because we deal here with probabilities of a two-dimensional random variable: number of doublets and that of triplets. One could think that calculating the probability of 2 *or more* doublets and 1 *or more* triplets, being $\sim 7 \cdot 10^{-3}$, is the right way. However, it is not, because it would not take into account several other possibilities with small probabilities. (The author thanks Peter Kiraly for pointing this to her.) Such a case is e.g. zero triplets and 5 or more doublets. Thus, we think that the proper approach is to sum up the probabilities of all multiplet combinations (events) with values smaller than that of the event which accually has occurred (2 doublets and 1 triplet,

$p = 4.9 \cdot 10^{-3}$). After adding this sum to p we obtain the probability of any of the events with individual probabilities equal or smaller than that of the event accually happened. Doing this we get $1.1 \cdot 10^{-2}$ i.e. about two times more. This value could be compared with confidence levels 0.1 or 0.05, usually assumed in the χ^2 test, and then the isotropy hypothesis would be definitely rejected.

A similar analysis can be performed for any other angular distribution assumed. As N_2 (N_3) depends on the square (cube) of the local angular density (integrated over Ω) it is clear that *any* non-isotropic distribution will predict larger numbers of multiplets (unless, by some strange chance, the distribution of the arriving showers, multiplied by the exposure, results in a flatter function on the sky than the exposure itself). E.g. the already mentioned hypothesis on CRs originating in luminous infrared galaxies (mostly colliding systems) predicts for the probability of 2 doublets and 1 triplet a value $\sim$ 13 times larger. This itself does not necessarily mean that the above distribution describes better the CR sky than isotropy.

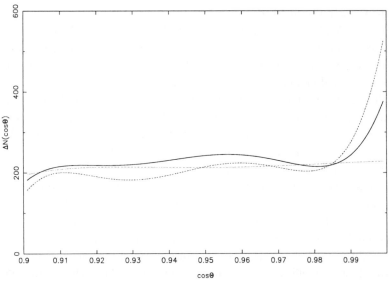

Figure 9.   Distribution of $\cos\Theta$, where Θ is the separation angle between AGASA showers (47 events with $(4 \div 8) \cdot 10^{19}$ eV) and 5000 directions drawn from isotropic (flat line) and LIRG (solid line) distributions. Dashed line is for separation of LIRG-LIRG simulated events.

An analysis which would indicate whether the LIRG (or any other objects) directions are actually correlated with the directions of the showers is to study distribution of $\cos \Theta$, where Θ is the angular distance between a registered shower from one side and "events" drawn from the distribution predicted by the particular hypothesis. Fig.9 shows the result of such analysis [26] for the AGASA showers with $E = (40 \div 80)$ EeV and the LIRG hypothesis. An excess of small separation angles can be seen for the AGASA-LIRGs line (solid). It should be confess, however, that the main contribution to this excess is provided by the triplet, coinciding with the Arp 299 source. The LIRG hypothesis, although plausible, must await its verification by more data.

Another analysis, strongly indicating an existence of the point sources against the isotropic distribution, has been done by Tinyakov and Tkachev [31]. These authors calculated the correlation function (difference between the observed number of pairs and that expect from isotropy, expressed in units of the standard deviation of the latter, as a function of the angular distance of the two showers in a pair) for the AGASA showers (with $E > 48$ EeV) and those from Yakutsk ($E > 24$ EeV). The result is presented in Fig.10 taken from their paper. It is seen that for both data sets there is a large, $(5-7)\sigma$, excess of shower pairs within the angular resolution of the experiments ($2.5°$ and $4°$ for AGASA and Yakutsk respectively). It is the only statistically significant deviation from isotropy of the correlation function in the whole range of Θ.

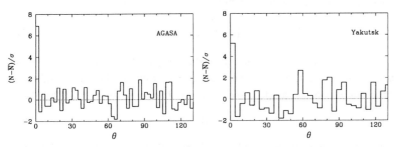

Figure 10. Angular correlation function for: left – AGASA ($E > 4.8 \cdot 10^{19}$ eV) and right – Yakutsk ($E > 2.4 \cdot 10^{19}$ eV). Binning angles correspond to the experiment angular resolution $2.5°$ and $4°$ correspondingly. (Figure taken from[31])

The nature of the possible astrophysical objects associated directionally with the UHE events, different than those discussed already, has been a subject of several papers. Farrar and Bierman [32] found that 5 events (two from AGASA, two from Haverah Park and one from the Fly's Eye experiment), correlate with positions of 5 quasars, with $z = 0.29 - 2.177$. The chance probability, $5 \cdot 10^{-3}$ eV, estimated by the authors seems rather small. However, $z = 0.3$ corresponds to a distance of about 1.5 Gpc, not to mention the higher z values! Neither a proton nor a nulceus can survive such a distance with $E \sim 10^{20}$ eV! Or these particles have much lower energies (below the cutoff), either it is a coincidence, otherwise one would have to call for new physics(!?).

3.3. Implication for the intergalactic matter

Particle trajectories from their sources to the observer undergo deflections in the ambient magnetic fields. It is usually assumed that there may be an irregular $\vec{B}$ in the intergalactic space. A study of this subject was done by Kronberg [33] who estimated that $B \leq 10^{-9}$ nG if the field correlation length is about 1 Mpc (strictly speaking the author speakes about "reversal scale"). On this basis one can calculate the expected deflection θ of a charged particle trajectory (i.e. of its velocity vector) and the resulting deviation η of its arrival direction from that towards the source (note that these angles are not the same [34,26]; see later).

For the simplest situation, it is usually assumed that the distance D between the source and the observer is divided into cells of equal length l, the magnetic field $\vec{B_i}$ in each cell is independent of those in other cells and all $\vec{B_i}$ are isotropic. If the final deflection angle is small, we have

$$\overline{\theta^2} = \frac{2}{3} \left(\frac{q}{E} \right)^2 \overline{B^2} l D \qquad (37)$$

where q is the particle charge and E – its energy. This angle, however, is not observable, since the initial velocity is not known. The angle of more interest is the deviation angle η (see above). It can be shown [34,26] that

$$\overline{\eta^2} = \overline{\theta^2}/3 \qquad (38)$$

It is the deviation η angle which is relevant if we want to correlate the particular objects on the sky with HE particle directions. From (37) and (38) we obtain

$$\eta_{rms} = 1.4° Z \cdot \frac{B_{rms}}{1 \text{ nG}} \cdot \frac{10^{20} \text{ eV}}{E} \left(\frac{l}{1 \text{ Mpc}} \cdot \frac{D}{30 \text{ Mpc}} \right)^{1/2} \qquad (39)$$

where Z in the particle charge (in e units).

The assumption about all cells being equal is certainly not realistic. A more accurate treatment should allow for a distribution of their lengths and a (possible) correlation of the cell length and the value of the magnetic field in it. This is automatically taken into account when one starts from a power spectrum of the intergalactic field $\vec{B}$ [35], instead of single cells. The correlation length l_c is defined then by the following relation

$$\int_{-\infty}^{+\infty} \left\langle \vec{B}(0) \cdot \vec{B}(\vec{l}) \right\rangle \mathrm{d}l \equiv l_c B_{rms}^2 \tag{40}$$

and the deflection angle η_{rms} (following the formula for θ_{rms} by Harari [35]) equals

$$\eta_{rms} = \frac{1}{\sqrt{6}} \frac{q}{E} B_{rms} \sqrt{l_c D} \tag{41}$$

This value is by factor $\sqrt{3/4} = 0.87$ times smaller than that for constant cell size l, if $l = l_c$. Thus, the difference is not large and, since the power spectrum is not known anyway, the constant cell approximation seems sufficient at this stage. (It is interesting to learn, however, that for the Kolmogorov power spectrum $l_c \simeq l_{max}/5$, where l_{max} is the maximum wavelength of the Fourier modes [35]).

So far, the CR sources are not identified but if the observed multiple members do come from a single source, their relative separation angle α can be used to restrict the parameters of the intergalactic medium. For the constant cell model we have that

$$\overline{\alpha^2} = \overline{\eta_1^2} + \overline{\eta_2^2} = \frac{2}{9} q^2 \overline{B^2} Dl \left(\frac{1}{E_1^2} + \frac{1}{E_2^2} \right) \tag{42}$$

where for simplicity we have assumed that both particles have the same charge q. E_1 and E_2 can be determined and if we adopt that the measured $\alpha \simeq \alpha_{rms}$ then the value $q B_{rms} \sqrt{Dl}$ could be evaluated.

However, as the deviation angles η are of the same order as shower direction resolution, one should take the latter into consideration. Then we have

$$\overline{\alpha^2} = \overline{\eta_1^2} + \sigma_1^2 + \overline{\eta_2^2} + \sigma_2^2 \tag{43}$$

Takeda et al.[36] gives $\sigma_{19} \simeq 2.8°$ and $\sigma_{20} \simeq 1.2°$ as resolutions for $E = 10^{19}$ eV and 10^{20} eV respectively. Taking as an example the AGASA doublet with $E_1 = 51$ EeV, and $E_2 = 2.13$ EeV, $\sigma_1 \simeq 1.7°$ and $\sigma_2 \simeq 1.56°$ we can estimate an upper limit for $q B \sqrt{Dl}$ such that a value of α smaller than that measured has a given (small) probability (say 0.1). As α has two dimensional Gaussian distribution, it can be calculated that

$$B(\mathrm{nG}) \left[(D/30 \text{ Mpc}) \cdot l(\mathrm{Mpc}) \right]^{1/2} < 0.9 \qquad (44)$$

if the particles were protons. (Without taking the direction resolutions into account, this upper limit would be over two times smaller). If we knew the source, hence its distance (and the particle charge!), we could obtain an estimation for $Bl^{1/2}$.

In the above considerations we tacitly assumed that the particle energy does not change on their way from a source to the observer. For large source distances often considered this assumption is no more valid. This effect can be taken into account analytically [26]. On the assumption that a particle loses its energy in an average way (no fluctuations in this process are allowed for) it can be derived that

$$\overline{\eta(E_f)^2} = 3\overline{\eta_0^2} \cdot \int_0^1 u^2 \left(\frac{E_f}{E(u)} \right)^2 du \qquad (45)$$

where E_f is the particle final energy, η_0 is the deviation angle without energy losses, and u measures the fraction of particle full trajectory passed from the start to the point where particle energy is $E(u)$ ($E(1) = E_f$). If a proton loses its energy on the microwave background then, after traversing 100 Mpc, its η_{rms} goes down by 0.96 and 0.9 for $E_f = 10^{19}$ eV and 10^{20} eV correspondingly, with respect to values $\eta_{0,rms}$. Thus, this effect is not large up to $D \leq 100$ Mpc.

The observation of doublets was first used by Cronin [37] to estimate particle charges, by assigning the angular separation of the doublet members as due to particle deflection in the *regular Galactic* magnetic field. Adopting a particular model of the field [38], he traced back the two particles of a doublet (of measured energies) to the outside of the Galaxy (with the halo size $\simeq 0.8$ kpc) and calculated the particle angular separation depending on their charge. For the doublet considered above, with $\alpha \simeq 1.6°$, the Galactic separation was 0.3° for protons, 0.9° for α particles and 5.9° for carbon nuclei. The latter is over 3 times larger than the observed α, hence carbon should be excluded.

Our knowledge of the large scale Galactic regular field is rather poor (particularly in the halo) and it would be interesting to calculate the same for another reasonable model of the field, just to know how sensitive the results are (we think that they are).

Returning to the effects of the intergalactic medium on the multiple events one could use another parameter to study particle propagation – the time delay τ (the difference between the particle propagation time from the source to the observer and D/c). It can only be useful, however, if one assumes that the multiplet members were emitted by the source at the same time (strictly speaking, if the emitting time difference was much shorter than the difference of the arriving times). The average particle time delay has been correctly derived by Achterberg et al [34] as

$$\overline{\tau} = \frac{1}{12}\overline{\theta^2} \cdot D/c$$

or $\qquad\qquad\qquad\qquad\qquad\qquad\qquad\qquad\qquad\qquad\qquad$ (46)

$$\overline{\tau} = 1.4 \cdot 10^4 \text{ yr} \cdot \left(Z \cdot \frac{B}{1 \text{ nG}} \cdot \frac{10^{20} \text{ eV}}{E} \cdot \frac{D}{30 \text{ Mpc}} \right)^2 \cdot \frac{l}{1 \text{ Mpc}}$$

for the model of small scattering angles and cells with equal lengths l. Further development has been made by Giller et al [39] where the dispersion of τ has been derived:

$$\delta_\tau = \frac{1}{\sqrt{360}} \cdot \overline{\theta^2} \cdot \frac{D}{c} \qquad\qquad (47)$$

so that $\delta_\tau/\tau = \sqrt{2/5} = 0.63$.

As it is seen, the relative dispersion of the time delay is always the same, i.e. does not depend on propagation conditions, nor on particle energy (if the energy losses are disregarded). The same ratio ($\sqrt{2/5}$) would be obtained if τ had the χ^2 distribution (gamma) with 5 degrees of freedom. (Our numerical calculations have shown that it really is a very good approximation to the actual τ distribution.)

The above result that the dispersion of the time delay is almost as large as the mean has an implication for drawing conclusions about the nature of CR ejection from the source, i.e. for distinguishing between a continous and explosive (instantaneous) emission. When considering this question for a multiple event, assuming common origin of its members, it is often argued that if the less energetic shower arrives earlier then the instantaneuos emission is impossible. This is not necessarily true: if the time delay (and its dispersion) is of the order of the difference of the two shower arrival times, it is quite possible that a less energetic particle hits the Earth earlier than that with higher energy.

Another thing to be noticed here is that for some multiplets the energy difference of two showers is not much larger than its uncertainty, which should also be taken into account when concluding about the same origin of the two showers (in space and time).

Conclusions

In view of the new cutting-off energy spectrum obtained by the HiRes experiment our conclusion are now different from those drawn during the Erice School. Then, the main weight was put on looking for models explaining the AGASA flat spectrum. It seemed that universal protons (with 10^{17}–10^{20} eV) were excluded but primary heavies (Fe) could find their way to us in a low level of the infrared background. Now, we are almost certain that at least one of the two experiments has a much larger error of the flux measurement above $\sim 3 \cdot 10^{19}$ eV, than quoted.

The observation of multiplets, with a rather small probability of their being statistical fluctuations, looks promising. They may indicate the directions towards the sources (the AGASA triplet!) and provide limits on the intergalactic irregular magnetic field characteristics.

In a few years time, there will be data from the Pierre Auger Observatory. We just remind here the main advantage of this experiment: it will be able to measure *the same shower* by the two methods, used separately by HiRes (fluorescence) and AGASA (ground particles). The two energies of a shower must agree, otherwise we miss something in our understanding of the shower development in the atmosphere (I assume here that the detectors are well understood).

In a more distant future there could be a quantum leap in the UHECR observations: the shower fluorescence and scattered Cherenkov emission could be observed from above, by telescopes on a satellite having a field of view orders of magnitude larger than that of the ground arrays (OWL experiment [41]).

Acknowledgments

I thank the Organizers of the School for hospitality and support. I thank also W.Michalak and A.Śmiałkowski for their collaboration. This paper was supported by the Polish Committee for Scientific Research (KBN) grant 2PO3D03723.

References

1. G. T. Zatsepin and V.A. Kuzmin, *Zh. Eksp. Teor. Fiz. Pisma Red.*, **4**, 144 (1966)
2. K. Greisen, *Phys. Rev. Lett.*, **16**, 748 (1996)
3. M. Giller and M. Zielińska, *Nuclear Physics*, **A663&664** 852c (2000)
4. Y. Uchihori *et al*, *Astroparticle Phys.*, **13**, 151 (2000)

5. M. Takeda *et al*, *Phys. Rev. Lett.*, **81**, 1163 (1998)

6. T. Abu-Zayyad *et al*, *astro-ph* 0208243 (2002)
   as above *astro-ph* 0208301

7. V. S. Berezinsky and S. I. Grigoreva, *Astron. Astrophys.*, **199**, 1 (1988)

8. A. Donnachie, *Proc. 1971 Int. Symp. Electron and Photon Interactions at High Energies*, ed. N.B.Mistry, Cornell University, Ithaca, NY (1972)

9. F. W. Stecker and M. H. Salamon, *ApJ*, **512**, 521 (1999)

10. J. L. Puget, F. W. Stecker and J. H. Bredekamp, *ApJ*, **205**, 638 (1976)

11. M. Ave *et al*, *astro-ph* 0112253 v2 (2002)

12. S. Yoshida and M. Teshima, *Prog. Theor. Phys.*, **89**, 833 (1993)

13. M. Nagano and A. A. Watson, *Rev. Mod. Phys.*, **72**, 689 (2000)

14. M. Takeda *et al*, *astro-ph* 0209422 (2002)

15. M. Teshima *et al*, *Nucl. Instrum. Methods Phys. Res. A*, **247**, 399 (1986)

16. A. M. Hillas, *Proc. ICRR Int. Symp* (Kofu) World Scientific, p.74 (1990) (and ref. therein)

17. J. N. Capdevielle *et al*, *Proc. Int. Cosmic Ray Conf.* (Hamburg), 479 (2001)

18. T. Abu-Zayyad T *et al*, *Proc. Int. Cosmic Ray Conf.* (Salt Lake City), **5**, 349 (1999)
   J. Boyer *et al*, *Nucl. Instrum. Methods*, **A482**, 457, (2002)

19. V. S. Berezinsky, A. Z. Gazizov and S. I. Grigoreva, *astro-ph* 0204357 (2002)

20. F. W. Stecker and M. H. Salamon, *ApJ*, **512**, 52 (1999)

21. J. Szabelski *et al*, *Astroparticle Phys.*, **17**, 125 (2002)

22. G. Medina-Tanco, *ApJ Lett*, **510**, L91 (1999)

23. N. Hayashida *et al*, *astro-ph* 0008102 (2000)

24. Haverah Park: M. A. Lawrence *et al*, *J. Phys. G: Nucl. Part. Phys.*, **17**, 733 (1991)
   Yakutsk: B. N. Afanasiev *et al*, *Proc. Int. Symp. Extremely High Energy Cosmic Rays* (Tokyo), ed. M.Nagano, p.32 (1996)
   Volcano Ranch: J. Linsley, *Catalogue of Highest Energy Cosmic Rays* ed. M.Wada, World Data Center of Cosmic Rays, Tokyo

25. T. Stanev *et al*, *Phys. Rev. Lett.*, **75**, 3056 (1995)

26. A. Śmiałkowski, M. Giller and W. Michalak, *J. Phys. G: Nucl. Part. Phys.*, **28**, 1359 (2002)

27. C. Cesarsky and V. Ptuskin, *Proc. Int. Cosmic Ray Conf.* (Calgary), **2**, 341 (1993)

28. S. S. Al-Dargazelli *et al*, *Proc. Int. Cosmic Ray Conf.* (Durban), **4**, 465 (1997) but see also
   F. Jones, AIP Conf. Proc. 433 *Workshop on Observing Giant Air Showers from Space*, (ed. Krizmanic *et al*), p.37 (1998)

29. N. J. Fisher *et al*, *Statistical Analysis of Spherical Data*, Cambridge University Press (1993)

30. G. Medina-Tanco, *ApJ*, **549**, 711 (2001)

31. P. G. Tinyakov and I. I. Tkachev, *astro-ph* 0102101 (2001), (to appear in Pisma v ZhETF) (2001)

32. G. R. Farrar and P. L. Biermann, *Phys. Rev. Lett.*, **81**, 3579 (1998)

33. P. P. Kronberg, *Rep. Prog. Phys.*, **57**, 325 (1994)

34. A. Achterberg *et al*, *astro-ph* 99070670 (1999)
35. D. Harari *et al*, *astro-ph* 0202362 (2002)
36. M. Takeda *et al*, *ApJ*, **522**, 225 (1999)
37. J. Cronin, *Proc. Int. Symp. Extremely High Energy Cosmic Rays* (Tokyo), ed. M.Nagano, p.2 (1996)
38. J. P. Vallée, *ApJ*, **366**, 450 (1991)
39. M. Giller, W. Michalak and A. Śmiałkowski, *in preparation*
40. M. Ave *et al*, *astro-ph* 0112253 (2001)
41. R. E. Streitmatter, AIP Conf. Proc. 433 *Workshop on Observing Giant Air Showers from Space* (ed. Krizmanic *et al*) p.95 (1998)

GAMMA RAY BURSTS, SUPERNOVAE, AND COSMIC RAY ORIGIN

CHARLES D. DERMER*

Code 7653, Naval Research Laboratory
4555 Overlook Ave. SW
Washington, DC 20375-5352
E-mail: dermer@gamma.nrl.navy.mil

Progress in the solution of the mystery of gamma ray bursts (GRBs) is leading to the solution of the mystery of cosmic ray origin. Successes and difficulties with the hypothesis that cosmic rays originate from galactic supernovae (SNe) are outlined. In particular, SNe satisfy power requirements, and first-order shock acceleration in supernova remnant (SNR) shocks explains the power-law spectrum of cosmic rays below the knee. The lack of detection of the π^0 bump and weakness of hadronic TeV emission from SNRs suggests, however, that cosmic rays are accelerated by a rare type of supernova. Statistical arguments show that the average kinetic power of GRBs and clean and dirty fireballs into an L* galaxy like the Milky Way is at the level of 10^{40} ergs s^{-1}, with ~ 1 event every 10^3-10^4 years. This power and rate is sufficient to explain the origin of hadronic cosmic rays observed locally. The rare SNe that give rise to GRBs are proposed to be those which accelerate cosmic rays.

1. Introduction

The strongest argument that hadronic cosmic rays (CRs) are powered by SNRs may be the claim that only SNe inject sufficient power into the Galaxy to provide the measured energy density of CRs.[1,2] The local energy density of CRs is $u_{CR} \sim 1$ eV cm$^{-3} \approx 10^{-12}$ ergs cm^{-3}. The required CR power is thus $L_{CR} \approx u_{CR} V_{gal}/t_{esc}$, where V_{gal} is the effective volume of the Galaxy from which CRs escape on a timescale t_{esc}. If CRs are produced in a disk of 15 kpc radius and 100 pc scale height, then $V_{gal} \approx 4 \times 10^{66}$ cm^3. Observations of the light elements Li, Be, and B that are formed through spallation of C, O, and N indicate that CRs with energies of a few GeV per nucleon, which carry the bulk of the CR power, pass through ≈ 10 gm

---

*Work supported by the Office of Naval Research and NASA DPR# S-15634-Y.

cm^{-2} before escaping from the disk of the Galaxy. A mean disk density of one H atom cm^{-3} gives $t_{esc} \approx 6 \times 10^6$ yr, implying that $L_{CR} \approx 2 \times 10^{40}$ ergs s^{-1}. Analysis of the composition of isotopic CR ^{10}Be yields a larger value of t_{esc}, implying a smaller mean matter density but a larger effective containment volume of the Galaxy, so that in either case

$$L_{CR} \approx \frac{u_{CR}V_{gal}}{t_{esc}} \approx 5 \times 10^{40} \text{ ergs s}^{-1} . \tag{1}$$

The galactic SN luminosity $L_{SN} \approx (1 \text{ SN}/30 \text{ yrs}) \times 10^{51}$ ergs/SN $\approx 10^{42}$ ergs s^{-1} which, given a 10% efficiency for converting the directed kinetic energy of SNe into CRs that seems feasible through the shock Fermi mechanism, is completely adequate to power the hadronic cosmic radiation.

Although γ-ray astronomy was supposed to solve the cosmic-ray origin problem, this has not happened. The predicted π^0 decay features at 70 MeV have not been detected from SNRs with EGRET,[3] and the Whipple imaging air Cherenkov telescope has not detected emission consistent with hadronic CR acceleration by SNRs.[4] Observations show that TeV electrons are accelerated by SNRs, but there is as yet no direct observational evidence for hadronic CR acceleration by SNRs except in the case of Cas A, though at an unexpectedly low level.[8] Moreover, it is becoming increasingly clear that the stochastic nature of explosive phenomena in the Galaxy is important for interpreting radio and gamma radiation emitted by Galactic CRs.[6,7] The shock Fermi mechanism explains the CR injection index due to strong nonrelativistic shocks.[9]

The EGRET observations of the diffuse galactic γ radiation[5] contradicts the assumption that hadronic CRs are uniformly distributed throughout the Galaxy, so that the required power could be significantly smaller if CR leptons emit most of the diffuse galactic radiation. Alternately, we could live in a region of enhanced CR hadron energy density compared to the Galactic average. This is more likely if CRs are produced by rare powerful events, in which case the power requirements are also reduced. Our location in the Gould belt shows that we live near a region of enhanced star formation activity.[10] These rare powerful events could be the stellar collapse events associated with GRBs.

The redshifts of over 20 GRBs are now measured, with the mean redshift near unity and the largest at $z = 4.5$. The corresponding distances imply apparent isotropic γ-ray energy releases in the range from $\approx 10^{51}$-10^{54} ergs. Recent results suggest that GRB emissions are strongly beamed,[11] so that the total energy release is actually in the neighborhood of 10^{51} ergs, corresponding to the typical total energies released in the kinetic outflow of a supernova. Delayed reddened enhancements detected in the optical light curves of a few GRBs could also be a consequence of a supernova emission component. If the beaming results are correct, then many more sources of GRBs exist than are implied through direct statistical studies of detected GRBs. The implied rate of both aligned and misdirected GRB sources begins to approach the expected rate of Type Ib/c supernovae (SNe). These and other lines of evidence indicate that GRBs are related to a rare type of supernova.

In this paper, I briefly review GRBs from the point of view that GRB sources accelerate the bulk of the hadronic GeV-TeV CRs,[13] as well as CRs above the knee, including the ultrahigh energy ($> 10^{19}$ eV) cosmic rays (UHECRs). The confirming prediction of this model is that one out of every 10-50 SNRs display intense hadronic emission and signatures of earlier GRBs (e.g., bipolar outflows, remnant black holes, and neutron-decay halos[13]). See reviews[14,15,12] for more detail.

2. Observations of GRBs

The integral size distribution of BATSE GRBs in terms of 50-300 keV peak flux ϕ_p is very flat below ~ 3 ph cm^{-2} s^{-1}, and becomes steeper than the $-3/2$ behavior expected from a Euclidean distribution of sources at $\phi_p \gtrsim 10$ ph cm^{-2} s^{-1}.[18,12] GRBs typically show a very hard spectrum in the hard X-ray to soft γ-ray regime, with a photon index breaking from ≈ -1 at photon energies $E_{ph} \lesssim 50$ keV to a -2 to -3 spectrum at $E_{ph} \gtrsim$ several hundred keV.[16] Consequently, the distribution of the peak photon energies E_{pk} of the time-averaged νF_ν spectra of BATSE GRBs are typically found in the 100 keV - several MeV range.[17]

The duration of a GRB is defined by the time during which the middle 50% (t_{50}) or 90% (t_{90}) of counts above background are measured. A bi-modal duration distribution is measured in both the t_{50} or t_{90} durations.[20] About two-thirds of BATSE GRBs are long-duration GRBs with $t_{90} \gtrsim 2$ s, with the remainder comprising the short-duration GRBs.

The Beppo-SAX GRB observations revealed that essentially all long-duration GRBs have fading X-ray afterglows.[21] The small X-ray error boxes allow deep optical and radio follow-up studies. GRB 970228 was the first GRB from which an optical counterpart was reliably identified,[22] and GRB 970508 was the first GRB for which a redshift was measured.[23,24] Redshifts of GRBs are inferred from host galaxy optical emission lines, X-ray spectral features, and absorption lines in the fading optical afterglow due to the presence of intervening gas.[25] No optical counterparts are detected from approximately one-half of GRBs with well-localized X-ray afterglows, and are termed "dark" bursts. These sources may be undetected in the optical band because of dusty media.[26] Approximately 40% of GRBs have radio counterparts. The transition from a scintillating to smooth behavior in the radio afterglow of GRB 980425 provides evidence for an expanding source.[27]

X-ray features have been detected in 6 GRBs, including variable Fe absorption during the γ-ray luminous phase of GRB 990705,[28] X-ray emission features in the afterglow spectra of GRB 991216,[29] low significance X-ray Fe K features observed in GRB 970508,[30] GRB 970828,[31] and GRB 000214,[32] and multiple high-ionization emission features detected in GRB 011211.[33]

3. GRB Source Models

Keeping in mind that only members of the class of long-duration GRBs have measured redshifts, considerable evidence connects GRBs to star-forming regions[25] and, consequently, to high-mass stars. For example, the associated host galaxies have blue colors, consistent with galaxy types that are undergoing active star formation. GRB counterparts are found within the optical radii and central regions of the host galaxies.[34] Lack of optical counterparts in some GRBs could be due to extreme reddening from large quantities of gas and dust in the host galaxy.[35] Supernovae-like emissions have been detected in the late-time optical decay curves of a few GRBs.[36,37]

The two leading scenarios to explain the origin of gamma-ray bursts are the collapsar and supranova models. The collapsar model[38,39] assumes that GRBs originate from the direct collapse of a massive star to a black hole. During the collapse process, a nuclear-density, several Solar-mass accretion disk forms and accretes at the rate of ~ 0.1-1 $M_\odot$ s^{-1} to drive a baryon-dilute, relativistic outflow through the surrounding stellar envelope. The duration of the accretion episode corresponds to the prompt gamma-ray luminous phase, which is commonly thought to involve internal shocks, though this interpretation is disputed.[40] A wide variety of collapsar models can be envisaged,[41] but their central feature is the one-step collapse of the core of a massive star to a black hole. A major difficulty of this model is to drive a baryon-dilute, relativistic outflow through the stellar envelope.[42] The delayed reddened optical excesses are claimed to be SN optical light curves, and the light curve of SN 1998bw is often used as a template to model these excesses.

The supranova model[43] involves a two-step collapse process of an evolved massive star to a black hole through the intermediate formation of a neutron star with mass exceeding several Solar masses. The neutron star is initially stabilized against collapse by rotation, but the loss of angular momentum support through magnetic dipole and gravitational radiation leads to collapse of the neutron star to a black hole after some weeks to years. The accretion-induced collapse of a neutron star in a binary system could also form a GRB.[44] A two-step collapse process means that the neutron star is surrounded by a SN shell of enriched material at distances of $\sim 10^{15}$-10^{17} cm from the central source. The earlier SN could yield ~ 0.1-$1\ M_\odot$ of Fe in the surrounding vicinity. The X-ray features in prompt and afterglow spectra can be explained in the context of the supranova model.

A pulsar wind and pulsar wind bubble consisting of a quasi-uniform, low density, highly magnetized pair-enriched medium within the SNR shell is formed by a highly magnetized neutron star during the period of activity preceding its collapse to a black hole.[45] The interaction of the pulsar wind with the shell material will fragment and accelerate the SNR shell, and the pulsar wind emission will be a source of ambient radiation that can be Comptonized to gamma-ray energies.[46] In the context of the supranova model, the delayed reddened excesses could be the cooling thermal emissions of a SNR shell heated by the pulsar wind.[47]

4. Cosmic Ray Production by GRBs

Several lines of evidence[14] indicate that GRBs are closely related to a subset of SNe that drive relativistic outflows in addition to the nonrelativistic ejecta expelled during the collapse of the massive core to a neutron star. The relativistic ejecta decelerate to nonrelativistic speeds at the Sedov radius by sweeping up matter from the external medium. Particle acceleration occurs at these shocks, just as in the nonrelativistic shocks from "normal" Type Ia and Type II supernova remnants (see lectures by Danziger), though with the addition of an earlier relativistic shock phase. The first-order shock Fermi mechanism is generally recognized as the mechanism that accelerates GeV-PeV cosmic rays in the converging flows formed by the forward shock. In addition, second-order Fermi acceleration of particles through gyroresonant particle-wave interactions with magnetic turbulence generated in the shocked fluid can also accelerate particles to ultra-high energies.[48] First-order Fermi processes cannot accelerate particles to ultra-high energies for particles accelerated from low energies for typical ISM conditions.[49]

4.1. UHECR Production by GRBs

The Larmor radius of a particle with energy $10^{20}E_{20}$ eV is $\approx$ $100E_{20}/(ZB_{\mu G})$ kpc. Unless UHECRs are heavy nuclei, they probably originate from outside our Galaxy. The energy density of UHECRs is comparable to the energy density that would be produced by GRB sources within the GZK radius, assuming that the γ-ray and UHECR power from GRBs are roughly equal.[50,51] The energy density of UHECRs observed near Earth is

$$u_{UH} \cong \zeta \, \frac{L_{GRB}t_{esc}}{V_{prod}} \tag{2}$$

where L_{GRB} is the power of GRBs throughout the production volume V_{prod} of the universe. UHECRs are produced with an efficiency ζ compared with the γ-ray power and "escape" from the universe primarily through photohadronic processes with an effective escape time t_{esc}.

The mean γ-ray fluence of BATSE GRBs is $F_\gamma \approx 3 \times 10^{-6}$ ergs cm^{-3} and their rate over the full sky is $\dot{N}_{GRB} \approx 2/$day. If most GRBs are at redshift $\langle z \rangle \sim 1$, then their mean distance is $\langle d \rangle \approx 2 \times 10^{28}$ cm, so that the average isotropic energy release of a typical GRB source is $\langle E_\gamma \rangle \approx 4\pi \langle d \rangle^2 F_\gamma /(1 + z) \cong 8 \times 10^{51}$ ergs, implying a mean GRB power into the universe of $L_{GRB} \approx 2 \times 10^{47}$ ergs s^{-1}. (This estimate is independent of the beaming fraction, because a smaller beaming fraction implies a proportionately larger number of sources.) UHECR protons lose energy due to photomeson processes with CMB photons in the reaction $p + \gamma \to p + \pi^0, n + \pi^+$. The effective distance for 10^{20} eV protons to lose 50% of their energy is 140 Mpc,[52] so that $t_{esc} \cong 1.5 \times 10^{16}$ s. This implies that the energy density observed locally is $u_{UH} \approx 10^{-22}\eta$ ergs cm^{-3}. This estimate is about equal to the energy density of super-GZK particles with energies exceeding 10^{20} eV, as measured with the High Res air fluorescence detector.

This coincidence is verified in a detailed estimate of GRB power in the context of the external shock model for GRBs.[53,13] A testable prediction of the hypothesis is that star-forming galaxies which host GRB activity will be surrounded by neutron-decay halos.

4.2. Rate and Power of GRBs in the Milky Way

The preceding estimates were made for extragalactic sources of cosmic rays. GRBs taking place in galaxies such as our own will also power cosmic rays with energies below the knee of the cosmic ray spectrum.

The local density of L^* galaxies can be derived from the Schechter luminosity function, and is $\approx 1/(200\text{-}500 \text{ Mpc}^3)$. The BATSE observations imply, as already noted, ~ 2 GRBs/day over the full sky. Due to beaming, this rate is increased by a factor of $500f_{500}$, where $f_{500} \sim 1$. Given that the volume of the universe is $\sim 4\pi(4000 \text{ Mpc})^3/3$, this implies a rate per L^* galaxy of

$$\rho_{L^*} \approx \frac{300 \text{ Mpc}^3/L^*}{\frac{4\pi}{3}(4000 \text{ Mpc})^3} \frac{2}{\text{day}} \frac{365}{\text{yr}} \times 500f_{500} \times SFR \times K_{FT}$$

$$\approx 2 \times 10^{-4} \left(\frac{SFR}{1/6}\right) \times \left(\frac{K_{FT}}{3}\right) \times f_{500} \text{ yr}^{-1} . \tag{3}$$

The star-formation rate factor SFR corrects for the star-formation activity at the present epoch $[SFR(z = 0) \cong (1/6)SFR(z = 1)]$, and the factor K_{FT} accounts for dirty and clean fireball transients that are not detected as GRBs. Thus a GRB occurs about once every several thousand years throughout the Milky Way.

Explosion Type	Outflow Speed (km s^{-1})	$\langle\beta_0\Gamma_0\rangle$	Rate (century^{-1})
SN Ia	$\lesssim 2 \times 10^4$	0.03	0.42
SN II	$\sim 10^3$-2×10^4	0.01	1.7
SN Ib/c	$\sim 1.5 \times 10^3$-2×10^5	0.2	0.28
Dirty	3×10^5	30	?
GRB	3×10^5	300	~ 0.02
Clean	3×10^5	3000	?

Table 1 gives the outflow speeds and rates of different types of SNe and GRBs in the Galaxy. The data for the outflow speeds of Types Ia, II, and Ib/c are from [54] and [55]. The mean values of the initial dimensionless momentum $\langle\beta_0\Gamma_0\rangle$ of the outflows are also given. The GRB rate is about 10% as frequent as Type Ib/c SNe, and about 1% as frequent as Type II SNe. The rates of dirty and clean fireballs are unknown, but if the X-ray rich γ-ray bursts comprise the dirty fireball limit of the BATSE

GRB population, then they could be even more frequent than the GRB population. Statistical fits[53] to the BATSE data suggest that the rate of clean fireballs is smaller than the GRB rate.

By weighting the GRB rate, equation (3), by the mean energy of $3 \times 10^{51}/(\eta_\gamma/0.2)$ ergs (η_γ is the efficiency to convert total energy to γ-ray energy), we see that the time-averaged power of GRBs throughout the Milky Way or other L_* galaxies is

$$L_{MW} \approx 2 \times 10^{40} \frac{f_{500}}{(\eta_\gamma/0.2)} \left(\frac{SFR}{1/6}\right) \left(\frac{K_{FT}}{3}\right) \text{ergs s}^{-1} . \tag{4}$$

Because each GRB has a total energy release of a few $\times 10^{51}$ ergs, comparable to SNe, but occur ~ 100 times less frequently, the available power from the sources of GRBs is only a few per cent of the power from Type Ia and Type II SNe. This would suggest that GRB sources can only make a minor contribution to cosmic ray production below the knee, but the efficiency for accelerating cosmic rays in the nonrelativistic SN outflows could be considerably less than in the relativistic outflows of GRBs.

In this regard, note that Cas A has recently been detected with HEGRA at TeV energies at the level of $\approx 8 \times 10^{-13}$ ergs cm^{-2} s^{-1}.[8] Spectral considerations suggest that the emission is hadronic. The expected level of 100 MeV- TeV γ-ray production can be estimated to be

$$F_\gamma \approx (4\pi d^2)^{-1} \times 10^{51} \text{ ergs } \eta_p \times c\sigma_{pp} n_0 \approx$$

$$7 \times 10^{-11} \left(\frac{\eta_p}{0.1}\right) n_0 \left(\frac{d}{3.4 \text{ kpc}}\right)^{-2} \text{ergs cm}^{-2} \text{ s}^{-1} , \tag{5}$$

where the strong interaction cross section $\sigma_{pp} \approx 30$ mb. This expression may underestimate the spectral energy flux measured at a fixed frequency range and the γ-ray production efficiency, but unless the density is unusually tenuous in the vicinity of Cas A and other SNRs examined with the Whipple telescope,[4] the TeV observations suggest that most SNRs do not accelerate cosmic rays with high efficiency.

We propose that cosmic rays are predominantly accelerated by the subset of SNe that eject relativistic outflows. Bright enhancements of hadronic emission from those SNe which host GRB events are implied. From the rate estimates shown in Table 1, we see that about 1 in 10 to 1 in 100 SNRs would exhibit this enhanced emission. The better imaging and sensitivity of the *GLAST* telescope and the next generation of ground-based air Cherenkov telescopes, including HESS, VERITAS, and MAGIC will test this hypothesis.

5. Conclusions

GRB sources release energy into an L^* galaxy such as the Milky Way at the time-averaged rate of $\sim 10^{40\pm1}$ ergs s^{-1}, which is nearly equal to the cosmic ray power into our Galaxy. Relativistic blast waves can accelerate particles well beyond the knee of the cosmic-ray spectrum through second-order processes,[48] overcoming difficulties that first-order Fermi acceleration faces to accelerate cosmic rays to energies above the knee of the cosmic ray spectrum in SNe. If the beaming results are correct, then GRBs occur in our Galaxy at the rate of once every several thousand years, suggesting that GRBs are associated with a subset of SNe.[13] GRB sources are probably related to an unusual type of supernova occuring at $\sim 1\%$ of the rate of Type II SNe, with energy releases several times greater than Type Ia or Type II SNe. The stochastic nature of these rather rare GRB-associated SNe is in accord with recent inferences from CR observations, including the hard spectrum in the diffuse galactic gamma-radiation.

Acknowledgments

This work is supported by the Office of Naval Research and the NASA Astrophysics Theory Program (DPR S-13756G).

References

1. Gaisser, T. K. 1990, Cosmic Rays and Particle Physics (New York: Cambridge University Press)
2. Ginzburg, V. L., and Syrovatskii, S. I. 1964, The Origin of Cosmic Rays (New York: MacMillan) (1964)
3. Esposito, J. A., Hunter, S. D., Kanbach, G., and Sreekumar, P. 1995, Astrophys. J., 461, 820
4. Buckley, J. H., et al. 1998, Astron. and Astrophys., 329, 639
5. Hunter, S. D., et al. 1997 Astrophys. J., 481, 205
6. Pohl, M., and Esposito, J. A. 1998, Astrophys. J., 507, 32
7. Strong, A. W. 2000, private communication
8. Aharonian, F. A. et al. 2001, Astron. Astrophys., 370, 112
9. Blandford, R. & Eichler, D. 1987, Phys. Rep., 154, 1
10. Grenier, I. A., and Perrot, C. 1999, in Proc. 26th ICRC, 3, 476
11. Frail, D. A. et al. 2001, Astrophys. J., 562, L55
12. van Paradijs, J., Kouveliotou, C., and Wijers, R. A. M. J. 2000, Ann. Rev. Astron. Astrophys. 38, 379
13. Dermer, C. D. 2002, Astrophys. J., 574, 65, and astro-ph/0005440 v. 1
14. Dermer, C. D. 2002, in the 27th ICRC (astro-ph/0202254)
15. Mészáros, P. 2002, Annual Rev. Astron. Astrophys., 40, 137

198

16. Band, D. et al. 1993, Astrophys. J., 413, 281
17. Mallozzi, R. S., et al. 1995, Astrophys. J., 454, 597
18. Meegan, C. A., et al., 1996, Astrophys. J. S., 106, 65
19. Mallozzi, R. S., et al. 1997, in 4th Huntsville Symposium on GRBs, p. 273
20. Kouveliotou, C., et al. 1993, Astrophys. J., 413, L101
21. Costa, E. 1999, Astron. Astrophys. S, 138, 425
22. van Paradijs, J., et al. 1997, Nature, 386, 686
23. Djorgovski, S. G., et al. 1997, Nature, 387, 876
24. Metzger, M. R., et al. 1997, Nature, 387, 878
25. Djorgovski, S. G., et al. 2001, in Gamma Ray Bursts in the Afterglow Era, ed. E. Costa et al. (Springer: Berlin), 218
26. Groot, P. J. et al. 1998, Astrophys. J.l, 493, L27
27. Frail, D. A., et al. 1997, Nature, 389, 261
28. Amati, L., et al. 2000, Science, 290, 953
29. Piro, L., et al. 2000, Science, 290, 955
30. Piro, L., et al. 1999, Astrophys. J., 514, L73
31. Yoshida, A. et al. 2001, Astrophys. J., 557, L27
32. Antonelli, L. A. et al. 2000, Astrophys. J., 545, L39
33. Reeves, J. N. et al. 2002, Nature, 416, 512
34. Bloom, J. S., et al. 1999, Astrophys. J., 518, L1
35. Galama, T. J. and Wijers, R. A. M. J. 2001, Astrophys. J., 549, L209
36. Reichart, D. E. 1999, Astrophys. J., 521, L111
37. Bloom, J. S. et al. 2002, Astrophys. J., 572, L45
38. Wang, W., & Woosley, S. E.. 2002, astro-ph/0209482
39. Woosley, S. E. 1993, Astrophys. J., 405, 273
40. Dermer, C. D. & Mitman, K. E. 1999, Astrophys. J., 513, L5
41. Fryer, C. L., Woosley, S. E., & Hartmann, D. 1999, Astrophys. J., 526, 152
42. Tan, J. C., Matzner, C. D., and McKee, C. F. 2001, Astrophys. J., 551, 946
43. Vietri, M. and Stella, L. 1998, Astrophys. J., 507, L45
44. Vietri, M. and Stella, L. 1999, Astrophys. J., 527, L43
45. Königl, A. & Granot, J. 2002, Astrophys. J., 574, 134
46. Inoue, S., Guetta, D., & Pacini, F. 2002, Astrophys. J., in press (astro-ph/0111591)
47. Dermer, C. D. 2002, to be submitted to Astrophys. J. Letters
48. Dermer, C. D., and Humi, M. 2001, Astrophys. J., 536, 479
49. Gallant, Y. A., and Achterberg, A. 1999, M. Not. Roy. Astron. Soc., 305, L6
50. Vietri, M., Astrophys. J. 1995, 453, 883
51. Waxman, E. 1995, Phys. Rev. Letters, 75, 386
52. Stanev, T., Engel, R., Mücke, A., Protheroe, R. J., and Rachen, J. P. 2000, Phys. Rev. D, 62, 93005
53. Böttcher, M., and Dermer, C. D., Astrophys. J. 2000, 529, 635
54. Lozinskaya, T. A. 1992, Supernovae and Stellar Wind in the Interstellar Medium (New York: AIP)
55. Weiler, K. W., et al. 2000, (astro-ph/0002501)

THE ALPHA MAGNETIC SPECTROMETER (AMS)

MARIA IONICA*

INFN Sezione di Perugia, Via A. Pascoli,Perugia, I- 06100
E-mail: maria.ionica@pg.infn.it

The Alpha Magnetic Spectrometer (AMS), once installed on the International Space Station will provide precise measurements of the cosmic ray spectra up to TeV energy range, and will search for cosmological antimatter and missing matter. A prototype version of the detector was operated successfully on the space shuttle Discovery in June 1998 (STS-91). Here we briefly report on the design of the AMS apparatus and present the results of the measurements of the fluxes of proton, electron, positron and helium from the STS-91 flight.

1. Introduction

The existence of the cosmological matter-antimatter asymmetry [1] is one of the puzzles of modern astroparticle physics. Due to the limited energy which can be reached at accelerators, this problem can only be studied by performing very accurate measurement of the composition of Cosmic Rays. To perform these measurements, a high energy particle physics experiment, the Alpha Magnetic Spectrometer (AMS)[2], is scheduled for installation on the International Space Station (ISS) in 2005 for an operational period of three years. The aims of the AMS experiment are twofold: to search for new physics in the form of antimatter and dark matter of cosmological origin and to measure with high precision the cosmic ray energy spectra in the rigidity range of 0.1GV to several TV. In preparation for the ISS mission, a protoype AMS experiment was flown for ten days on the space shuttle Discovery, in June 1998 (STS-91). During this flight, at low earth orbital altitude, the protoype AMS apparatus collected almost 100 millions of Cosmic Rays. Here we present the AMS apparatus and some important results obtained by AMS during the STS-91 mission.

---

*Permanent address: I.M.T. - Bucharest, Erou Iancu Nicolae Street 32, R-72996, Bucharest, Romania

2. The AMS detector on the STS-91 mission

The AMS apparatus in STS-91 configuration (Figure 1) consists in a permanent magnet with silicon microstrip tracker planes, scintillation counters hodoscopes above and below the magnet, two layers of Aerogel Threshold Cerenkov and scintilator veto counters covering the inner surface of the magnet.

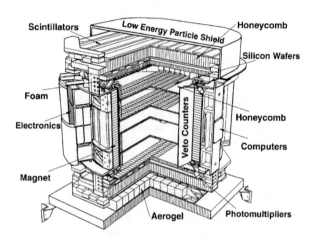

Figure 1. The AMS apparatus on the space shuttle during STS-91 precursor flight

The AMS magnet has a cylindrical shape of $800mm$ length with an inner diameter of $1115mm$ and an outer diameter of $1298mm$. The magnet is made of blocks of high-grade Nd-Fe-B with $BL^2 = 0.14Tm^2$ and a total weight about of $1.9tons$. The magnetic field vector is orthogonal to the cylinder longitudinal axis.

The Time of Flight system, was composed of four planes of coincidence scintillation counters. It is used to measure the absolute charge, the direction (upward or downward) of charged particles and the velocity of the particles traversing the spectrometer, with a resolution of $120ps$ over a distance of almost $1.4m$. It also provides the first level trigger by coincidence.

Pattern recognition and tracking is performed using large area, high accuracy silicon tracker [3,4] that consist of six planes of double sided silicon microstrip detectors, mounted on light-weight carbon fiber support planes. The silicon tracker measures both position and energy loss of particles. Four layers of silicon sensors are mounted on $1m$ diameter support planes within the magnet. Two additional layers each mounted on a $1.25m$ diameter support plane are placed at the entrance and exit of the magnet. The total sensitive area of the silicon was $2m^2$ for the STS-91 mission.

The momentum resolution for AMS on the STS-91 mission was about ($\frac{\Delta p}{p} \approx 7\%$) at 10GV, reaching ($\frac{\Delta p}{p} \approx 100\%$) at about 400GV.

By combining various measurements it is possible to determine the type of each particle traversing the spectrometer.

3. Physics results from the AMS experiment on the STS-91 flight

During the ten days STS-91 flight, June 2nd to June 12th, the shuttle Discovery performed 154 orbits at an inclination 51.7^0 and at an altitude between 320Km to 390Km, across all geomagnetic longitudes. More then 100 millions of triggers have been collected at various attitudes. The experiment was successfully producing the first high quality CR data collected with a magnetic spectrometer located outside the atmosphere [5].

3.1. Search for Anti-helium

Within $2.8millions$ He events collected by AMS during the STS-91 flight no anti-helium have been found up to a rigidity of $140GV$. Assuming identical He and anti-He spectra we obtain a superior limit on the ratio anti-He/He of $1.1 \cdot 10^{-6}$ over the rigidity interval 1 to $140GV$ [6].

3.2. Measurements of the Cosmic Ray spectrum

In this section we review the result obtained from data collected with AMS during STS-91 to study cosmic ray proton spectrum from energies of 0.1 to $200GeV$ [7,8], the helium spectrum over the kinetic energy range 0.1 to $100GeV/nucleon$ [9] and the spectra of electrons and positrons over the respective energy ranges of 0.2 to $30GeV$ and 0.2 to $3GeV$ [10], the later spectra being limited by proton background. Data taken near the South Atlantic Anomaly were excluded.

The high statistics, 10^7 protons and 10^5 electrons, available allow the variation of the spectrum with position to be measured both above and below the geomagnetic cutoff.

The differential proton flux as a function of geomagnetic latitude (Figure 2) shows that in addition to the primary CR spectrum visible above the geomagnetic cutoff, there is a substantial second spectrum, below the cutoff.

202

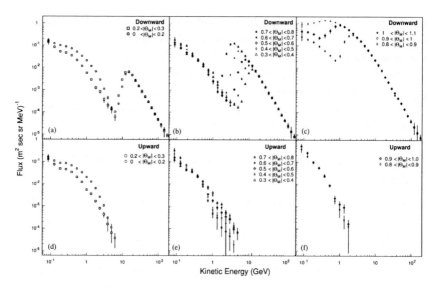

Figure 2.   Proton spectra measured by AMS for different geomagnetic latitudes intervals

Figure 3 shows the results for electrons and positrons for three intervals of geomagnetic latitudes at which they were observed.

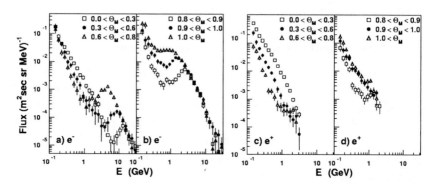

Figure 3.   Flux spectra for downward going electrons (a,b) and positrons (c,d), separated according to the geomagnetic latitude at which they were detected

Adding all data collected above the geomagnetic cutoff is possible to obtain a precise estimate of the primary CR differential flux. The AMS data was taken during a period of maximum solar activity, therefore it

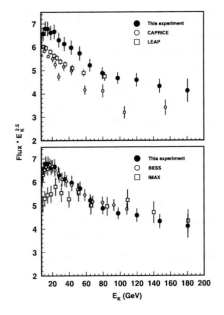

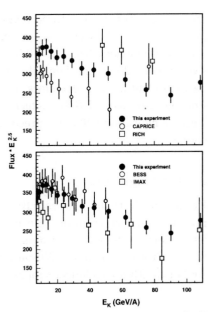

Figure 4. Primary proton flux measured by AMS and compared with existing balloons measurements

Figure 5. Primary He flux measured by AMS and compared with existing balloons measurements

is meaningful to compare the spectra at high energies $(R > 20GV)$ with previous balloons experiments [11−16]. In Figure 4 and Figure 5 are shown the primary spectra for protons and helium compared with existing balloons measurements.

The primary spectrum has been parameterized by a power law function in rigidity of type $\Phi(R) = \Phi_0 \times R^{-\gamma}$. Fitting the measured proton spectra over the rigidity range $10 < R < 200GV$ yields $\gamma = 2.78 \pm 0.009(fit) \pm 0.019(syst)$ and $\Phi_0 = 17.1 \pm 0.15(fit) \pm 1.3(syst) \pm 1.5(\gamma)GV^{2.78}(m^2.sr.sec.MeV)^{-1}$. For the helium primary spectra, fitting over the rigidity range $20 < R < 200GV$ we obtained $\gamma = 2.740 \pm 0.010(fit) \pm 0.016(syst)$ and $\Phi_0 = 2.52 \pm 0.09(fit) \pm 0.13(syst) \pm 0.14(\gamma)GV^{2.78}(m^2.sr.sec.MeV)^{-1}$.

4. Conclusions and outlook

The performance and results from the first version of the AMS experiment on STS-91 were extremely encouraging. In 2005 an improved AMS experiment is scheduled to be installed on the International Space Station for an operational period of three years. The new apparatus [17] will have improved particle identification by including a superconducting magnet with a stronger magnetic field ($B = 0.8T$), a fully equipped silicon tracker (sensitive area of $7.2m^2$), the Time of Flight system together with three powerful particle identification detectors, a Transition Radiation Detector, a Ring Imaging Cerenkov (RICH) detector and an Electromagnetic Calorimeter. This configuration will allow precise particle identification up to O(TeV) of energy.

Acknowledgments

I would like to thank Professor Roberto Battiston for his help in the preparation of this presentation. Sincere thanks to Professor M. Shapiro, Professor John Wefel for encouraging me to give this talk at the 13th ISCRA, and to Professor Thomas Wilson for his warm encouragement and to Dr. Arthur Smith for technical support.

References

1. Steigmann, G. Ann. Rev. Astron. Atroph. **14** , 339(1976)
2. Ahlen, S. et al Nucl. Instr. Meth. **A350**, 351(1994)
3. Battiston, R. Nucl. Instr. Meth (Proc.Suppl), **B44**, 274(1995)
4. Alcaraz,J. et al Il Nuovo Cimento **112A**, 1325(1999)
5. AMS Collaboration: M. Aguillar et al. Physics Report **366**, 331(2002)
6. AMS Collaboration, Alcaraz, J. et al Physics Letter B **461**, 4, 387(1999)
7. AMS Collaboration, Alcaraz, J. et al Physics Letter B **472**, 1-2, 215(2000)
8. AMS Collaboration, Alcaraz, J. et al Physics Letter B **490**, 27(2000)
9. AMS Collaboration, Alcaraz, J. et al Physics Letter B **494**, 193(2000)
10. AMS Collaboration, Alcaraz, J. et al Physics Letter B **484**, 10(2000)
11. Sanuki, T. et al. astro-ph/002481, 2000
12. Boezio, M. et al. Ap.J **518**, 457(1999)
13. Menn, W. et al. The Astrophys. J **533**, 281(2000)
14. Bellotti, R et al. Phys. Rev. **D60**, 052002(1999)
15. Seo, E. S. et al. Ap.J **378**, 763(1991)
16. Buckley, J. et al. Astrophys. J.**429**, 736(1994)
17. Alpat, B. Nucl. Instr. Meth. **A461**, 1-3, 272(2001)

THE DECONVOLUTION OF THE ENERGY SPECTRUM FOR THE TRACER EXPERIMENT

A. A. RADU, D. MÜLLER AND F. GAHBAUER

The Enrico Fermi Institute, The University of Chicago
Chicago, IL 60637, USA
E-mail: andrei@radu.uchicago.edu

TRACER (Transition Radiation Array for Cosmic Energetic Radiation) is a large area detector built at the University of Chicago for direct measurements of heavy cosmic ray nuclei up to about 10 TeV/amu. The deconvolution of the energy spectra for different nuclei, from the data collected by TRACER, is one component of our data analysis efforts. Two methods used to estimate the spectra will be discussed and the deconvoluted spectrum for iron will be presented.

1. Introduction

TRACER is a balloon-borne, large area detector system built at the University of Chicago to study high energy cosmic rays. A description of TRACER is given by Hörandel et al.[1] The observational goals of TRACER include the measurement of the energy spectra of O, Ne, Mg, Si and Fe nuclei up to 10 TeV/amu using a combination of proportional tubes and transition radiation detectors. In the following sections, we will discuss two methods for the deconvolution of the energy spectra and present the estimated spectrum for iron. TRACER was designed for a circumglobal, long duration flight, but the following discussion refers to the data collected during the one-day test flight which took place in September 1999 from Ft. Sumner, NM. Due to the short duration of this flight not many events above 1 TeV/amu, which would produce transition radiation, were recorded. Therefore, the present work attempts to determine the energy spectra at lower energy, just utilizing the logarithmic dependence of the specific ionization in gas on the Lorentz factor of the particle. The energy resolution in this region is much inferior than that expected for the transition radiation region above 1 TeV/amu.

2. The Deconvolution of the Energy Spectrum

Estimating the incident spectrum from the measured data with the response function of the instrument, is a nontrivial deconvolution problem. The following describes the deconvolution methods used for the TRACER data.

2.1. *Method A*

The true flux N_i, in the energy interval i, is derived from the measured flux M_j, in the apparent energy interval j, by correcting for inefficiencies and finite energy resolution (overlap corrections) according to Buckley et al.[2]:

$$N_i = \frac{1}{\varepsilon_i} \cdot \sum_{j=1}^{n} A_{ij} \cdot M_j \qquad (1)$$

The energy overlap matrix A_{ij} and the efficiency ε_i are determined from a complete Monte Carlo simulation of TRACER's response. Figure 1 illustrates the simulated response for iron nuclei (Z=26) by displaying the "measured" signals (intervals j) versus the true energies (intervals i). The relativistic increase is clearly visible, but the fluctuations in response are significant. The simulation assumes a power law differential spectrum ($E^{-\gamma}$) of primary particles above 10 GeV/amu with a power law index γ. The number α_{ij} gives the total count of events in cell (i,j). Then:

$$A_{ij} = \frac{\alpha_{ij}}{\sum\limits_{i=1}^{n} \alpha_{ij}} \qquad (2)$$

The efficiency ε_i is the ratio of the number of events which pass the data cuts in the energy bin i to the total number of events in the same energy bin.

The estimated spectrum obtained with this method may depend on the spectral index of the power law assumed when the detector's response (fig.1) was generated.

To investigate this question we generated a reference data set from a primary spectrum with the index $\gamma_0 = 2.5$. Then we generated a set of response functions for spectral indexes ranging from $\gamma = 1.7$ to $\gamma = 3.0$. With these, we attempted to reconstruct the primary spectrum of the reference data set according to equation (1). The reconstructed primary spectra were then characterized by spectral indexes γ_R which may be expected to be different from γ_0. However, as figure 2 illustrates, γ_R approximate the true value γ_0 fairly well, even if the assumed response function has been derived with an index γ which is quite different from γ_0. Thus, the dependence of the deconvolution technique on the choice of γ for the energy

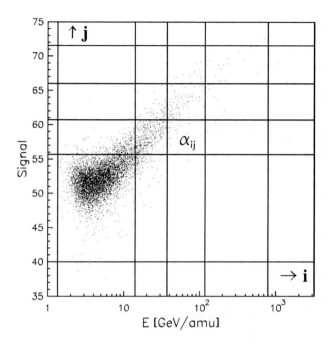

Figure 1. Scatter plot of TRACER's response in the relativistic rise region. Each event is simulated with GEANT4. A single power law spectrum of cosmic rays was assumed above 10 GeV/amu.

overlap matrix A_{ij} is not very strong in this case.

Based on these observations we analysed the measured iron flight data with response functions corresponding to several different spectral indexes. The result is shown in table 1 and in figure 3.

Table 1. (γ) - the spectral index of the response function, (γ_R) - the spectral index of the reconstructed spectrum

Simulations (γ)	Flight Data (γ_R)
2.5	2.564
2.2	2.572
2.7	2.510

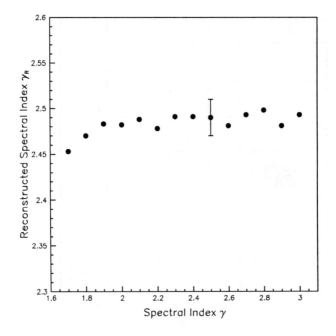

Figure 2. The spectral index (γ_R) of the reconstructed primary spectrum of the reference data set versus the spectral indexes of the response functions (γ) used to reconstruct this spectrum. A typical error bar is also shown.

TRACER's differential energy spectrum of the iron nuclei above 10 GeV/amu is characterized by a power law index $\gamma = 2.56 \pm 0.09$, well in agreement with the results of previous measurements by the CRN (Müller et al.[3]) and the HEAO 3 C-2 (Engelmann et al.[4]) experiments. The data points of TRACER (fig. 3) still include an arbitrary normalization to agree with the CRN flux at $\sim$ 50 GeV/amu, and thus are to be regarded as preliminary.

2.2. Method B

This method was developed by Loredo & Epstein[5] and uses a more general algorithm which does not depend on the spectral index used to generate the response of the instrument.

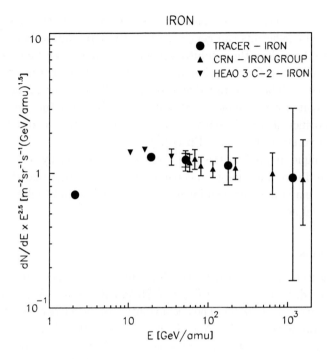

Figure 3. Preliminary results on the energy spectrum of the iron nuclei from the data collected by TRACER (circles), CRN (triangles pointing up) and HEAO 3 C-2 (triangles pointing down) experiments. The TRACER points were normalized to the CRN flux at ~ 50 GeV/amu

The incident spectrum is estimated with a linear combination of data - M_j:

$$\hat{s}(E_k) = \sum_{j=1}^{n} a_j(E_k)M_j = \int \sum_{i=1}^{N} a_i(E_j)R_i(E)s(E)dE \qquad (3)$$

E_k is the energy where the spectrum is estimated, s(E) is the true spectrum function and $R_j(E)$ is the energy distribution of all events in a certain signal bin j. If the resolution function

$$\delta(E, E_k) = \sum_{j=1}^{n} a_j(E_k)R_j(E) \qquad (4)$$

is a Dirac function, then the estimate of the spectrum is perfect. However, no set of $a_j(E_k)$ coefficients can generate ideal, Dirac resolution functions. In order to see how close the resolution function is to a Dirac function, a spread of the resolution function should be defined. On the other hand, any spectral estimate will be uncertain due to statistical fluctuations in the data and the propagated variance of the estimate should be taken into account. The $a_j(E_k)$ coefficients are chosen so that they minimize the weighted sum of spread and variance (Loredo & Epstein[5]).

This technique applied to the iron flight data collected by TRACER estimated the differential energy spectrum only with a large uncertainty. This situation may improve when longer duration flights lead to data with better statistical accuracy.

3. Conclusions

Two methods were used to estimate the differential energy spectrum in the case of the iron data from the first flight of TRACER. Method A generated a spectral index of -2.56 ± 0.09. The deconvolution of the energy spectra for other cosmic ray nuclei (O, Ne, Mg, Si) using this method is in progress. Method B, very sensitive to the response of the instrument and to the statistical fluctuations of the analyzed data, estimated the spectral index with a large uncertainty. However, both methods should provide higher quality estimates when the statistical fluctuations of the data are reduced. Also, studies are currently in progress to determine the optimum trade-off between energy intervals (bin-sizes) and statistical fluctuations in the data.

References

1. J. R. Hörandel et al., *Nucl. Phys. B (Proc. Suppl.)* **97**, 142-145 (2001).
2. J. H. Buckley, *Astrophys. J.* **429** 736-747 (1994).
3. D. Müller et al., *Astrophys. J.* **374** 356-365 (1991).
4. J. J. Engelmann et al., *A&A* **233** 96 (1990).
5. T. J. Loredo and R. I. Epstein, *Astrophys. J.* **336**, 896-919 (1989).

COMPTON SCATTER TRANSITION RADIATION DETECTORS FOR ACCESS

GARY L. CASE AND MICHAEL L. CHERRY

Dept. of Physics and Astronomy,
Louisiana State University,
Baton Rouge, LA 70803, USA
Email: case@phunds.phys.lsu.edu

The detection of transition radiation x-rays can provide a direct, non-destructive measurement of a particle's Lorentz factor. Standard transition radiation detectors (TRDs) typically incorporate thin plastic foil radiators and gas-filled x-ray detectors, and are sensitive up to $\gamma \sim 10^4$. To reach higher Lorentz factors (up to $\gamma \sim 10^5$), thicker, denser radiators can be used, which consequently produce x-rays of harder energies (≥ 100 keV). At these energies, scintillator detectors are more efficient in detecting the hard x-rays, and Compton scattering of the x-rays out of the path of the particle becomes an important effect. The Compton scattering can be utilized to separate the transition radiation from the ionization background spatially. We have designed and built a Compton Scatter TRD optimized for high Lorentz factors and exposed it to the electron beam at the CERN SPS. In this paper, we discuss the design principles for a high energy TRD; present preliminary results of the accelerator tests, demonstrating the effectiveness of the Compton Scatter TRD approach; and finally, discuss the application of this technique to the ACCESS cosmic-ray mission.

1. Introduction

Transition radiation (TR) is produced when a charged particle crosses the interface between two materials with different dielectric constants, resulting in the rapid rearrangement of the particle's electric field as it passes from one material to the next. Detailed calculations of the TR phenomenon are given in the literature,[1,2,3] but a brief description is given here. The emitted TR intensity can be derived from the Lienard–Wiechart potentials.[4] We will limit ourselves to the case of high frequencies and assume that the incident particle is highly relativistic ($\gamma = E/mc^2 \gg 1$). The formation zone, the distance in the medium over which the electric field of the particle changes and hence the transition radiation is generated, is then defined as

$$Z_{1,2} = \frac{4c}{\omega} \left(\frac{1}{\gamma^2} + \frac{\omega_{1,2}^2}{\omega^2} + \theta^2 \right)^{-1}, \tag{1}$$

where $\omega_{1,2}$ is the plasma frequency of medium 1 and 2, respectively, θ is the angle the radiation makes with respect to the direction of the incident particle and ω is the frequency of the radiation. For definiteness, material 1 is assumed to be a solid and material 2 is assumed to be a gas or vacuum, such that $\omega \gg \omega_1 \gg \omega_2$. The radiated intensity is then

$$\frac{d^2 S}{d\Omega d\omega} = \frac{1}{c} \left(\frac{q e \omega \theta}{4\pi c} \right)^2 (Z_1 - Z_2)^2, \tag{2}$$

where q is the charge of the incident particle. At frequencies above $\gamma\omega_1$, $Z_1 \approx Z_2$ and the intensity vanishes. The total intensity emitted from a single interface is proportional to $q^2\gamma$, i.e. the total intensity increases with γ as the spectrum extends to higher frequencies $\gamma\omega_1$.

2. Optimizing TRDs for high energies

For a highly relativistic particle the TR is emitted at x-ray frequencies, and the spectrum produced depends on the plasma frequencies and thicknesses of the two materials as well as the energy of the particle. The TR intensity emitted from a single interface is weak, and in practical applications, a radiator is constructed with a large number N of thin foils of thickness l_1 separated by a distance l_2, with TR produced at each of the $2N$ interfaces. Interference effects from the superposition of the amplitudes produced at each interface give rise to pronounced minima and maxima in the spectrum, with the last (highest frequency) maximum near

$$\omega_{\max} = \frac{l_1 \omega_1^2}{2\pi c}(1 + \rho), \tag{3}$$

where ρ is 1 for a metal and 0 for a nonconductor.[5] As the particle energy increases, the total radiated intensity increases up to a Lorentz factor

$$\gamma_s \approx \frac{0.6\omega_1}{c} \sqrt{l_1 l_2 (1 + \rho)}, \tag{4}$$

above which saturation sets in due to the interference. The saturation energy and characteristic frequency can be tuned by varying the radiator foil material, thickness, and separation.

   An x-ray detector appropriate for absorbing the TR x-rays must be placed after the radiator. Since the TR is strongly beamed forward ($\theta \sim$

$1/\gamma$), the x-rays are coincident with the ionization energy deposited in the detector by the particle itself. Therefore, in conventional applications, the detector must be made thin in order to minimize the ionization signal, yet with sufficient stopping power to absorb the x-rays. For ω_{max} less than about 40 keV, gaseous detectors (e.g. Xenon-filled wire chambers) are typically employed. In order to improve statistics and for redundancy, a complete transition radiation detector (TRD) consists of multiple layers of radiators and x-ray detectors. Such TRDs have been used successfully[a] at accelerators, underground, in ground-based cosmic-ray experiments, on balloons, and in space, where the energies of cosmic ray nuclei with $\gamma \geq 3 \times 10^3$ have been measured by the Space Shuttle CRN experiment.[8,9]

In order to increase the maximum particle energy γ_s that the TRD can measure, one must increase the plasma frequency (or equivalently, density), thickness, and/or spacing of the foils (Eq. 4). In a space instrument, the overall length will be constrained, putting a limit on Nl_2 (assuming $l_2 \gg l_1$). Increasing ω_1 and/or l_1 results in a hardening of the x-ray spectrum produced (Eq. 3). With $\gamma_s \approx 10^5$ and a typical spacing $l_2 = 0.1 - 1$ cm, $\omega_{max} \approx 0.4\gamma_s^2 c/l_2$ can be in excess of several hundred keV. Gas detectors are then no longer efficient in detecting these hard x-rays. Although the potential use of gas TRDs near $\gamma_s \approx 10^5$ by optimizing the radiator design has been described recently[10] and preliminary results of accelerator tests of such TRDs have been discussed at this meeting,[11] scintillators such as NaI or CsI provide an efficient alternative at these high Lorentz factors and the corresponding high x-ray energies. The high density of the scintillators leads to an increase in the ionization energy deposited by the particle as it traverses the detector. However, as the TR spectrum hardens, Compton scattering in the radiators becomes important. A significant portion of the x-rays produced are scattered out of the path of the incident particle. Thus, a detector that is segmented or positioned outside of the beam can efficiently detect the TR signal spatially separated from the ionization.

3. Accelerator test of Compton Scatter TRD

A scintillator-based Compton Scatter TRD was designed to investigate the predicted increase in saturation energy obtained by using thick, dense radiator materials. The instrument consisted of 6 identical modules, each consisting of a stack of radiators (see Table 1) and viewed by three NaI

---

[a]These applications of TRDs have been described in recent reviews[6,7]

Table 1. Parameters of radiator configurations tested

Radiator	ω_1 (eV)	l_1 (μm)	l_2 (mm)	N	ω_{max} (keV)	γ_s
Thin Mylar	24.4	125	3.4	50	61	4.9×10^4
Thick Mylar	24.4	250	3.4	50	122	6.9×10^4
Thin Teflon	28.5	125	3.4	50	83	5.8×10^4
Thick Teflon	28.5	200	3.4	50	133	7.2×10^4
Aluminum	32.7	135	3.8	48	230	9.9×10^4

detectors positioned outside of and parallel to the trajectory of the incident particles. The instrument was exposed to high energy electrons at the CERN SPS in August/September 1999 and again in August/September 2001. Beam energies ranged from 7 to 150 GeV, covering the range of Lorentz factors $\gamma = 1.4 \times 10^4 - 2.9 \times 10^5$. Materials investigated included Mylar ($\rho = 1.4$ g/cm^3), Teflon ($\rho = 2.0$ g/cm^3) and aluminum ($\rho = 2.7$ g/cm^3). The parameters of each configuration tested are given in Table 1.

Spectra produced by 150 GeV electrons passing through aluminum honeycomb radiators and detected in a downstream scintillator are shown in Fig. 1. Two things are immediately evident: first, Compton scattered TR is being detected away from the path of the incident electron; and second, the detected TR x-ray spectrum peaks about 120 keV, with some x-rays detected at energies > 300 keV. In order to compare the detected yields of the different radiator configurations, the total number of photons detected per NaI detector was summed over the energy range $35 - 500$ keV. Figure 2

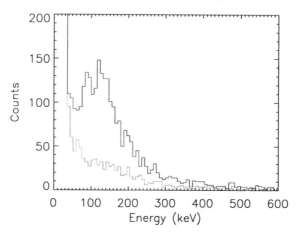

Figure 1. Spectra measured in a single downstream detector using 150 GeV electrons with aluminum honeycomb radiators (upper curve) and solid aluminum background plates (lower curve) in place.

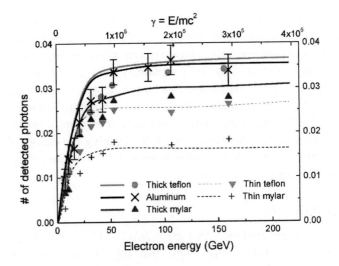

Figure 2. Average number of photons detected in the energy range 35 – 500 keV per detector per event in the first module as a function of electron energy for various radiator configurations. The symbols represent experimental data and solid lines represent results of calculations. The error bars represent statistical errors.

shows the detected yield as a function of electron energy. The observed saturation Lorentz factors are $\approx 10^5$, as expected from the calculated values in Table 1 and from simulations.

4. Application to ACCESS

The Advanced Cosmic-Ray Composition Experiment for Space Sciences (ACCESS) is a proposed cosmic-ray mission for either the space station or a satellite.[12,13,14] The goal of this mission is to measure the composition and energy spectra of individual elements up to energies of ~ 100 TeV per nucleon. ACCESS will consist of a calorimeter, TRD and charge detector. As stated above, standard TRDs will need to be modified in order to achieve the required sensitivity at $\gamma \sim 10^5$ for $Z > 3$. The Compton Scatter TRD described here meets these requirements, and work has been undertaken to design an appropriate high energy TRD for ACCESS using this technique.

5. Conclusion

TRDs can be used to make direct, nondestructive measurements of the Lorentz factor of charged particles. Increasing the saturation threshold of the TRD by using thicker, denser foils results in x-rays of harder energies being produced, suggesting the use of scintillators as the x-ray detector. At these hard x-ray energies, Compton scattering becomes important, and the TR signal can be detected spatially separated from the ionization signal of the incident particle if a segmented detector or detector outside the particle beam is employed. A prototype high energy Compton Scatter TRD was tested at CERN using high energy electrons. The results verify that the technique provides an effective means of determining the energy of charged particles with $\gamma \sim 10^5$. A Compton Scatter TRD has been developed for the ACCESS cosmic-ray mission.

Acknowledgments

This work was supported by NASA grant NAG5-5177 and NASA/Louisiana Board of Regents grant NASA/LEQSF-IMP-02. The authors wish to thank P. Altice, J. Isbert, D. Patterson and especially J. Mitchell for their invaluable help in running the experiment at CERN; A. Aranas, S. Apewokin, T. Brown and O. Shertukde for their many hours of work during the construction and testing of the radiators and detectors; the staff at CERN for their superb cooperation and assistance, particularly D. Lazic and G. Bencze; and J. Anderson, J. Marsh and C. Welch for assisting with the data analysis.

References

1. M. L. Ter-Mikaelian, *High Energy Electromagnetic Processes in Condensed Media*, (Wiley, New York, 1972).
2. M. L. Cherry, *Phys. Rev. D*, **17**, 2245 (1978).
3. X. Artu, G. Yodh and G. Mennessier, *Phys. Rev. D*, **12**, 1289 (1975).
4. J. D. Jackson, *Classical Electrodynamics*, (Wiley, New York, 1975).
5. M. L. Cherry and G. L. Case, *Astroparticle Physics*, in press (2002).
6. C. Favuzzi, N. Giglietto, N. M. Mazziotta and P. Spinelli, *La Rivista del Nuovo Cimento*, **24**, 1 (2001).
7. M. L. Cherry and J. P. Wefel, in *Proc. TRDs for the 3rd Milleniumem*, eds. N. Giglietto and P. Spinelli, **25**, 151 (2002).
8. J. M. Grunsfeld et al., *Ap. J. L.*, **327**, L31 (1988).
9. S. P. Swordy et al., *Phys. Rev. D*, **42**, 3197 (1990).
10. S. P. Wakely, *Astroparticle Physics*, **18**, 67 (2002).
11. P. J. Boyle, in *Proc. 13th Intl. School of Cosmic-Ray Astrophysics*, Erice (2002).
12. J. P. Wefel and T. L. Wilson, *Proc. 26th Intl. Cosmic-Ray Conf.*, **5**, 84 (1999).
13. T. L. Wilson and J. P. Wefel, NASA report TP-1999-209202 (1999).
14. M. Israel et al., NASA GSFC report NP-2000-05-056-GSFC (2000).

A NEW MEASUREMENT OF THE μ^+ AND μ^- SPECTRA AT SEVERAL ATMOSPHERIC DEPTHS WITH CAPRICE98

PATRICIA HANSEN

Royal Institute of Technology
10691 Stockholm,
Sweden
E-mail: hansen@particle.kth.se

FOR THE NMSU-WIZARD/CAPRICE98 COLLABORATION

A new measurements of the μ^+ and μ^- spectra at several atmospheric depths in the momentum range 0.3-20 GeV/c was made by the balloon-borne experiment CAPRICE98. The data were collected during the ascent of the payload on the 28 May 1998 from Fort Summer, New Mexico, USA. This apparatus consists of a magnet spectrometer, with a superconducting magnet and a drift-chamber tracking device, a time of flight scintillator system, a silicon-tungsten imaging calorimeter and a gas ring imaging Cherenkov detector. This gas RICH radiator made it possible to safely identify μ^+ up to 20 GeV/c. This is the first time that the μ^+ component has been measurement over a wide momentum range.

1. Introduction

Atmospheric neutrinos are produced from the decay of muons, pions and kaons. The production of electron and muon neutrinos is dominated by the processes $\pi^+ \to \mu^+ + \nu_\mu$ followed by $\mu^+ \to e^+ + \overline{\nu_\mu} + \nu_e$ (and their charge conjugates). Therefore there will be two muon neutrinos for each electron neutrino and it will give an expected ratio ν_μ/ν_e of the flux of $\nu_\mu + \overline{\nu_\mu}$ to the flux of $\nu_e + \overline{\nu_e}$ of about 2. The ν_μ/ν_e flux ratio has been measured in deep underground experiments. The measurements are reported as $R = (\nu_\mu/\nu_e)_{data} / (\nu_\mu/\nu_e)_{MC}$ where $(\nu_\mu/\nu_e)_{data}$ is the flux ratio observed in the detector and $(\nu_\mu/\nu_e)_{MC}$ is obtained by the Monte Carlo simulation. This ratio cancels many experimental and theoretical uncertainties, especially the uncertainty in the absolute flux. A value of $R = 1$ is expected if the physics and the models used in the simulation are correct.

The deep underground experiments have obtained significantly smaller values of R . This disagreement with theoretical predictions has been interpreted in terms of possible neutrino oscillations.

The correct interpretation of the results requires a detailed understanding of neutrino production in the atmosphere. This can be accomplished by comparing the simulations with other detectable particles, like muons (M. Boezio et al. [1])

We present in this paper results of the muon spectra in the atmosphere obtained with the CAPRICE98 instrument. We describe the experiment and the detector system in section 2, the data analysis in section 3 and give the results in section 4.

It is worth pointing out that the primary cosmic ray hydrogen and helium spectra (M. Boezio et al. [2]) were also measured with the CAPRICE98 instrument. These can be used as the input spectra for the cascade simulation in order to reduce the overall systematic uncertainties associated with the comparison of observed and calculated muon fluxes.

2. The experiment

The experiment was launched from Ft. Summer, New Mexico, USA on the 28th of May 1998. The flight lasted 24 hour. It was more than 20 hours above 35000 meters. At this altitude, which corresponds to approximately 5 g/cm^2 of residual atmosphere above the experiment, the spectrometer collected data from more than 5.3 million cosmic ray events.

The CAPRICE98 apparatus (M. Ambriola, et al. [3]), from top to bottom, consisted of a gas-RICH detector, a time-of-flight (TOF) device, a magnet spectrometer, and a silicon-tungsten imaging calorimeter (see figure 1).

2.1. Rich

The solid radiator Rich, in the CAPRICE94 flight was changed to a 1 m long gas RICH with a proton threshold at 18 GeV/c in the flights of 1997 and 1998. The CAPRICE94 spectrometer were able to identify positive muons up to 2 GeV/c, whereas the gas rich radiator of CAPRICE98 made it possible to safely identify μ^+ up to 20 GeV/c. The CAPRICE98 Rich detector (see figure 2) consists of a 1 m tall aluminum radiator box filled with a volume of about 800 l with C_4F_{10} gas (at an average pressure of 911 $mbar$), a Multi Wire Proportional Chamber (MWPC) with pad readout and a spherical mirror. (T. Francke et al. [4]) (D. Bergström et al. [5])

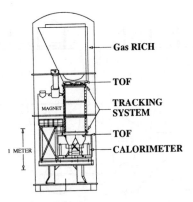

Figure 1. A schematic view of the CAPRICE98 detector

A charged particle entering the RICH traverses the MWPC and gives an ionization signal in the pad plane.

If $\beta > 1/n$, ($\beta = v/c$ where v is the particle velocity). Cherenkov light is emitted in the gas (C_4F_{10}) radiator, along the trajectory of the particle.

The particle exits through the bottom of the radiator box but the produced light is reflected and focused back up onto the MWPC by a spherical mirror on the bottom of the radiator box. In the MWPC the photons are converted to photoelectrons by the photo-electric effect. For $\beta \simeq 1$ charge one particles, an average of 12 photoelectrons per event were detected in the pad plane. The Cherenkov angle was calculated using a Gaussian potential method. The reconstructed Cherenkov angle of 50 $mrad$ for a $\beta \simeq 1$ particle had a resolution of 1.2 $mrad$ (D. Bergström et al. [5]).

2.2. The time of flight

The time of flight (TOF) system consists of four 25×50 cm^2 plastic scintillating paddles with a thickness of 1 cm, two above the tracking system and two below. They were placed in pairs to cover the detector area of 50×50 cm^2. The distance between the two scintillator layers was 1.19 m. This system gave the time-of-flight measurement an accuracy of approximately 230 ps. The purpose of the TOF system is:

- The TOF provides the read out trigger for the data recorded on board.
- The TOF gives velocity of the particle and distinguishes upward moving particle from downward moving particles. Consequently albedo particles can be rejected.

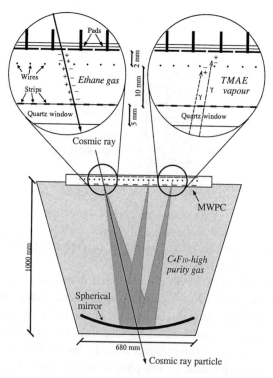

Figure 2. A schematic view of the CAPRICE98 RICH detector. The left upper expanded view illustrate how the gas is ionized by the cosmic ray particle. The location of the pads, wires and strips are indicated. The '-' and '+' in the figures refer to electron and ions liberated in the ethane gas. In the lower part the Cherenkov light emitted in the C_4F_{10} gas close to the particle track is indicated by the shaded area.The light is reflected by a spherical mirror back up onto the MWPC. In the right upper expanded part two Cherenkov photons (γ) enter the MWPC through the quartz window

- Separation of particles of different charge can be done by the energy loss in the scintillators.

2.3. The tracking system

The tracking system of CAPRICE98 consists of three drift chambers modules (M. Hof *et al.* [6]) and a superconducting magnetic field. The outer diameter of the magnet is 61 *cm* and the inner 36 *cm*. The operating current is 120 *A* which produces a field of 4 *T* in the center of the coil.

The drift chambers provide 12 measurements of the particle position in the non-bending view and 18 measurements in the bending view.

The rigidity is calculated from the measured deflection of the particle in the magnetic field. This deflection is derived from the fitted trajectory in the drift chambers. The fitting of the track is done with an iterative least square method in which a particle is described by five parameters: x and y coordinates, two direction angles ($\partial x/\partial z$ and $\partial y/\partial z$), and the particle deflection.

2.4. *The calorimeter*

The CAPRICE98 calorimeter configuration (M. Bocciolini *et al.* [7], M. Ricci *et al.* [8]) was the same as for CAPRICE94 and CAPRICE97. It was designed to distinguish between minimum ionizing particles, hadronic and electromagnetic showers.

3. Data Analysis

3.1. *Particle selection*

The CAPRICE98 instrument was well suited to measure the muon spectra and charge ratio in the atmosphere against a background of electrons, protons and heavier particles. The background in the muon sample depended strongly on the atmospheric depth. At float the dominant background for positive muons was protons. With increasing atmospheric depths, the abundance of the proton component decreased, becoming a few percent of the positive muon component at ground level.

It is important to stress that the spectrometer accepted particles with a zenith angle less than 20 degrees around the vertical and the average zenith angle was about 9 degrees.

3.1.1. *Tracking*

To achieve a reliable estimation of the rigidity, a set of conditions were imposed on the fitted tracks:

At least 10 (out of the 18) position measurements in the x direction and 6 (out of the 12) in the y direction were used in the fit.

There should be an acceptable chi-square for the fitted track in both directions with stronger requirements on the x-direction.

3.1.2. *Scintillators and time-of-flight*

The ionization loss dE/dx in the TOF scintillators was used to select minimum ionizing single charged particles by requiring a measured dE/dx of less than 1.8 times the most probable energy loss by a minimum ionizing particle (mip).

Multiple track events were also rejected by requiring that not more than one of the two top scintillator paddles was hit.

With the TOF information, the particle velocity can be estimated. By requiring $\beta \geq 0$, downward moving particles can be selected and they assure that no contamination of albedo particles remained in the selected sample.

The velocity β of the particle from the TOF was compared to the β obtained from the fitted deflection assuming that the particle had the mass of a muon. An event is accepted as a muon if it has a β greater than the β obtained from the fitted deflection minus one standard deviation of the resolution.

3.1.3. The Rich

The RICH was used to measure the Cherenkov angle of the particle and thereby its velocity. Muons started to produce Cherenkov photons in the C_4F_{10} gas at about 2 GeV/c and protons at about 18 GeV/c while electrons were above threshold in the whole momentum range of interest. Hence, below about 2 GeV/c muons were selected requiring no light while above it muons were selected according to their reconstructed Cherenkov angle. To accept a particle as a muon it was required that the measured Cherenkov angle agrees within three standard deviation of the resolution from the expected Cherenkov angle of muon.

The RICH was also used to reject events with multiple charged particles. Since ionization from charged particles produced significantly higher signals than converting Cherenkov photons, we required that an event contained only one cluster of pads with high signal.

It is important to stress that in our selection we used a condition to reject pions. The reconstructed Cherenkov angle was required to be more than 3 $mrad$ (about 1 standard deviation for pions below 5 GeV/c) away from the expected Cherenkov angle for pions. This was done for momenta smaller than 5 GeV/c because at higher rigidities the difference between the Cherenkov angle for muons and pions is too small.

3.1.4. The Calorimeter

The calorimeter was used to separate non-interacting particles from showers. It was very powerful to reject electrons or interacting hadrons but it could no distinguish muons from non-interacting pions or protons except at low momenta where the particles could be separated using their energy losses in the silicon detector.

3.2. Contamination

We estimated the contamination of protons in the muon sample with the data at float, because in that way it was easy to select a clean and large sample of protons. Below 1.5 GeV/c the TOF system was efficient in rejection protons and the RICH started to reject them about 2.1 GeV/c. The overall proton rejection factor was better than 10^3 and, hence, proton contamination was essentially negligible in all momentum intervals of this analysis with the exception of the interval 1.5 to 2.1 GeV/c where μ^+ could not be separated from protons. Therefore, no results on μ^+ are presented in this momentum interval. Electrons were efficiently rejected by the calorimeter in the whole momentum range and by the RICH below about 5 GeV/c , hence the remaining electron contamination was negligible.

To calculate the contamination of pions we select interacting hadrons with the calorimeter and by a simulation we calculate the efficiency and contamination in this selection. This have been done with a simulation because it is not possible to select a clean sample of pions with the others detectors, only within a small interval of rigidity with the RICH. This was used to check that the result obtained with the simulation is correct. So selecting interacting pions with the calorimeter and knowing the efficiency and contamination of this selection it was possible to determine the pion contamination in the muon sample.

At higher altitudes ($25g/cm^2$) in the atmosphere the contamination of pions was very large at low momentum and for this reason we exclude the data from momentum 0.3 GeV/c till 2 GeV/c.

3.3. Efficiency determination

For the efficiency study, a large sample of negative muons collected at ground prior the flight was used. Then, the results were cross-checked with smaller samples of negative muons from flight data (P. Hansen et al. [9]) To determine the efficiency of a given detector, a data set of muons was selected by the remaining detectors. The number of muons correctly identified by the detector under test divided by the number of events in the data set provided a measure of the efficiency. This procedure was repeated for each detector. The efficiency of each detector was determined as a function of rigidity in a number of discrete bins and then parameterized to allow an interpolation between bins.

4. Results

The absolute particle fluxes were calculated from the number of observed muons taking into account the spectrometer geometrical factor and live time as well as selection efficiencies. The Figure 3 shows the muon flux as function of the momentum for positive and negative muons at different atmospheric depth. The data are compare with data of CAPRICE94 (M. Boezio et al. [10]) and in the case of negative muons also with MASS89 (R. Bellotti et al. [11]) and MASS91 (A. Codino et al. [12]). For negative muons there is a good agreement with the data of CAPRICE94 and also with MASS data. In the case of positive muons there is a good agreement between the CAPRICE data from 1998 and 1994 for muon momenta below 2 GeV/c. It is worth pointing out that this is the first time that atmospheric positive muons were measured up to 20 GeV/c in a balloon-borne experiment. It was possible because the CAPRICE98 apparatus used a gas RICH detector with an average threshold Lorentz factor of about 19 at float (corresponding to a muon momentum of about 2 GeV/c). It is also possible to see in figure 3 that below 1 GeV/c, the CAPRICE98 results are lower than the results from CAPRICE94 which could indicate geomagnetic effects. It is important to stress that for CAPRICE98 and CAPRICE94 the solar modulation was close to the minimum. For momenta above 2 GeV/c the CAPRICE98 data follows an approximate exponential decrease as a function of momentum, similar to the negative muon data. This measurement of the flux of muons is a powerful tool to check and/or calibrate air shower simulation programs. There is a new project discussed for the Wizard Collaboration to re fly CAPRICE98 stopping at different atmospheric depths to sample the muon flux. This new experiment will improve the muon statistic and will also get rid of the problem of averaging the value of atmospheric depth of measurements, since the data will be collected at approximately fixed altitudes (M. Circella et al. [13]).

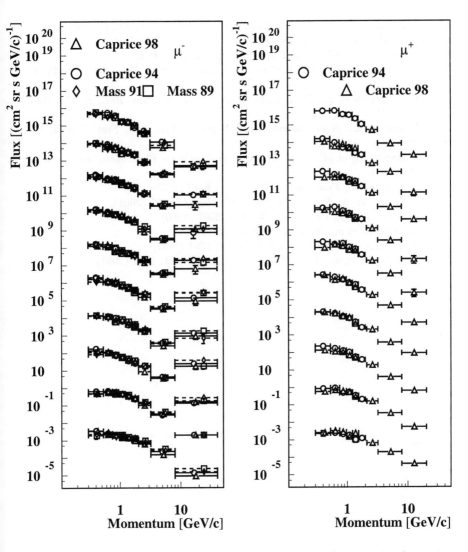

Figure 3. Different curves represent different altitudes (in g/cm^2): (1) $25g/cm^2$, scaled by 10^{18}. (2) $50g/cm^2$, scaled by 10^{16}. (3) $77.5g/cm^2$, scaled by 10^{14}. (4) $105g/cm^2$, scaled by 10^{12}. (5) $135g/cm^2$, scaled by 10^{10}. (6) $170g/cm^2$, scaled by 10^8. (7) $220g/cm^2$, scaled by 10^6. (8) $315g/cm^2$, scaled by 10^4. (9) $480g/cm^2$, scaled by 10^2. (10) $735g/cm^2$, scaled by 1.

References

1. M. Boezio et al., *Phys. Rev. Lett.* **82**, 4757 (1999).

2. M. Boezio et al., The Cosmic-Ray Proton and Helium Spectra measured with the CAPRICE98 balloon experiment (submitted to publish)
3. M. Ambriola, et al., *Nucl. Phys. B* (Proc. Suppl), 78, 32
4. T. Francke, et al., *Nucl. Instrum. and Methods Phys. Res., Sect. A* **433**, 87 (1999).
5. D. Bergström et al., *Nucl. Instrum. and Methods. Phys. Res., Sect. A* **463**, 161 (2001).
6. M. Hof, et al., *Nucl. Instrum. and Methods Phys. Res., Sect. A* **345**, 561 (1994).
7. M. Bocciolini, et al., *Nucl. Instrum. and Methods Phys. Res., Sect. A* **370**, 403 (1996).
8. M. Ricci, et al., *Proc. 26th ICRC (Utah)* **5** 49 (1999).
9. P. Hansen, et al., *Proc. 27th ICRC (Hamburg)* **1**, 921 (2001).
10. M. Boezio et al., *Phys. Rev. D.* **62**, 032007 (2000).
11. R. Bellotti, et al., *Phys. Rev. D.* **53**, 35 (1996).
12. A. Codino, et al., *Nucl. Part. Phys.* 23, 1751 (1997)
13. M. Circella, et al., *Proc. 27th ICRC (Hamburg)* **1**, 1251 (2001).

EXTENSIVE AIR SHOWERS

ULTRA HIGH ENERGY COSMIC RAYS: PRESENT STATUS AND FUTURE PROSPECTS

A A WATSON

Department of Physics and Astronomy
University of Leeds
Leeds LS2 9JT, UK

Reasons for the current interest in cosmic rays above 10^{19} eV are described. The latest results on the energy spectrum, arrival direction distribution and mass composition of cosmic rays are reviewed. The enigma set by the existence of ultra high-energy cosmic rays remains. Ideas proposed to explain it are discussed and progress with the construction of the Pierre Auger Observatory is outlined.

1. Introduction

For the purposes of this paper I define ultra high-energy cosmic rays (UHE-CRs) as those cosmic rays having energies above 10^{19} eV. There is currently great interest in them, partly because we have little idea as to how Nature creates particles or photons of these energies. Also we know enough about their energy spectrum and arrival direction distribution to believe that we have an additional problem: their sources must be reasonably nearby (within 100 Mpc) but there is no evidence of the anisotropies anticipated if the galactic and inter-galactic magnetic fields are as weak as astronomers tell us.

The distance limit comes from a combination of well-understood particle physics and the universality of the 2.7 K radiation. Interactions of protons and heavier nuclei with this, and other, radiation fields degrade the energy of particles rather rapidly. In the case of protons, the reaction is photopion production, while heavier nuclei are photodisintegrated by the 2.7 K radiation and the diffuse infrared background. These effects were first recognised by Greisen [1], and by Zatsepin and Kuzmin [2], and lead to the expectation that the energy spectrum of cosmic rays should terminate rather sharply above 4×10^{19} eV (the GZK cut-off). Above 4×10^{19} eV about 50% of particles must come from within 130 Mpc, while at 10^{20} eV the corresponding distance is 20 Mpc.

The most recent data suggest that particles do exist with energies beyond the GZK cut-off and that the arrival direction distribution is isotropic. The mass of the cosmic rays above 10^{19} eV is not known, although there are experimental limits on the fraction of photons that constrain one of the models proposed to resolve the enigma. A relatively recent, and detailed, review can be found in [3].

2. Measurement of UHECR

The properties of UHECRs are obtained by studying the cascades, or extensive air showers (EAS), they create in the atmosphere. Many methods of observing these cascades have been explored. Currently two approaches seem to be most effective. In one, the density pattern of particles striking an array of detectors laid out on the ground is used to infer the primary energy. At 10^{19} eV the footprint of the EAS on the ground is several square kilometres so detectors can be spaced many hundreds of metres apart. Alternatively, on clear moonless nights, the fluorescence light emitted when shower particles excite nitrogen molecules in the atmosphere can be observed by massive photomultiplier cameras. This technique, uniquely, allows the rise and fall of the cascade in the atmosphere to be inferred.

The primary energy of the initiating particle or photon is deduced in different ways. For the detector arrays, Monte Carlo calculations have shown that the particle density at distances from 400 – 1200 m is closely proportional to the primary energy and that the fluctuations expected from the stochastic nature of shower development should be small [4, 5]. Such a density can be measured accurately (usually to around 20%) and the primary energy inferred from parameters found by calculation. The estimate of the energy depends on the realism of the representation of features of particle interactions within the Monte Carlo model, at energies well above accelerator energies. The currently favoured model (QGSjet) is based on QCD and is matched to accelerator measurements. Although this model appears to describe a variety of data from TeV energies up to 10^{20} eV [6] one cannot be certain of the systematic error in the energy estimates. Another model, Sibyll 2.1 [7] predicts quite different numbers of muons for the same primary and a much higher cross-section for proton or pion collisions with air nuclei.

For the fluorescence detectors, the primary energy is found by integrating the number of electrons in the cascade curve and assuming that their rate of energy loss is close to that at the minimum of the dE/dx curve for electrons, ~2.2 MeV per g cm^{-2} in the case of air. A model-dependent correction, at the 10% level, must be made to account for the energy carried by muons and neutrinos into the ground [8]. Ideally, one wants to compare estimates of the primary energy made in the same shower by the two techniques operating simultaneously, but this has yet to be done at the energies of interest. So far, all that has been possible is to compare estimates of the fluxes at nominally the same energy, but the dual measurement is a strong feature of the hybrid Pierre Auger Observatory now under construction and is expected to be decisive.

3. The Energy Spectrum, Arrival Direction Distribution and Mass of UHECRs

3.1. *Energy Spectrum*

A spectrum summarising data from Fly's Eye (the earliest fluorescence experiment), Haverah Park (a ground array that used water-Cherenkov detectors), HiRes and AGASA is shown in figure 1. This set of spectra is very different from the summary plot given in [3]. There are several points to note in understanding the changes. Firstly, the HiRes data have been reported in detail for the first time [9, 10] and are very different from preliminary results presented at the ICRC in Utah in 1999. Secondly, the Haverah Park energy estimates have been re-assessed [11] using the QGSjet98 model. In the range 3×10^{17} to 3×10^{18} eV there is very good agreement between the Fly's Eye, Haverah Park and HiRes results. A recent analysis of Haverah Park data on the lateral distribution in showers [12] suggests that protons and iron are in the ratio 35:65 in this energy range. With this mixture the intensity agreement is remarkable. This implies that the QGSjet98 provides an adequate description of important features of showers up to 10^{18} eV. However, the AGASA energies have also been estimated with a QGSjet model under the assumption that the primaries are protons at energies above 3×10^{18} eV, the lowest AGASA energy plotted. There is no evidence as to what mass species is appropriate at the highest energies but the method used (measurement of the scintillator density at 600m) would lead to an estimate lower by only about 20% if iron nuclei were assumed. This change would be sufficient to reconcile the AGASA-HiRes differences, particularly with regard to the point at which the spectrum slope flattens above 10^{18} eV. A combination of a change in the QGSjet model and iron primaries (for which there is no evidence) might go some way to aligning the different results at the highest energies. This

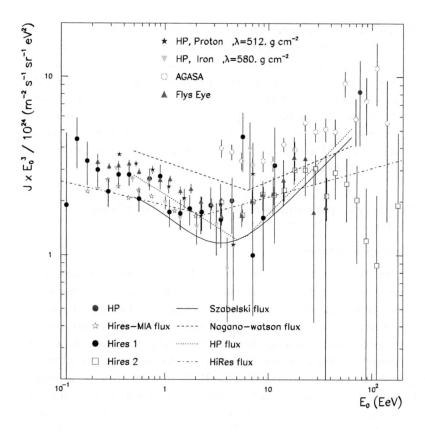

Figure 1. A composite energy spectrum from AGASA Fly's Eye, Haverah Park and HiRes. This plot was prepared with the help of Maximo Ave.

change would be insufficient to reconcile the AGASA-HiRes differences, particularly with regard to the point at which the spectrum slope flattens above 10^{18} eV. Here uncertainties in the aperture of the AGASA instrument may be important.

There are also unanswered questions about the HiRes data. The 'disappearance' of the events reported as being above 10^{20} eV in 1999 is attributed to a better understanding of the atmosphere which is now claimed to be clearer than had previously been supposed. The Hamburg results were prepared using an 'average atmosphere' so presumably some events will subsequently be assigned larger energies and some smaller ones. Two further issues need resolving. Firstly, an accelerator-based calibration of the fluorescence yield [13] claimed 'that the fluorescence yield of air between 300 and 400 nm is proportional to the electron dE/dx.' This claim is not consistent with information tabulated in the paper, where it is shown that the yield from 50 keV electrons is very similar to that from 1.4 MeV electrons, or with the dE/dx curve plotted there, normalised to the 1.4 MeV measurements, which does not fit the accelerator data for 300, 650 and 1000 MeV electrons. The latter discrepancy is about 15 – 20% and in such a direction as would increase the HiRes energies. Secondly, Nagano et al. [14] has described a new measurement of the yield in air from 1.4 MeV electrons. In what seems to be a very careful study, they find that the earlier results give a higher yield at 356.3 nm and 391.9 nm than is found now. Nagano attributes the absence of background corrections as being responsible for at least some of the discrepancies [15]. These long wavelengths become increasingly important when showers are observed at the large distances needed at the highest energies because of Rayleigh scattering. The magnitude of the adjustments that need to be made to the HiRes data are presently unclear and further fluorescence yield measurements are certainly required.

At the Hamburg meeting, the HiRes group also reported data from their stereo system. In 20% of the monocular exposure they found one event with an energy estimated as being close to 3×10^{20} eV, the energy of the largest event found with the Fly's Eye detector [16]. The significance of this event was stressed strongly at a meeting in April 2002 [17]. My opinion is that the spectra from AGASA and HiRes will come together as further understanding is gained of the models and of the atmosphere. Knowledge of the mass composition will also help considerably. For now it seems that trans-GZK events do exist but that the flux of them is less certain than appeared a few years ago.

3.2. Arrival Direction Distribution

The angular resolution of shower arrays and of fluorescence detectors is typically $2 - 3°$. The arrival direction of the 59 events with energy above

4×10^{19} eV registered by the AGASA group is shown in figure 2[18]. It is evident that the distribution is isotropic and that there is no preference for events to come from close to the galactic or the super-galactic planes. The AGASA group draw attention to a number of clusters, where a cluster is defined as a grouping of 2 or more events within 2.5°. It is claimed that the number of doublets (5) and triplets (1) could have arisen by chance, with probabilities of 0.1% and 1%. The implications of such clusters would be profound but my view is that the case for them is not proven. The angular bin was not defined a priori and the data set used to make the claim for clusters is being used in what has become the 'hypothesis testing' phase. Furthermore, I note that the directions of the 7 most energetic events observed by Fly's Eye, Haverah Park, Yakutsk and Volcano Ranch do not line up with any of the 6 cluster directions.

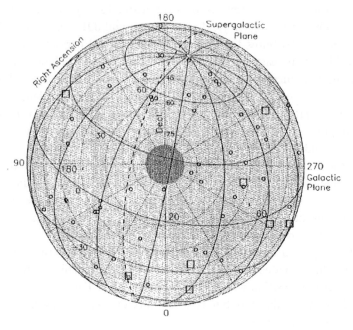

Figure 2. AGASA arrival direction distribution for 59 events above 4×10^{19} eV. The most energetic events ($> 10^{20}$ eV) are shown by squares [18].

It is hard to understand the isotropy if the local extragalactic magnetic field is really 10^{-9} gauss. A proton of 10^{20} eV would be deflected by only about 2° over a distance of 20 Mpc if the field has a 1 Mpc correlation length [19]. If the fields were much higher, as has been suggested [20], then the lack of anisotropy might be understood, but more energy is then stored in the magnetic field and this might create other difficulties. Similarly, if the charge of the particles initiating the showers was much higher than Z=1, the isotropy could be explained.

3.3. Mass Composition

Interpretation of the data on UHECRs is hampered by our lack of knowledge of the mass of the incoming particles. There are data from several experiments which can be interpreted as showing a change from a dominantly iron beam near 3×10^{17} eV to a dominantly proton beam at 10^{19} eV, but with another model a quite different interpretation can be made. The situation is very unclear (and highly unsatisfactory) and is quite open at higher energies: the data are just too limited and the interpretations are ambiguous. A difficulty is that both the fluorescence detectors and ground arrays must rely on shower models to deduce composition information.

Some progress has been made at constraining the photon flux. It is unlikely that the majority of the events claimed to be near 10^{20} eV have photons as parents as some of the showers seem to have normal numbers of muons (the tracers of primaries that are nuclei) and the cascade profile of the most energetic fluorescence event is inconsistent with that of a photon primary [21]. Furthermore, there is now evidence that less than 40% of the events at 10^{19} eV are photons. This limit has been set in two ways. Taking the energy spectrum as measured by Fly's Eye as being independent of the mass of the incoming particles, the rate of showers coming at large angles to the vertical can be calculated. Using Haverah Park data, it has been found that the observed rate of inclined showers is much higher than would be expected if the primary particles were mainly photons [22]. For the Haverah Park data, it was shown that the energy of the primaries could be estimated with reasonable precision and an energy spectrum could be derived. This has led to a demonstration that the photon flux at 10^{19} eV is less than 40% of the baryonic primary component. In addition to this novel approach, a more traditional attack on the problem by the AGASA group, searching for showers, which have significantly fewer muons than normal, has given the same answer [23]. These experimental limits are in contrast to the predictions of large photon fluxes from the decay of super-heavy relic particles, discussed below. It is also unlikely that the majority of events are created by neutrinos as the distribution of zenith angles would be different from that observed. Indeed, in all aspects so far measured, events of 10^{20} eV look like events of 10^{19} eV, but ten times larger, and this statement can be reiterated as we go to lower and lower energies were nuclei seem certain to be the progenitors of showers.

4. Theoretical Interpretations

The UHECR enigma is attracting significant theoretical attention. Many of the ideas suppose a form of electromagnetic acceleration while others invoke processes that demand new physics.

4.1. *Electromagnetic Processes*

Currently, it is popularly believed that cosmic rays with energies up to about 10^{15} eV are energised by a process known as 'diffusive shock acceleration'. Supernovae explosions are identified as the likely sites, although so far there is no direct evidence of acceleration of nuclei by supernova remnants at any energy. The diffusive shock process, which has its roots in some early ideas of Fermi, has been extensively studied since its conception in the late 1970s. In [24] it is shown that the maximum energy attainable is given by $E = kZeBR\beta c$, where B is the magnetic field in the region of the shock, R is the size of the shock region and k is a constant less than 1. The same result has been obtained by a number of people and most authors agree upon it. However, some claim that the diffusive shock acceleration process can be modified to give much higher energies than indicated by the equation and that radio galaxy lobes, in particular, are probable acceleration sites. It is difficult to see how an energy of 3×10^{20} eV can be accounted for if the size of the shock region is 10 kpc and the magnetic field is 10 μG (values thought typical of lobes of radio galaxies), as even the optimum estimate of the energy is lower by a factor of 3 than the observational upper limit. It could be that the magnetic fields are less well known than is usually supposed, a line of argument that also comes from the arrival direction work mentioned above.

4.2. *Non-electromagnetic Processes*

A number of proposals have been made which dispense with the need for electromagnetic acceleration. In general, attention has been focused on the very highest energy events ($> 10^{20}$ eV). However, it is my view that proposers of some of the more exotic mechanisms often overlook one or more important points. Any mechanism able to explain the very highest energy events must also explain those above about 3×10^{18} eV, where the galactic component possibly disappears. The spectrum above this point is probably too smooth to imagine that there are two or more radically different components — although this might be seen by some as an almost philosophical argument, particularly in view of the scatter seen in figure 1! In addition, the solutions proposed must produce particles at the top of the atmosphere that can generate showers of the type we see, and now understand rather well. Finally, source energetics cannot be ignored: there seems little point in inventing a mechanism to 'solve' the GZK cut-off problem that requires a source region that is unrealistically energetic.

An overview of the various mechanisms proposed can be found in [3] and I will only discuss one of these here. It has been suggested that UHECR arise from the decay of super-heavy relic particles. In this picture, the cold dark matter is supposed to contain a small admixture of long-lived super-heavy particles with a mass $> 10^{12}$ GeV and a lifetime greater than the age of the Universe [25]. It is argued that such particles can be produced during reheating following inflation or through the decay of hybrid topological defects such as monopoles connected by strings. I find it hard to judge how realistic these ideas are but the decay cascade from a particular candidate [26] has been studied in some detail [27, 28]. A feature of the decay cascade is that an accompanying flux of photons and neutrinos is predicted which may be detectable with a large enough installation. The anisotropy question has been examined by several authors and specific predictions for the anisotropy that would be seen by a Southern Hemisphere observatory have been made. The observation of the predicted anisotropy, plus the identification of appropriate numbers of neutrinos and photons, would be suggestive of a super-heavy relic origin. While the super-heavy relic idea has received much attention, the experimental result on the photon/proton ratio at 10^{19} eV appears not to support it (figure 3). Predictions of the photon/proton ratio are in the range 2 to 10.

5. Detectors of the Future

5.1. *The Pierre Auger Observatory*

The Pierre Auger Observatory was conceived to measure the properties of the highest energy cosmic rays with unprecedented statistical precision. When completed, it will consist of two instruments, constructed in the Northern and Southern Hemispheres, each covering an area of 3000 km^2.

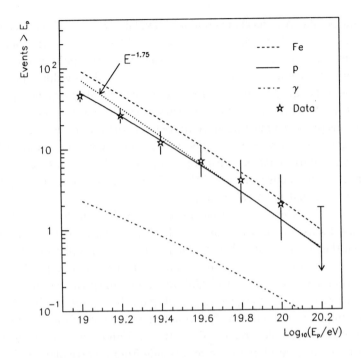

Figure 3.   The measured spectrum of inclined showers ($60° < \theta < 80°$) compared with predictions made for protons, iron nuclei and photons [22].

The design calls for a hybrid detector system with 1600 particle detector elements and three or four fluorescence detectors at each of the sites. The particle detectors will be deep water-Cherenkov tanks arranged on a 1.5 km hexagonal grid. These detectors have been selected because water acts as a very effective absorber of the multitude of low energy electrons and photons found at distances of about 1 km from the shower axis.

At the Southern site (see figure 4) fluorescence detectors will be set up at four locations, on small promontories at the array edge: the site is close to the town of Malargue. During clear moonless nights, signals will be recorded in both the fluorescence detectors and the particle detectors, while for roughly 90% of the time only particle detector data will be available. Construction of an engineering array in Mendoza Province, Argentina, containing 40 water tanks and a section of a fluorescence detector has been completed (September 2001) and all of the sub-systems of the Observatory have been demonstrated. The array includes many novel features in its design [29]. The first 'hybrid' events were recorded in December 2001 and there is great confidence that the observatory will work as designed. When the Auger Observatory at Malargue has operated for 10 years, we would expect to have recorded over 300 events above 10^{20} eV.

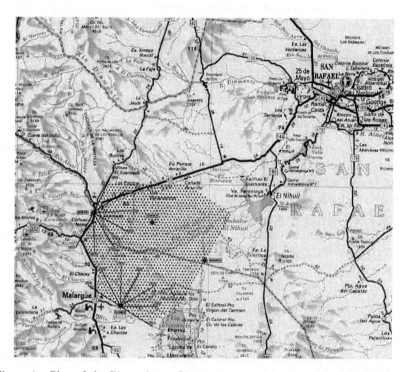

Figure 4. Plan of the Pierre Auger Observatory near Malargue, Mendoza Province, Argentina. Most of the water tanks will be located on the Pampa to the north east of the town of Malargue, which is about 200 km south of the city of San Rafael. Each dot, within the area to the left of route 40, marks the planned position of a water tank. The tanks are separated by 1.5 km. Fluorescence detectors will be established at 4 sites (from Auger plot library).

6. EUSO and OWL

Achieving an exposure greater than that targeted by the Auger Observatory is a formidable challenge. A promising line is the development of an idea due to Linsley [30]. The concept is to observe fluorescence light produced by showers from space with satellite-borne equipment. It is proposed to monitor $\sim 10^5$ km^2 sr (after allowing for an estimated 8% on-time). Preliminary design studies have been carried out in Italy and the USA. An Italian-led collaboration has proposed a design that is under study for flight in the International Space Station and is now (September 2002) in Phase A. This is known as EUSO (the Extreme Universe Space Observatory), and has the potential to detect neutrinos in large numbers as well as UHECRs. Observations are scheduled to start in 2008: the twin satellite OWL project will follow sometime later. These projects require considerable technological development but may be the only way to push to energies beyond whatever energy limits are found with the Auger instruments.

Acknowledgements

I am very grateful to Maury Shapiro for inviting me to his excellent Erice School. It was a great pleasure to meet so many bright young students who, I hope, will find much challenge and pleasure in studying astroparticle physics. I thank Maximo Ave for his help in preparing figure 1. Work on Ultra High Energy Cosmic Rays at the University of Leeds is supported by the UK Particle Physics and Astronomy Research Council under grants PPA/G/S/1998/00453 and PPA/Y/S/1999/00276 and this help is warmly acknowledged.

[1] Greisen, K., *Phys Rev Letters*, **16**, 748 (1966).
[2] Zatsepin, G.T. and Kuzmin, V.A., *Sov. Phys. JETP Letters*, **4**, 78 (1966).
[3] Nagano, M. and A. A. Watson, *Rev Mod Phys*, **27**, 689 (2000).
[4] Hillas, A.M., *Acta. Phys. Acad. Sci.*, **29**, Suppl.3, 355 (1970).
Hillas, A.M., et al., *Proc 12th Int. Conference on Cosmic Rays (Hobart)*, 1971, **3**, 1001.
[5] Dai, H. Y., et al., *J. Phys. G.*, **14**, 793 (1988).
[6] Knapp, J., et al. *Astroparticle Physics* (in press) and astro-ph/0206414 (2002).
[7] Alvarez-Muñiz, J., et al., *Phys. Rev. D* (in press) and astro-ph/0205302 (2002).
[8] Song, C., et al. *Astroparticle Physics*, **14**, 7 (2000).
[9] Jui, C.H., et al., *Proc. 27th Int. Conf. on Cosmic Rays (Hamburg)*, 2001, **1**, p354.
[10] Abu-Zayyadd et al., papers submitted to *Phys Rev Letters* (astro-

ph/0208243) and *Astroparticle Physics* (astro-ph/0208301) (September, 2002).

[11] Ave, M., et al., *Astroparticle Physics* (in press) and astro-ph/0112253 (2001).

[12] Ave, M., et al., *Astroparticle Physics* (in press) and astro-ph/0203150 (2002).

[13] Kakimoto, F., et al., *Nucl Inst and Methods*, **A372**, 527 (1996).

[14] Nagano, M., et al., *Proc 27th International Conference on Cosmic Rays (Hamburg)*, 2001, **2**, p675.

[15] Nagano, M., private communication, September 2001.

[16] Bird, D., et al., *Phys. Rev. Lett.*, **71**, 3401 (1993).

[17] Loh, G., *NEEDS Workshop*:
http:// www-ik.fzk.de/~needs/needs_talks.htm, 2002.

[18] Takeda, M, et al., *Proc 27th International Conference on Cosmic Rays (Hamburg)* 2001, **1**, p341.

[19] Kronberg, P.P., *Rep Prog Phys*, **57**, 325 (1994).

[20] Farrar, G.R. and Piran, T., *Phys Rev Letters*, **84**, 3527 (2000).

[21] Halzen, F., et al., *Astroparticle Physics*, **3**, 151 (1995).

[22] Ave, M et al., *Phys. Rev. Lett.*, **85**, 2244 - 2247 (2000).
Ave, M et al., *Phys. Rev. D* **65**, 063007 (2002).

[23] Shinosaki, K., et al., *Astrophysical Journal*, **571**, L117 (2002).

[24] Drury, L. O'C., *Contemporary Physics*, **35**, 232 (1994).

[25] Berezinsky, V., Kachelreiss, M. and Vilenkin, A., *Phys Rev Lett*, **22**, 4302 (1997).

[26] Benakli, K., Ellis, J. and Nanopolous, D.V., *Phys Rev D*, **59**, 047301 (1999).

[27] Birkel, M and Sarkar, S., *Astroparticle Physics*, **9**, 297 (1998).
Sarkar, S., and Toldra, R., Nuclear Physics B, 621, 495 (2002).

[28] Rubin, N. A., *M Phil Thesis*, University of Cambridge, 1999.

[29] Cronin, J.W., *Proc 27th International Conference on Cosmic Rays (Hamburg)* 2001 Invited, Rapporteur and Highlight papers www.auger.org/

[30] Linsley, J., in Field Committee Report *"Call for projects and ideas in High Energy Astrophysics in the 1980s"* National Science Foundation (unpublished) www.ifcai.pa.cnr.it/~EEUSO/

MEASUREMENT AND RECONSTRUCTION OF EXTENSIVE AIR SHOWERS WITH THE KASCADE FIELD ARRAY

G. MAIER[B]*, T. ANTONI[A], W.D. APEL[B], F. BADEA[A,1], K. BEKK[B],
A. BERCUCI[B,1], H. BLÜMER[B,A], H. BOZDOG[B], I.M. BRANCUS[C],
C. BÜTTNER[A], A. CHILINGARIAN[D], K. DAUMILLER[A], P. DOLL[B],
J. ENGLER[B], F. FEßLER[A], H.J. GILS[B], R. GLASSTETTER[A],
R. HAEUSLER[A], A. HAUNGS[B], D. HECK[B], J.R. HÖRANDEL[A],
A. IWAN[A,2], K.-H. KAMPERT[A,B], H.O. KLAGES[B], H.J. MATHES[B],
H.J. MAYER[B], J. MILKE[A], M. MÜLLER[B], R. OBENLAND[B],
J. OEHLSCHLÄGER[B], S. OSTAPCHENKO[A,3], M. PETCU[C], H. REBEL[B],
M. RISSE[B], M. ROTH[B], G. SCHATZ[B], H. SCHIELER[B], J. SCHOLZ[B],
T. THOUW[B], H. ULRICH[A], J.H. WEBER[A], A. WEINDL[B], J. WENTZ[B],
J. WOCHELE[B], J. ZABIEROWSKI[E]

[a] Institut für Experimentelle Kernphysik, Universität Karlsruhe, 76021
Karlsruhe, Germany,
[b] Institut für Kernphysik, Forschungszentrum Karlsruhe, 76021 Karlsruhe,
Germany
[c] National Institute of Physics and Nuclear Engineering, 7690 Bucharest,
Romania
[d] Cosmic Ray Division, Yerevan Physics Institute, Yerevan 36, Armenia
[e] Soltan Institute for Nuclear Studies, 90950 Lodz, Poland

[1] on leave of absence from National Institute of Physics and Nuclear
Engineering, 7690 Bucharest, Romania
[2] on leave of absence from University of Lodz, 90236 Lodz, Poland
[3] on leave of absence from Moscow State University 119899 Moscow, Russia

The main aim of the KASCADE experiment is the determination of the primary
energy spectrum and chemical composition of cosmic rays through the measurement of extensive air showers at energies around $4 \cdot 10^{15}$eV, the so called knee
region. An overview over the measurement and reconstruction procedures of the
large field array is presented.

*corresponding author; gernot.maier@ik.fzk.de, phone +49 7247 824174

1. Introduction

The energy spectrum of cosmic rays (CR) extends over at least 12 decades in energy and 30 decades in intensity. It follows a power law with only few features, the most prominent is a change of the spectral index at about 4 PeV, the so called *knee*. Little is known about the origin of the knee. This is mainly due to the very low fluxes of CR at these energies of about 1 particle per square meter and year. Direct measurements, usually with balloons and satellites, are impossible at these low fluxes. CR are therefore studied at ground level with large detector systems via their secondary particles in extensive air showers (EAS).

An EAS induced by a primary particle with an energy of e.g. 1 PeV consists at sea level typically of about 10^6 particles, mainly photons and electrons but also muons and hadrons. The secondary particles form a shower disk with a thickness of typically 1 m in the center with a lateral extension of up to few hundred meters. In a simplified picture, the number of muons of an EAS is an estimator of the energy of the primary particles, the number of electrons is more sensitive to their mass. Detailed measurements of many different components of individual EAS allow the reconstruction of the primary energy spectra and chemical compositions of cosmic rays.

2. The KASCADE experiment

The KASCADE[a] (Karlsruhe Shower Core and Array Detector) experiment[1] is located at the Forschungszentrum Karlsruhe, Germany (110 m a.s.l.). It consists of three major detector systems (see Fig. 1), the 200x200 m² field array with electron and muon counters, a central detector with a large hadronic calorimeter and muon detectors with different threshold energies and a muon tracking detector in an underground tunnel. Lateral distributions and time profiles for all particle types as well as incident angles of individual high energetic muons or hadrons and the energies of the hadrons are reconstructed. This gives a very complete picture of single air showers and in combination with detailed EAS simulations, answers to the main scientific questions of KASCADE, the energy spectra and chemical compositions of cosmic rays can be expected.

This paper will focus on the field array. More details about the other components of KASCADE can be found in Ref.[1].

---

[a]WWW pages of KASCADE: http://www-ik.fzk.de/KASCADE_home.html

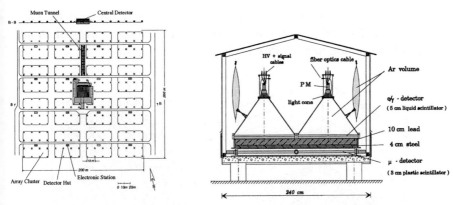

Figure 1. Schematic layout of the KASCADE experiment (left) and KASCADE field array detector station (right).

3. Field Array

The purpose of the field array is the reconstruction of the incident angle, shower core, and electron and muon numbers of individual air showers.

The 252 detector stations of the field array are arranged on a rectangular grid of 13 m spacing (Fig. 1, left). The stations are organized in 16 electronically independent clusters each with 16 stations in the 12 outer clusters and 15 stations in the four inner clusters.

The stations (Fig. 1, right) contain four detectors in the four inner resp. two in the 12 outer clusters filled with 5 cm thick custom made liquid scintillator (PMP) with an area of 0.8 m^2 each for the electromagnetic component. The energy resolution of the detectors at 12 MeV (energy deposit of m.i.p) is about 12%, the time resolution 0.8 ns. Energy deposits equivalent to 1250 m.i.p. are detected linearly with a threshold of 0.25 m.i.p. (about 3 MeV). The 192 stations in the 12 outer clusters are additionally equipped with four sheets of plastic scintillator detectors, 90 x 90 x 3 cm^3 each. These detectors, mounted below an absorber of 10 cm lead and 4 cm iron which corresponds to more than 20 radiation lengths and to a muon threshold of 0.3 GeV, are used to measure the muonic component of an air shower. The total detector area is about 490 m^2 for the electromagnetic and 622 m^2 for muon detectors. The energy sum, the arrival time of the first shower particle in a station and the hit detectors are read out individually for the muon and e/γ detectors. In addition, 16 Flash ADCs are installed in one cluster to measure more detailed the time profile of electrons and muons in the shower disk.

The present trigger conditions are a detector multiplicity of 10/20 out of 32/60 e/γ detectors fired in at least one of the outer/inner clusters. This results in a 3 Hz trigger rate and a threshold in primary energy of about 0.6 PeV for proton and 0.8 PeV for iron showers.

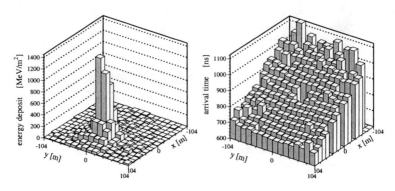

Figure 2. Energy deposits (left) and arrival times (right) of the particles in an air shower measured by the e/γ detectors. The reconstructed electron number is about 258000 electrons, the shower core is at x = 40 m, y = 0.6 m and the zenith angle 32 degrees.

4. Reconstruction of extensive air showers

A typical shower event seen by the KASCADE electron detectors in the field array is shown in Fig. 2. The reconstruction of the shower parameters is done by an threefold iteration procedure.

In the reconstruction, detector signals are corrected for 'wrong' particle contributions, i.e. the electron detectors are corrected for contributions of particles others than electrons, the muon detectors for the punch-through of other particles than muons. The expected contamination of wrong particles is estimated from their fitted lateral distributions of previous reconstruction levels. These corrections are obtained from CORSIKA[2] air shower simulations followed by a detailed detector simulation based on GEANT[3].

In the first reconstruction level the shower core is obtained by a center of gravity of the signals in the electron detectors, the shower direction by applying a gradient method using the arrival times in these detectors. Electron and muon shower sizes are obtained by weighted sums of the relevant detector signals. No minimizations are used in this level, its main purpose is giving good starting parameters for the following levels.

In level 2 and 3, the shower direction fit takes into account expectation values and spreads of the arrival time distributions dependent on the dis-

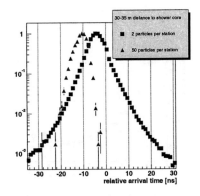 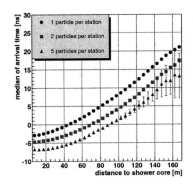

Figure 3. Two examples of arrival time distributions (left) for two and 50 particles per station and 30-35 m shower core distance. The right figure shows three examples of the median of these distributions vs. the shower core distance for 1, 2 and 5 particles per station. The arrival times are relative to a plane fitted to the shower front.

tance to the shower core and the particle numbers in the stations[4]. Fig. 3 shows as an example the arrival time distributions for 2 and 50 particles per station for a distance of 30-35 m from the shower core. Stations with more particles have arrival time distributions with expectation values shifted to earlier times and a smaller spread compared to stations with less particles. The latter gives them a larger weight in the direction fit. The distance dependence of the median of the arrival time distributions can also be seen for some particle numbers in Fig. 3. A parameterization of the median and spread of these measured distributions is used in a robust absolute value minimization to determine the shower direction.

A fit of the Nishimura-Kamata-Greisen (NKG) function to the particle densities gives the core position and the electron and muon numbers. The NKG-function is an approximation of an analytical calculation of electromagnetic showers and describes the particle density at a distance r from the shower core for an vertical EAS with particle number N and form parameter (age) s:

$$\rho_{NKG}(r, s, N) = \frac{N}{r_M^2} \frac{\Gamma(4.5 - s)}{2\pi\Gamma(s)\Gamma(4.5 - 2s)} \left(\frac{r}{r_M}\right)^{s-2} \left(1 + \frac{r}{r_M}\right)^{s-4.5}$$

The Moliere radius r_M is fixed because of its strong correlation to the shower age (see Fig. 4 in Ref.[5]) to 89 m respectively 420 m for the fit to the electron resp. muon lateral distribution.

The accuracy of the direction and core reconstruction can be seen in Fig. 4, the accuracy of the electron and muon reconstruction in Fig. 5.

248

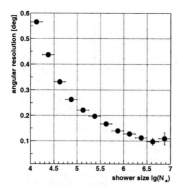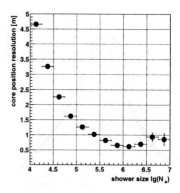

Figure 4. Angular and core position resolution obtained by the chessboard method for KASCADE array data. The stated values correspond to 68% confidence levels of the associated distributions. The core position resolution worsens due to detector overflows at very large shower sizes.

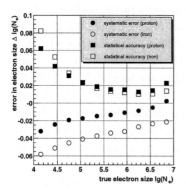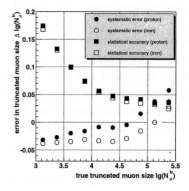

Figure 5. Systematic error and statistical accuracy of reconstructed electron (left) and muon (right) number obtained from Monte Carlo simulations. The truncated muon number is the number of muons reconstructed between 40 and 200m.

5. Summary and Conclusion

The KASCADE array is in operation since December 1995, more than 650 millions events have been measured. The measured electron and muon number spectra (Fig. 6) of KASCADE at various zenith angles show the changing of the power law index due to the knee. The more prominent knee in the electron size spectra attenuates as expected with zenith angle of the EAS due to the increasing thickness of the passed atmosphere[6].

Several analysis of the electron and muon size spectra have been carried out to obtain the primary energy spectra[7,8,9,10]. An overview over recent results of KASCADE is given in Ref.[11].

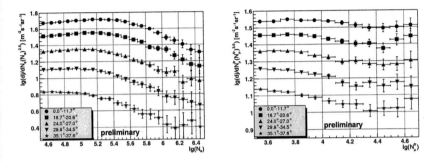

Figure 6. Electron (left) and muon (right) size spectra for different angular bins.

References

1. T.Antoni et al.,(KASCADE Collaboration), *Nucl.Instr.and Meth.in Phys.Res. A*, to be published, 2003.

2. D.Heck et al., Report FZKA 6019, Forschungszentrum Karlsruhe, 1998.

3. GEANT, CERN Program Library Long Writeup W5013, Geneva, 1993.

4. H.Krawczynski et al., *Nucl.Instr. and Meth. in Phys.Res. A* **383**, 431-440, 1996.

5. T.Antoni et al., (KASCADE Collaboration), *Astroparticle Physics* **14**, 245-260, 2001.

6. G.Maier et al., (KASCADE Collaboration), Proc. 27th ICRC, Hamburg, 161, 2001.

7. R.Glasstetter et al., (KASCADE Collaboration), Proc. 26th ICRC, Salt Lake City, **2**, 222, 1999.

8. J.Weber et al., (KASCADE Collaboration), Proc. 26th ICRC, Salt Lake City, **1**, 347, 1999.

9. T.Antoni et al., (KASCADE Collaboration), *Astroparticle Physics* **16**, 245-263, 2001.

10. H.Ulrich et al., (KASCADE Collaboration), Proc. 27th ICRC, Hamburg, 97, 2001.

11. K.-H.Kampert et al., (KASCADE Collaboration), Highlight talk, 27th ICRC, Hamburg, 2001.

ASPECTS OF THE RECONSTRUCTION CHAIN FOR THE FLUORESCENCE TELESCOPES OF THE PIERRE AUGER OBSERVATORY

F. NERLING

FOR THE

AUGER COLLABORATION

Institut für Kernphysik
Forschungszentrum Karlsruhe GmbH,
POB 3640 76021 Karlsruhe
Germany
E-mail: nerling@ik.fzk.de

The Fluorescence Technique is one of the two methods used by the Pierre Auger Observatory to detect Extensive Air Showers. An overview of the reconstruction chain developed in our group as well as some analyzed real data concerning the geometrical reconstruction is shown. Finally an outlook of future work due to the reconstruction of the longitudinal shower development is presented.

1. Introduction

The Fluorescence Detector of the Pierre Auger Observatory (PAO) measures the longitudinal profile of Extensive Air Showers (EAS). An EAS is initiated by Cosmic Rays (CR) arriving on Earth and interacting with the molecules of the atmosphere. Mainly it consists of three parts, the hadronic, the muonic and the electromagnetic component. One main goal of the PAO is to measure the energy spectrum of CR at the very high end of the energy spectrum exceeding 10^{18} eV to clarify whether there is the predicted GZK cutoff or any directional anisotropy as well as what is the composition. Furthermore aims are to investigate how such energies are reached (so-called Bottom-Up or Top-Down scenarios), what the sources are and the question how particles of such energies propagate to the Earth[a]. By the PAO, with two complementary detector types, the longitudinal profile and the particle distribution on ground are measured.

2. Experimental Setup

The PAO is situated near Marlagüe in Argentina and contains of two different detectors. On the one hand there is the Surface Detector (SD) consisting of 1600 water Čerenkov tanks on a 1.5 km grid covering an area of 3000 km^2 and on the other hand the Fluorescence Detector (FD) observing the atmosphere above. The FD contains of 24 fluorescence telescopes ($30° \times 30°$ each) distributed over 4 so-called 'eyes' ($180° \times 30°$) and sensitive in the wavelength band of $300 - 400$ nm. Each camera of these telescopes is built of 440 photomultipliertubes (PMT). The SD measures the Čerenkov yield of ground particles in water (lateral distribution). As the charged particles excite the nitrogen of the atmosphere, which de-excites by emitting - with small efficiency - fluorescence light, the longitudinal profile is detected by the FD, for more detailed technical information see Ref.[1] . For predictions about the shower geometry, the primary energy and the composition, one has to reconstruct the CR from the raw data in the form of ADC counts.

3. Reconstruction Chain

There are different possibilities concerning reconstructing data, particularly at which level to compare the data with simulations. In this article the way for a comparison of photons at the diaphragm is described. After introducing the basic parameters concerning the shower geometry, an overview of the whole reconstruction chain is given. The geometrical reconstruction is discussed in more detail with reference to examples of real data. Later, a couple of improvements are introduced.

3.1. Shower Geometry

As it is also needed for the energy and composition the geometrical reconstruction is the most basic step. Considering the EAS being detected at large distance of the detector and the fluorescence light being emitted isotropically the shower can be assumed as a point source with speed of light. Fig.1 shows the Shower Geometry within the so-called Shower Detector Plane (SDP), which is the plane spanned by the Shower Axis (SA)

---

[a]for more details, see also the contribution of A. Watson in these proceedings

and the detector. The unit normal to this plane is determined by fitting a plane to all the direction vectors of PMT which detect scintillation light. The SA is determined within the SDP by the angle χ_0, the impact parameter R_p and the time T_0 related to this parameter, see Ref.[9].

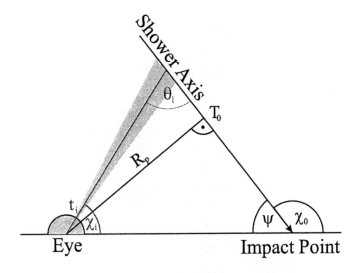

Figure 1.   Shower Geometry

The trigger times t_i of each pixel are connected with these parameters as given by the following equation.

$$t_i = T_0 + \frac{R_p}{c} tan(\frac{\chi_0 - \chi_i}{2}) \tag{1}$$

Thus after fitting Eq. (1) the shower geometry is determined.

3.2. Overview

The raw data of the FD consist in ADC counts as a function of time. Taking into account the detector calibration the ADC counts are directly connected to the number of photons at the diaphragm as a function of altitude and time. As result of the geometrical reconstruction the shower axis, i.e. its zenith and azimuth angle as well as the impact point are known (see Fig. 2). Having determined the SDP and the SA, an algorithm using the brightest pixel is employed to make an energy estimate, for details see Ref.[3]. In the case of hybrid events, i.e. showers which are detected by both the SD and the FD, additional information coming from the SD as e.g.

254

the SD core position is used to improve the minimization concerning the geometry. This provides a first geometry and energy guess which is given as input to CORSIKA[b], see Ref.[4]. As output CORSIKA gives the number of charged particles along the shower track. Taking into account the energy

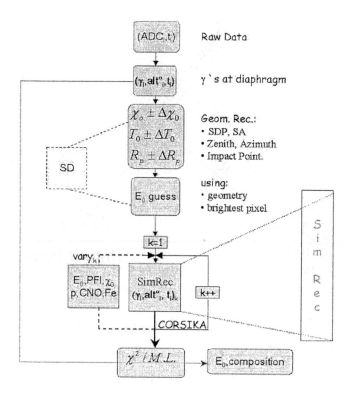

Figure 2.   Example of a reconstruction scheme

deposit as well as the flourescence yield and an atmospheric model[c], the number of photons reaching the detector is calculated. Knowing the field of view of each PMT one can loop over the triggered (here simulated) pixels and get finally the simulated number of photons at the diaphragm[d] as a

---

[b]*Cosmic Ray Simulation for Kascade and Auger*
[c]These items are not yet known with satisfying accuracy, e.g. the fluorescence yield has uncertainties up to about 10%, see Ref.[6].
[d]in units of 370 nm equivalent photons per 100 ns time bin, as described in Ref.[10]

function of time and altitude[e], which is the level of comparison to the data. Varying the primary energy and the composition within the simulations[f] implies different (simulated) photons as a function of time profiles at the telescope diaphragm. A maximum likelihood or χ^2 is then used to extract the primary energy and composition.

3.3. Some Results

From December 2001 until the end of the prototype phase, at the end of March 2002, the so-called Engineering Array consisting of about 30 water-tanks and two telescopes were already taking data in hybrid mode succesfully. Fig.3 and Fig.4 show exemplary the geometrical reconstruction of an event.

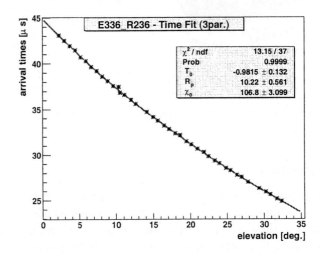

Figure 3. Time Fit

In Fig.3 the time-fit corresponding to Eq.1 is given for this particular hybrid event. All the information about the SA geometry is contained within this curve.

---

[e]This is done in the box "Sim Rec" in Fig. 2, for detailed description see Ref.[3].
[f]Of course finally this should be done more carefully using some unfolding to get rid of large fluctuations in the energy spectrum as well as using as many observables as possible measured by the SD.

Fig.4 shows a three dimensional view of this event. The small spots are the triggered pixels within the SDP, the larger ones on the ground stand for

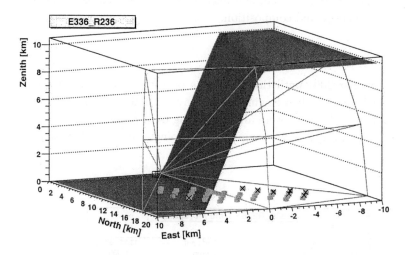

Figure 4. 3D-view of E336R236

the tanks of the Engineering Array while the crossed ones were not working that time.

3.4. *Pixel Pointing*

As the shower geometry is used to calculate the primary energy and the composition, it is important to determine it as precisely as possible. Since the optical axis of the cameras are not aligned perfectly and due to optical aberration as well as other optical unalignments in general the pixels are not pointing as theoretically planned and constructed. Using the star background one can calculate the real pointings, Ref.[8]. Therefore the predicted positions of bright stars within the wavelength band of 300 − 400 nm are compared with the measured ones, i.e. the pointing of these pixels viewing that star are determined. Fig.5 shows a star traversing a FD camera. Plotted are the variances of the outputs of those pixels pointing towards the star at a certain time. When the maximum of a light curve is reached and the surrounding pixels detect no signal, the star is close to the real center of the field of view of this pixel.

By averaging over a large number of stars the effective pointings of each pixel are available more precisely by using this method.

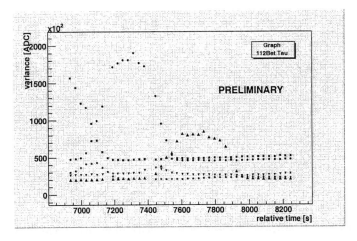

Figure 5. Star traversing FD Camera

3.5. Čerenkov Contamination

The FD measures fluorescence photons produced by charged particles in the atmosphere. Since these particles have velocities close to the speed of light, additionally Čerenkov photons are produced in this medium also in the fluorescence band. After traversing the atmosphere, a certain ratio of scattered (Mie and Rayleigh) and where appropriate of direct Čerenkov photons reach the cameras. As by doing the reconstruction it is calculated from the raw data over photons to energy deposit of charged particles in the atmosphere these Čerenkov photons can disturb the data interpretation. Thus, the amount of Čerenkov photons within the raw data has to be subtracted. In Fig.6 and 7 the number of Čerenkov photons per solid angle at three different atmospheric depths is plotted versus the angle with respect to the shower axis. The calculation is done by CORSIKA for vertical showers induced by an iron and a proton of 10^{19} eV each. Also plotted is the amount of fluorescence photons. These curves are horizontal as the fluorescence light is emitted isotropically. The ratio of Čerenkov to fluorescence light is not negligible especially for smaller angles to the shower axis, see Fig.6 and 7. While around 20° the contribution of different kinds of light is comparable at least for larger atmospheric depths, the contamination decreases with increasing angles to the shower axis.

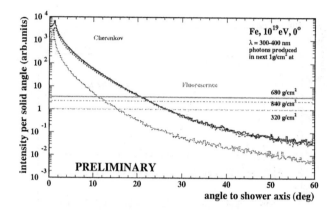

Figure 6. Emission of Čerenkov photons compared to Fluorescence for Iron

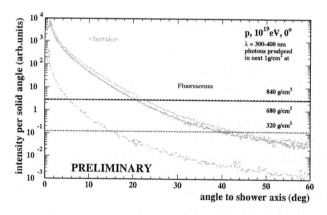

Figure 7. Emission of Čerenkov photons compared to Fluorescence for Proton

4. Conclusions & outlook

The Engineering Array of the PAO is working. Having detected nearly one hundred hybrid events the prototype phase was succesful. The geometrical reconstruction is also working. Not yet taken into account is the uncertainty of the pixel pointing but a possibility of improvement was presented and further calculations are under progress. An example of a reconstruction chain concerning energy and composition has been presented, in which particularly three aspects were not yet included. First the Čerenkov Contamination, which was shown to be not negligible. It is planned to connect the Čerenkov output given by CORSIKA to the simulation within the reconstruction for substracting the appropriate amount of photons. Secondly the important factor of fluorescence yield, which is up to now not known precisely enough. Some new measurements are already planned, see Ref.[5]. Finally since the fluorescence yield as well as the Mie and Rayleigh scattering depend on the atmosphere's conditions (mainly pressure and temperature), one needs a very good knowledge of this at the site in Argentina, see Ref.[7]. Therefore ballon measurements with radio sondes have been performed and are planned also for future to get a more precise atmospheric model. Additionally a good monitoring of the atmosphere is needed, that is why some groups are working on Lidar systems as well, see e.g. Ref.[2].

Acknowledgments

Many thanks to M. Boháčová for Fig.4 and 3, to C.K. Guérard for Fig.2 as well as creative discussion and to M. Risse for preparing Fig.6 and 7 as well as for critical reading.

References

1. Auger Collaboration, The Pierre Auger Observatory Design Report, 1997; http:www.auger.org/admin/DesignReport.
2. A. Biral et al., Auger technical notes GAP-2002-023 (2002).
3. C.K. Guérard and L. Perrone, Auger technical note, in preparation (2002).
4. D. Heck, et al., CORSIKA: A Monte Carlo Code to Simulate Extensive Air Showers, FZKA 6019, Forschungszentrum Karlsruhe, 1998.
5. International workshops in Utah (USA) and Bad Liebenzell (Germany); see http://www.physics.utah.edu/~fiwaf and http://www.auger.de/events
6. F. Kakimoto et al., *Nucl. Instr. and Meth.* **A372**, 527 (1996).
7. B. Keilhauer, H. Blümer, H.O. Klages and M. Risse, Auger technical notes GAP-2002-022 (2002).
8. H.O. Klages, personal communication (2002).
9. P. Sokolsky, Introduction to Ultrahigh Energy Cosmic Ray Physics, Addison-Wesley 1989.
10. P. Sommers et al., 'Life of an Event'; http://www.physics.utah.edu/ sommers/hybrid/specs/index.html

SIMULATIONS OF EXTENSIVE AIR SHOWERS:
A HYBRID METHOD

J. ALVAREZ-MUÑIZ[1], R. ENGEL[1], T.K. GAISSER[1],
JEFERSON A. ORTIZ[1,2] AND T. STANEV[1]

1 *Bartol Research Institute, University of Delaware,*
Newark, DE 19716, U.S.A.

2 *Instituto de Física "Gleb Wataghin", Universidade Estadual de Campinas,*
13083-970 Campinas-SP, BRAZIL

We have developed a fast one-dimensional hybrid method to calculate the development of extensive air showers. Based on precalculated pion induced showers and a bootstrap technique, this method allow us to simulate ultra-high energy showers with the collection of sufficiently high Monte Carlo statistics. It predicts the average shower profile, the number of muons at detector level above several energy thresholds as well as the fluctuations of the electromagnetic and muon components of the shower. As an application of this code we investigate the main characteristics of of proton-induced air showers up to ultra-high energy, as predict by four hadronic interaction models: SIBYLL 1.7, SIBYLL 2.1, QGSjet98 and QGSjet01.

1. Introduction

Extensive air showers (EAS) generated by cosmic rays in the Earth's atmosphere are the only way to study cosmic rays of energies above 10^{15} eV. Air shower experiments are either ground arrays of detectors that trigger in coincidence when the shower passes through them, or optical detectors that observe the longitudinal development of EAS. The most commonly observed EAS parameters are the number of charged particles at ground level for the shower arrays, or at shower maximum (S_{max}) for the optical detectors; the depth of shower maximum (X_{max}) itself, and the number of muons (N_μ) above different energy thresholds.

The analysis of air shower data relies on simulations that use the current knowledge of hadronic interactions to predict the observable shower parameters. Different hadronic models predict in fact different, and at times contradicting distributions for the values of important observable quantities for a set of showers generated by primary particles of a fixed type, energy and zenith angle. The main source of these differences arises from the uncertainties in the extrapolations of the hadronic interaction properties, performed by different models over a wide range of energy. This is due to the gap between the shower energy and the energy range studied in accelerator experiments, which increases with cosmic ray energy.

Once a hadronic interaction model has been chosen, it remains a technical but nonetheless very difficult problem: the calculation of the gigantic showers that corresponds to the cosmic ray energy spectrum ($>10^{18}$ eV). This is due to the huge number of charged particles that have to be followed in the Monte Carlo scheme, in which is proportional to the shower energy. For instance, highest energy cosmic ray showers[1,2] can have more than 10^{11} charged particles at X_{max}. As a consequence the direct simulation of the shower following each individual particle becomes practically impossible.

The widely used solution to this problem is the simulation of EAS using the thinning technique as suggested by Hillas[3]. This method is extremely useful to estimate detectable signals [4,5]. To keep the complexity of the problem at an affordable level, interactions, propagation and decay are simulated only for a representative number of EAS particles which are given higher weights. Due to this, artificial fluctuations are introduced even when small thinning thresholds are used. Various methods of reducing artificial fluctuations have been proposed recently[6,7], optimizing the compromise between time-consuming simulations and fluctuation-enhancing thinning.

In this paper we explore a different way of calculating the air shower development - an efficient one-dimensional hybrid calculation of the electromagnetic and muon component of EAS. Here we follow the approach of Gaisser et al[8] and treat the subthreshold particles with a library of shower profiles based on presimulated pion-initiated showers. This idea can be combined in a bootstrap procedure[9] to extend the shower library to high energy. The novelty of this work is that we extend the method[9,8] by accounting for fluctuations in the subshowers generated with the shower library, and also calculate the number of muons at detector level above several energy thresholds.

Some of the results obtainede with the method have been already published[10]. We briefly introduce the method and present results for more interaction methods.

2. A Hybrid Simulation Technique

The hybrid method used in this work consists of calculating shower observables by a direct simulation of the initial part of the shower. We track all particles with $E > f E_0$, where E_0 is the primary energy and f is an appropriate fraction of it (in the following we will use $f = 0.01$). Then presimulated showers for all subthreshold particles are superimposed after their first interaction point is simulated. The subshowers are described with parametrizations that give the correct average behavior and at the same time describe the fluctuations in shower development. This method allows naturally to "bootstrap" itself to higher and higher energy, because the results for showers of energy E_0 can then be used to calculate the development of showers of higher energy $E_1 > E_0$, and so on recursively. Although we do not need to account for the sub-PeV pion shower fluctuations for the calculation of 100 EeV showers, we parametrize the fluctuations in the whole energy range to be able to calculate correctly showers that could be simulated with direct Monte Carlo, and check and normalize the bootstrap procedure.

We build a library of presimulated showers by injecting pions of fixed energy E_π, at fixed zenith angle θ and depth X measured along the shower axis. The atmospheric density adopted here corresponds to Shibata's fit of the US Standard Atmosphere[16,17], very similar to Linsley's parametrization. We limit the injection zenith angles to $\theta < 45°$. Nucleons are followed explicitly in the Monte Carlo down to the particle production threshold.

Photon and electron/positron induced cascades are treated with a full screening electromagnetic Monte Carlo in combination with a modified Greisen parametrization. The electromagnetic branch of the Monte Carlo includes photoproduction of hadrons. For energies above 1 EeV, the Landau-Pomeranchuk-Migdal (LPM) effect[18,19,20] is taken into account using an implementation by Vankov[21].

When building the library we have accounted for fluctuations in the main observables describing the behavior of the longitudinal development of the subthreshold showers namely: S_{max}, the maximum number of $e^- + e^+$ in the shower; X_{max}, the depth at which the maximum of the shower occurs and N_μ^{obs}, the number of muons above the threshold energies of 0.3, 1, 3, 10 and 30 GeV both at sea level and at a depth of 400 g/cm^2 above the sea level measured along the shower axis.

We have simulated primary pions of energies between 10 GeV and 3 EeV with a step in energy of half a decade, interacting at fixed atmospheric depths $X_0 = 5,\ 50,\ 100,\ 200,\ 500$ and 800 g/cm^2. $10,000$ ($5,000$) pion showers were fully simulated for each interaction depth at energies E_0 from 10 GeV to 300 GeV (from 1 TeV to 316 EeV), 2 energies per decade. For each energy, injection zenith angle and depth (i.e. a point in the library) we obtain the distributions of $X_{\max}$ and $S_{\max}$, the correlation between them, the distributions of N_μ^{obs} with energies above the thresholds indicated above, and the slope of the muon longitudinal profile between sea level and a slant depth of 400 g/cm^2 above sea level.

Although it is unlikely to produce a high energy pion deep in the atmosphere, we also calculate their interactions at depths as large as 500 and 800 g/cm^2 to obtain an accurate description of the muon numbers at sea level and a better description of the late developing electromagnetic showers. For showers initiated after 500 g/cm^2 the atmosphere has been artificially extended beyond ground level. The distributions of muons are easily extended to other depths (corresponding to the observation level of different experiments) by extrapolation. For this task we use the slope of the muon longitudinal profile.

The longitudinal development of subthreshold meson induced showers is parametrized using a slightly modified version of the well-known Gaisser-Hillas function[22] that gives the number of charged particles at atmospheric depth X:

$$S_{\mathrm{GH}}(X) = S_{\max} \left(\frac{X - X_0}{X_{\max} - X_0} \right)^{(X_{\max} - X_0)/\lambda(X)} \times \exp\left[-\frac{(X - X_{\max})}{\lambda(X)} \right] \quad (1)$$

Here $\lambda(X) = \lambda_0 + bX + cX^2$ where λ_0, b and c are treated as free parameters. X_0 is the depth at which the first interaction occurs. The parameters b and c are assumed to be the same for all showers initiated at a given depth, angle, and energy. They are determined by fitting the mean shower profile of the parametrized showers to that obtained from the simulated shower profiles.

Instead of using the average values of $X_{\max}$ and $S_{\max}$ to generate subthreshold meson showers of a certain energy, we sample their values (as well as the number of muons) from their corresponding presimulated distributions, taking into account the correlation between them. This procedure accounts for the fluctuations in the subshower development. A technical remark is that we sample the observables directly from their precalculated histograms, i.e. we do not assume any functional form for the distribution. In this way our code is very flexible - it allows the study of hadronic models that predict distributions of observables not easily fitted by analytical functions.

We sample meson subshowers at a zenith angle, depth and/or primary energy different from those we have presimulated by interpolating between the relevant parameters of the shower development: $X_{\max}, S_{\max}, X_0, b, c, N_\mu$.

To ensure the consistency of our simulation approach, we have compared full simulations of pion showers to hybrid simulations for the same initial energy and depth using several energy thresholds. We find a very good agreement between the average values of the different observables and their fluctuations in the direct and hybrid simulations. Table 1 compares the direct simulations and the hybrid method for 5,000 vertical pion showers with fixed first interaction point at $X_0=5$ g/cm^2, energy 10^{16} eV for the four hadronic models. It is very important to note that the differences between the two methods of calculation are much smaller than those introduced by the different hadronic interaction models, i.e. by using the hybrid approach we do not lose sensitivity to the models we are considering. At primary energy 10^{16} eV, the saving in CPU time over direct simulations is a factor about 25 for the nominal energy threshold of $0.01E$. All CPU times illustrated in this work refer to a 1 GHz AMD Athlon processor.

The relative differences in the average numbers (Tab. 1) are less than 0.5% for all hadronic interaction models. The same comparison for showers generated by primary pions with incident zenith angle of 45° shows larger differences between direct and hybrid simulations, but they are smaller than 2%. We believe these relatively small errors come mostly from the representation of the intrinsic fluctuations in the shower development and from the interpolation in energy and atmospheric depth that the code performs.

Our results also show a remarkable stability under changes of the energy threshold, from which we conclude that the primary to threshold energy ratio we have used ($E_{\text{thr}} = E/100$) is sufficient to achieve a very good description of the average values and fluctuations of observables in nucleon and pion initiated showers.

3. Results for proton-initiated showers

In this section we apply the hybrid approach described above to simulate proton-initiated showers at fixed energy. In the following we consider the hadronic interaction models SIBYLL, and QGSjet. We have created libraries for the model versions SIBYLL 1.7[23], SIBYLL 2.1[24,25], QGSjet98[26] and QGSjet01[27]. QGSjet and SIBYLL are sufficiently different to illustrate various important points of how properties of hadronic interactions are re-

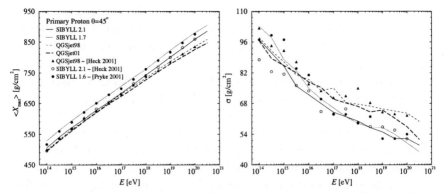

Figure 1. Average depth of maximum (left panel) and its fluctuation (right panel) of proton showers as function of primary energy. The lines represent 5,000 events generated by our hybrid method, at $\theta = 45°$, using SIBYLL 1.7 (dotted), SIBYLL 2.1 (solid), QGSjet98 (dashed) and QGSjet01 (thick dashed). The symbols show the values of $\langle X_{max} \rangle$ averaged over 500 showers obtained with CORSIKA using the thinning procedure.

flected in shower observables. In addition they are commonly used for the analysis of air shower measurements.

3.1. Depth of Maximum Development

X_{max} is a typical shower parameter measured by fluorescence and Cherenkov light detectors in several experiments. Knowing the shower energy, the mean depth of shower maximum and its fluctuations can be used to infer the primary cosmic ray composition. In Fig. 1 we compare our predictions for proton showers to those obtained in the framework of the COR-SIKA code[28] using similar (or identical) hadronic interaction models[29,30]. Each of the points generated with CORSIKA represents the mean value of X_{max} over 500 showers using the thinning procedure. The values of $\langle X_{max} \rangle$ and σ calculated by both codes for the same models are in very good agreement[31], within the larger statistical uncertainty of this particular CORSIKA calculation. This provides us a further check on the validity of the hybrid simulation method.

Fig. 1 shows the average values of X_{max} and σ as function of primary energy for proton showers injected at a zenith angle $\theta = 45°$. The lines were produced averaging X_{max} over 5,000 showers. The predictions of SIBYLL 1.7, SIBYLL 2.1, QGSjet98 and QGSjet01 are shown. The first important feature is that SIBYLL 2.1 predicts smaller $\langle X_{max} \rangle$ values than SIBYLL 1.7 by about 20 g/cm^2 from 10^{14} to 3×10^{20} eV. The predictions of SIBYLL 2.1

are closer to the values produced by QGSjet98. In fact, at energies below about 3×10^{17} eV the difference is smaller than 10 g/cm^2 and it increases with energy up to a maximum of 27 g/cm^2 at 3×10^{20} eV. QGSjet01 predicts smaller values of $\langle X_{max} \rangle$ than QGSjet98 by ~ 7 g/cm^2 at 10^{17} eV and up to 12 g/cm^2 at 3×10^{20} eV. QGSjet predicts values of $\langle X_{max} \rangle$ systematically smaller than the ones produced by SIBYLL. This is due to the much higher average particle multiplicity generated by both versions of QGSjet and ther lower elasticity compared to SIBYLL. These two features are responsible for the accelerated shower development in QGSjet. The width of the X_{max} distribution is a measure of the fluctuations (σ) of the position of the shower maximum. As shown in Fig. 1, the fluctuations become less important at very high energy. First of all, the fluctuations due to the position of the first interaction point are smaller at high energy due to the large cross section (small mean free path). Secondly, the large multiplicity of secondary particles produces a correspondingly larger number of subshowers.

3.2. Number of Muons

The number of muons in a shower is an important observable which depends strongly on the mass of the primary particle and is used in the studies of the elemental composition of cosmic rays. It also directly reflects the hadronic component of the shower and hence it is a sensitive probe of the hadronic interactions.

Fig. 2 depicts the distribution in number of muons for five energy thresholds obtained for 5,000 vertical primary proton showers at 10^{18}eV. The shape of the distributions for a fixed muon energy threshold is very similar for the four models. One could see that the average numbers of muons are higher, and their distributions wider, when calculated with QGSjet98 and QGSjet01 than in the distributions calculated with SIBYLL 1.7 and SIBYLL 2.1, for all energy thresholds illustrated in Fig. 2. This is a consequence of the higher multiplicity and multiplicity fluctuations in both versions of QGSjet. The difference in the averages (widths) between QGSjet98 and SIBYLL 2.1 is $\sim 17\%$ ($\sim 7\%$) for all energy thresholds while it becomes larger between QGSjet98 and SIBYLL 1.7, increasing from $\sim 28\%$ ($\sim 17\%$) at E_{μ}^{thr}=0.3 GeV to $\sim 35\%$ ($\sim 27\%$) at E_{μ}^{thr}=30 GeV. QGSjet01 produces more muons than QGSjet98. The difference in the averages (widths) between these versions of QGSjet is $\sim 8\%$ ($\sim 5\%$) for all energy thresholds.

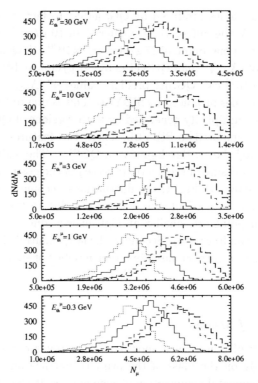

Figure 2. Shower distribution in number of muons at sea level. The results are obtained for 5,000 vertical primary proton showers at 10^{18} eV for different muon energy threshold, using different hadronic interaction models. The solid (dotted) line represents the values predicted by SIBYLL 2.1 (SIBYLL 1.7), while the dashed (thick dashed) line illustrates the values for QGSjet98 (QGSjet01).

4. Summary

We have discussed an efficient, one-dimensional hybrid method to simulate the development of extensive air showers. The combination of Monte Carlo techniques for the interactions of the shower particles above a certain hybrid energy threshold with a presimulated library of pion-induced showers, allows us to simulate the development of large statistical samples of air showers up to the highest energies observed. This technique accounts for fluctuations in the shower development as well as the correlations between the different parameters describing the electromagnetic and muon components of EAS. Showers simulated in this way can be used as input to simulations for experiments measuring the longitudinal development of the shower. Besides this, hybrid simulations are very helpful for comparing shower parameters predicted by different hadronic interaction models and to aid the interpretation of the experimental results in this way. We have shown the influence of different hadronic interaction models, namely SIBYLL 1.7, SIBYLL 2.1, QGSjet98 and QGSjet01, on some shower observables which are relevant for the determination of the energy and chemical composition of the primary cosmic ray flux. We also have presented average values of X_{max} and the number of muons above 0.3, 1, 3, 10 and 30 GeV at sea level, as well as the fluctuations of these quantities.

Acknowledgments

J.A. Ortiz is supported by CAPES "Bolsista da CAPES - Brasília/Brasil" and acknowledges Bartol Research Institute and Ettore Majorana Foundation and Centre for Scientific Culture for their hospitality and support. We are indebted to H.P. Vankov for making his LPM code available to us and for many discussions.This research is supported in part by NASA Grant NAG5-10919. RE, TKG & TS are also supported by the US Department of Energy contract DE-FG02 91ER 40626. We thank D. Heck for providing us with the values of X_{max} for CORSIKA. We also acknowledge fruitful discussions with D. Heck, P. Lipari, S. Ostapchenko, and T. Thouw.The simulations presented here were performed on Beowulf clusters funded by NSF grant ATM-9977692.

References

1. D. J. Bird *et al.*, *Astrophys. J.* **441**, 144 (1995).
2. M. Takeda *et al.*, *Phys. Rev. Lett.* **81**, 1163 (1998).
3. M. Hillas, *Nucl. Phys. B (Proc. Suppl.)* **52B**, 29 (1997).
4. Y. Shirasaki and F. Kakimoto, *Astropart. Phys.* **15**, 241 (2001).
5. J. N. Capdevielle, C. Le Gall and K. N. Sanosian *Astropart. Phys.* **13**, 259 (2000).
6. M. Kobal, *Astropart. Phys.* **15**, 259 (2001).
7. M. Risse, D. Heck, J. Knapp and S. S. Ostapchenko, Proc. 27th ICRC (Hamburg), 522 (2001).
8. T. K. Gaisser, P. Lipari and T. Stanev, Proc. 25th ICRC (Durban) **6**, 281 (1997).
9. T. Stanev and H. P. Vankov, *Astropart. Phys.* **2**, 35 (1994).
10. J. Alvarez-Muñiz et al., *Phys. Rev.* **D66**, 033011 (2002).
11. T. Abu-Zayyad *et al.*, *Nucl. Instrum. Methods Phys. Res. A* **450**, 253 (2000).
12. The Pierre Auger Project, Design Report **2nd ed.**, (1995).
13. O. Catalano, *Nuevo Cimento Soc. Ital. Fis. C* **24C**, 445 (2001).

14. R. E. Streitmatter (OWL), in Proceedings of the Workshop on Observing Giant Cosmic Ray Air Showers from $>10^{20}$ ev Particles from Space, College Park, MD, edited by J. F. Krizmanic, J. F. Ormes and R. E. Streitmatter, AIP Conf. Proc. **433**, 95 (1997).

15. M. Sasaki, in Proceedings of the EHECR 2001, International Workshop on Extremely High Energy Cosmic Rays, ICRR, Kashiwa, Japan, 2001 [*J. Phys. Soc. Jpn. Suppl. B* **70** (2001)].

16. S. E. Forsythe, Smithsonian Physical Tables (Smithsonian Institution Press, Washington, D.C. 1969).

17. T. K. Gaisser, M Shibata, and J. A. Wrotniak, Bartol Report **81-82**, (1981).

18. L. D. Landau and I. J. Pomeranchuk, *Dokl. Akad. Nauk SSSR* **92**, 535 (1953).

19. L. D. Landau and I. J. Pomeranchuk, *Dokl. Akad. Nauk SSSR* **92**, 735 (1953).

20. A. B. Migdal, *Phys. Rev.* **103**, 1811 (1956).

21. H. P. Vankov, (private communication).

22. T. K. Gaisser, Cosmic Rays and Particle Physics (Cambridge University Press, Cambridge, England, 1990).

23. R. S. Fletcher, T. K. Gaisser, P. Lipari and T. Stanev, *Phys. Rev. D* **50**, 5710 (1994).

24. R. Engel, T. K. Gaisser, P. Lipari and T. Stanev, Proc. 26th ICRC (Salt Lake City) **1**, 415 (1999).

25. R. Engel, T. K. Gaisser and T. Stanev, Proc. 27th ICRC (Hamburg), 431 (2001).

26. N. N. Kalmykov, S. S. Ostapchenko and A. I. Pavlov, *Nucl. Phys. B (Proc. Suppl.)* **52B**, 17 (1997).

27. Heck, D. et al., Proc. 27th ICRC (Hamburg) , 233, (2001).

28. D. Heck, J. Knapp, J. N. Capdevielle, G. Schatz and T. Thouw, Report FZKA 6019, (1998).

29. D. Heck, astro-ph/0103073 and private communication.

30. C. L. Pryke, *Astropart. Phys.* **14**, 4 (2001).

31. S. P. Swordy *et al.*, *Astropart. Phys.* (to be published).

Table 1. Average values of different observables and standard deviation of their distributions obtained by direct and hybrid simulations of 5,000 vertical pion showers with fixed interaction point $X_0=5$ g/cm^2, and primary energy $E = 10^{16}$ eV. The predictions of SIBYLL 1.7, SIBYLL 2.1, QGSjet98 and QGSjet01 are presented. The energy threshold in the hybrid calculation is 0.01 E=10^{14} eV.

Model	SIBYLL 1.7		SIBYLL 2.1	
	Direct	Hybrid	Direct	Hybrid
$\langle X_{max} \rangle$ [g/cm^2]	603	602	587	586
$\sigma\,(X_{max})$ [g/cm^2]	49	50	51	49
$\langle S_{max} \rangle /E$ [GeV^{-1}]	0.75	0.76	0.75	0.75
$\sigma\,(S_{max}/E)$ [GeV^{-1}]	6.8×10^{-2}	6.8×10^{-2}	6.3×10^{-2}	6.2×10^{-2}
$\langle N_\mu \rangle\,(> 0.3$ GeV)	5.39×10^4	5.41×10^4	6.10×10^4	6.13×10^4
$\sigma\,(N_\mu)$	1.79×10^4	1.81×10^4	1.86×10^4	1.87×10^4
CPU Time [min]	935	33	1091	41
Model	QGSjet98		QGSjet01	
	Direct	Hybrid	Direct	Hybrid
$\langle X_{max} \rangle$ [g/cm^2]	574	576	570	568
$\sigma\,(X_{max})$ [g/cm^2]	55	56	56	57
$\langle S_{max} \rangle /E$ [GeV^{-1}]	0.75	0.75	0.74	0.74
$\sigma\,(S_{max}/E)$ [GeV^{-1}]	6.5×10^{-2}	6.5×10^{-2}	6.3×10^{-2}	6.3×10^{-2}
$\langle N_\mu \rangle\,(> 0.3$ GeV)	6.87×10^4	6.91×10^4	7.14×10^4	7.12×10^4
$\sigma\,(N_\mu)$	2.25×10^4	2.28×10^4	2.26×10^4	2.24×10^4
CPU Time [min]	1398	79	1512	99

DELAYED SIGNALS – NEW METHOD OF HADRON STUDIES

KAROL JĘDRZEJCZAK

Andrzej Sołtan Institute for Nuclear Studies
Cosmic Ray Laboratory
Box 447, 90–950 Łódź, Poland
E-mail: kj@zpk.u.lodz.pl

In the Łódź array data we observe signals over a very long time ($\sim 600\mu s$) after the EAS front. The possibility of applying this phenomenon to building a new type of hadron detector is considered.

1. Introduction

A few years ago two teams reported signals registered by neutron monitors, which occurred several hundred microseconds after passing of the front of an extensive air shower (EAS) [1,2,3]. It is a very long time, considering the fact that an EAS signal in detectors lasts 10000 times shorter (most of particles are recorded within 50 ns). Such long delays, and the fact of their registration by specialised detectors, may suggest that they may be due thermal neutrons produced by a shower and diffusing in the matter around the detectors. If there neutrons come from EAS hadrons interaction with the detectors or arround them, there is the chance of constructing a hadron detector, using this phenomenon. Because of its simplicity and the possibility of building it in a large size, it may be a useful tool for research mass composition of primary cosmic rays. As this interesting phenomenon has not been much investigated so far, we have undertaken to register "delayed signals" (DS) using the Łódź array.

2. The Array

The Lodz EAS array is located in Łódź, Poland, close to the building of Cosmic Ray Laboratory of the A. Sołtan Institute for Nuclear Studies (http://ipj.u.lodz.pl). It consists of several parts (figure 2):

(1) Three muon detectors (marked as black rectangles in figure 2) containing altogether 104 units of Geiger - Mueller (GM) counters, covered by a layer of iron and lead for muon energy threshold 0.5 GeV.
(2) Four electron trays (marked as open rectangles) containing 72 GM counters covered only by a wooden roof.
(3) Three scintillation counters with a surface of 0.5 m^2 each and three with a surface of 1 m^2 each (marked s1 - s6). They produce trigger for the array.
(4) A block of counters designed to register "delayed signals" (DS), (marked as dashe rectangle).

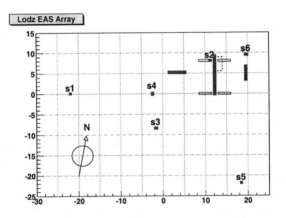

Figure 1. Plan of Łódź Shower Array, dimensions in meters

3. Delayed signals

Information about DS is collected in two ways. Firstly the sums of counts of 56 GM counters are recorded separately during twelve periods of 200 microseconds, started 5 μs after the EAS trigger (we can also switch the system to shorter 100 microsecond periods). The measurements using both electron and muon GM counters have shown DS. They will be discussed later. Secondly we use the set of counters, designed to register DS (shown in figure 3). It now consists of three $0.5m^2$ scintillation counters, a $0.004m^2$ stilben counter (a neutron sensitive organic scintillator) and a boron gas counter.

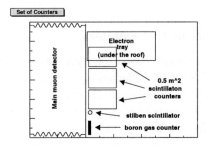

Figure 2. Set of counters, designed to register DS

The counters are connected to a special FADC (fast analogue to digital converter), which enables to investigate the time distribution of signals, similarly to digital oscilloscope. When the array is working the signal amplitude is measured every 100 ns. The result is recorded in the internal memory of the FADC in a cyclic way, i.e. recording the latest measurement removes ("pushes out") the oldest one from the memory. This way, at every moment, the memory stores 32 000 measurements (figure 3).

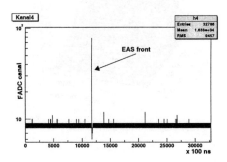

Figure 3. EAS signal in $0.5m^2$ scintillation counter

After the trigger the FADC still works for an allotted time, and then the memory content is moved to the central computer. This way we receive a continuous record of a counter's work before, during and after an EAS, 3.2 miliseconds in total (figure 4). The described above working method makes our array different from standard shower detectors which make only one measurement at the moment of a trigger signal. The necessity of a radical change of the equipment operating method makes it difficult to register DS in the presently existing EAS arrays.

4. Results

4.1. Delayed signals in GM counters

We present here the analysis based on registration of 173267 EAS. The time distribution of DS for all collected showers does not show any differences from the expected background. However, the situation will change completely if we choose events of high energy EAS i.e. a number of muons, registered while passing the EAS front, $N_\mu > 40$.

Figure 4 shows the time distribution of the number of DS for different EAS size. Figure 5 shows the same normalise to one EAS. The counters first register the remnants from the EAS front, then there is a break (minimum about $300\mu s$), after with they start to register delayed signals (maximum about $600\mu s$). The fact, that the DSs appear at nearly the same time, constitutes an argument for a physical, not random, character of the signals.

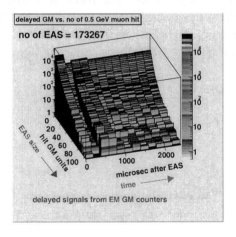

Figure 4.   Time distribution of the number of DS for different EAS size

4.2. Delayed signals in a set of counters

We do not have fully reliable results from the scintillation counters yet, as they are still in the process of calibration. On the other hand, the signals from the boron gas counter were observedonly after EASs and this is an argument for assumption that DS come from neutrons.

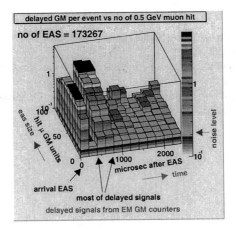

Figure 5. Time distribution of the number of DS for different EAS size, normalised to one EAS

5. Hypotheses on DS phenomenon

According to a current opinion "delayed signals" come from neutrons. There are two main hypotheses as to their origin:

(1) They are created during the EAS development, and for some unknown reason they arrive so late - the simulations seem to contradict it, because the lateral distribution of neutrons would be so flat, that DS would be registered with showers falling far from the array.

(2) They are created as a result of EAS hadron interactions in the lead detector, ground and ground water - if this were true, and were it possible to find the correlation between the number of DS and the number, or the total energy, of EAS hadrons, we could build a large and cheap hadron calorimeter. As the number of hadrons in EAS depends on on the nature of primary particle,it would be a step towards discovering mass composition of high energy cosmic rays.

6. Conclusion

Using the Lodz EAS array it has been possible to observe signals over a very long time ($\sim 600\mu s$) after the passing of the EAS front. The results are in agreement with earlier publications [1,2,3]. The possibility of applying the phenomenon of "delayed signals" to building a new type of hadron detector is considered. The work is in progress.

References

1. V. M. Aushev et al. *Izwiestia Rosyjskiej Akademii Nauk, Serija Fiziczeskaja* **61** (1997) 488–492
2. V. I. Antonova et al. *Izwiestia Rosyjskiej Akademii Nauk, Serija Fiziczeskaja* **63** (1999) 525–529
3. Yu. V. Stenkin, J. F. Valdes-Galicia, A. Hurtado, O. Musalam *Astroparticle Physics* **16** (2001) 157–168

THE ROLAND MAZE PROJECT

KAROL JĘDRZEJCZAK

Andrzej Sołtan Institute for Nuclear Studies
Cosmic Ray Laboratory
Box 447, 90–950 Łódź, Poland
E-mail: kj@zpk.u.lodz.pl

We are going to build in Łódź, Poland, EAS array for energy higher than $10^{18}eV$. The detection stations will placed on the roofs of high schools all over the city. We call our project with the name of Profesor Roland Maze, EAS codiscoverer.

1. Introduction

The Roland Maze Project is a project of an experiment for investigation of Cosmic Rays from GeV energy to the highest energy showers of above $10^{18}eV$. We are going to build detection stations on the roofs of high shools in Łódź, Poland. This way, we will use the existing infrastructure of a big city, in particular the power and telecommunication networks, to decrease the cost of our enterprise. More over the detection stations will be constructed from mass produced elements. Our array will be comparable to the largest experiments in the world, but cheaper (see fig. 1). School student would be involed to popularise science, especially cosmic ray studies.

2. The main aim of the experiment

The main aim of the experiment is to investigate the Ultra High Energy Cosmic Ray (UHECR) of energy over $10^{18}eV$. This is the only possibility to observe interactions between particles of energy 100 mln. times higher than those studied in laboratory experiments. In addition it is very interesting what astrophysical mechanizm is the source of so highly energetic particles. Because the flux of UHECR is very small (of the order of $1 particle/m^2 year$, see fig. 2) we need a very big array to register them with sufficient frequency.

280

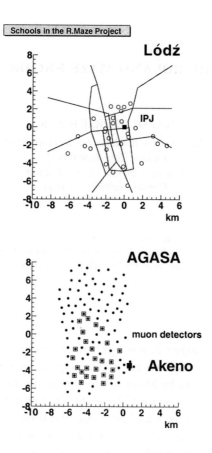

Figure 1. Sizes of Maze and AGASA

3. The idea of the experiment

As mentioned before, the array will consist of detection stations on the roofs of high schools all over the city. The stations will be connected via Internet to form one large experiment. Figure 3 shows the plan of a single station. This is an independent micro-array. It consist of 4 scintillation counters of an area $1m^2$, separated from each other by about 10 m. The electronic system should enable us to measure the relative times (with accuracy of ~ $5ns$) and the signal amplitudes from all the detectors. The results will be stored in a local PC computer.

Thanks to a GPS receiver, the time of an EAS arrival will be known with accuracy $\pm 300ns$. Which makes it possible to synchronise events at

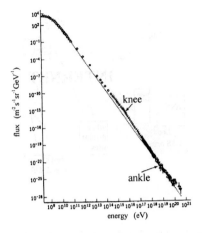

Figure 2.   Cosmic rays spectrum

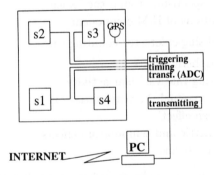

Figure 3.   The plan of a single station

different stations. Once a day the stored data will be transmitted via Internet to the main server, for storing and making them accessible to all interested (see fig. 3)

4. The additional subject

The realisation of the main aim of the project will require analising data from all over array, but the construction of a single station will allow to investigate also smaller EAS ($\sim 10^{15} eV$). The following list contains examples of phenomena, that can be examined using a single station:

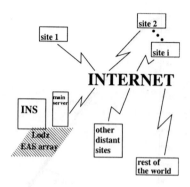

Figure 4. The idea of Maze Project network

(i) Tasks related to energies around $10^{15}eV$ (for one school array):
 (a) energy spectrum around the "knee"
 (b) fluctuations of E-M component

(ii) Tasks related single detector counting rate:
 (a) atmospheric effects
 (b) counting rate vs. solar activity
 (c) daily, seasonal, 11-years variation
 (d) Forbush effect
 (e) barometric and temperature effects
 (f) thunderstorms

Another, important or maybe more important than the physics aspect, is the educational of our project. The possibility of participating in the real experiment is going to popularise cosmic ray physics among high school students. Although the detectors will work automatically, a group of several students will be able to investigate the real physical problems using their "own" array.

5. Why Roland Maze ?

In the year 1938 Pierre Auger and Roland Maze discoverd the EAS phenomenon. The first EAS array was built by Maze on the roof of Ecole Normale Superieure in Paris. In the early fifties when the Łódź cosmic ray laboratory was established, Roland Maze closely collaborated with Polish physicists. At present when a great Pierre Auger experiment is being build, we have decided to call our project with the name of Roland Maze.

6. Summary

We are going to build in Lódź, Poland, EAS array for energy higher than $10^{18}eV$. Using the existing city infrastructure, and dividing the array in to independent stations (micro-arrays) we will considerably lower the costs. In addition, by placing the station on the roofs of high school buildings, we will achive an important educational effect. 30 secundary schools (students of ages 16 – 19) from the Lódź expressed their will to join the project.

We keep current information about the experiment status on our web page: "http://ipj.u.lodz.pl".

GAMMA RAY AND

NEUTRINO ASTRONOMY

TEV OBSERVATIONS OF EXTRAGALACTIC SOURCES AT THE WHIPPLE OBSERVATORY

DEIRDRE HORAN[1]

1. Harvard-Smithsonian Center for Astrophysics
P.O. Box 97, Amado, Arizona, 85645-0097
E-mail: dhoran@cfa.harvard.edu

After a brief introduction to the imaging atmospheric Čerenkov technique, the method by which very high energy γ-rays are detected, the most recent results from observations of extragalactic sources at the Whipple Observatory are described.

1. Detecting Very High Energy (VHE) γ-rays From The Ground

In most branches of astronomy, the Earth's atmosphere places a serious limit on the earth-bound astronomer. It is opaque to γ-rays, x-rays and ultraviolet photons and, although transparent to optical photons, resolution is seriously limited by atmospheric seeing. Observations of infrared photons can only be carried out in wave bands that are free of water vapour absorption. To very high energy (VHE) astronomers however, the Earth's atmosphere is essential to their technique, being an integral part of the detector[1].

This technique, the imaging atmospheric Čerenkov technique, is currently used to detect photons in the range 10^{11} to 10^{13} eV; the VHE regime. The VHE γ-rays are not detected directly, but rather through their interactions in the Earth's atmosphere, which is essentially the calorimeter in this technique. VHE γ-rays initiate extensive air showers in the Earth's atmosphere, and the charged particles subsequently generated, induce Čerenkov radiation which is detectable on Earth with large optical reflectors. These instruments are not wide-field survey instruments and so, like a conventional optical or radio telescope, they require a predetermined target which, for some reason is predicted to be an emitter of TeV γ-rays (Section 2.2).

The vast majority (>99%) of extensive air showers that propagate through the Earth's atmosphere are initiated by cosmic rays. Therefore, essential to performing ground based VHE astronomy is a means to extract the γ-ray signal from this very intense isotropic background of hadronic cosmic rays - one such approach is the imaging atmospheric Čerenkov technique.

First proposed in 1977[2], the imaging atmospheric Čerenkov technique utilises differences in the distribution of the Čerenkov light induced by cosmic rays and γ-rays, to effectively reject far more than 99% of the background, while retaining greater than 50% of the γ-ray signal.

The Whipple IACT, is located at an altitude of 2.3 km above sea level, on Mount Hopkins in Southern Arizona. The telescope consists of a large reflecting surface with the imaging camera at the focal plane. The Čerenkov images are recorded and processed using fast electronics. Currently, it is sensitive in the energy range ~250 GeV to 10 TeV with a maximum sensitivity at ~ 400 GeV. Its evolution and current status are described elsewhere in these proceedings[3].

2. Extragalactic Sources of VHE γ-rays

Active Galactic Nuclei (AGN) emit an unusually large amount of energy from a very compact central source and are among the most violent regions, embodying some of the most extreme conditions, to be found anywhere in the universe. Although much of the detailed physics is literally hidden from view because of their strongly anisotropic radiation patterns, there is general agreement on the basic components which comprise AGN. At the centre is a supermassive black hole whose gravitational potential energy is the ultimate source of the AGN luminosity.

AGN fall into two main categories: radio-loud AGN, which are powerful radio emitters with their radio-power output lying in the range 10^{35} to 10^{38} watts; and radio-quiet AGN, which emit most of their radiation in the infrared, and are not powerful at radio wavelengths. In radio-loud objects, jets emanate from the region near the black hole, initially at relativistic speeds.

It is now well established that the appearance of an AGN depends strongly on its orientation relative to the observer's the line of sight. Classes of apparently different AGN might actually be intrinsically similar, only viewed at different angles[4]. All of the AGN that have been detected at γ-ray energies, both with satellites and ground-based instruments, are members the *Blazar* subclass.

2.1. Blazars

Among AGN, blazars are the most extreme and powerful sources known and are believed to have their jets more aligned with the line of sight than any other class of radio loud AGN. They are high luminosity objects, characterized by large, rapid, irregular amplitude variability in all accessible spectral bands. Their overall spectral energy distribution (SED) is characterized by two broad emission peaks, the first located in the IR - extreme UV and sometimes the x-ray band, the second in the MeV - GeV band[5].

Several models, still in competition, have been proposed to explain the γ-ray emission from blazars. Depending on the dominant mechanism considered for the production of γ-rays, these models can be divided into two main groups: those in which the γ-rays are of leptonic origin (high energy electrons upscatter x-rays to γ-ray energies) and those in which the γ-rays are of hadronic origin (high energy protons produce neutral pions which subsequently decay into high-energy γ-rays).

Although united in a single class because of their similar properties, blazars come in different flavours - flat spectrum radio quasars (FSRQs) and BL Lacertae objects (BL Lacs). The main difference between the two subclasses lies in their emission lines, which are strong in FSRQs and are weak or non-existent in BL Lacs. Thus, contrary to most other astronomical sources, BL Lacs have been almost exclusively discovered at either x-ray or radio frequencies. Because BL Lacs found in x-ray and radio surveys had quite different properties, they were further designated as either x-ray or radio selected BL Lacs (XBLs and RBLs). Later, the terminology, 'High-frequency peaked BL Lacs' (HBLs) for those blazars that were strong x-ray emitters, and 'Low-frequency peaked BL Lacs' (LBLs) for those that were brighter in the radio band, was introduced[6]. This re-classification really only represented a change in nomenclature because objects tended to be discovered in the waveband at which they were the brightest. Recently however, deeper BL Lac surveys are revealing evidence which implies that rather than being separate subclasses, these objects may represent the extremes of a sequence of progressively different BL Lacs.

2.2. The Blazar Sequence

It has recently been proposed that blazars can be unified in terms of a luminosity sequence from HBLs through LBLs to FSRQs[7]. This model predicts that blazars form a well defined sequence and that knowledge of the properties of a particular object in one frequency band can be used to predict its emission characteristics in other regions of the spectrum. This unification scheme has had important implications for the search for TeV emission from blazars because it makes specific predictions about the x-ray properties that TeV emitters should have.

The $BeppoSAX$ observations of the TeV blazars, Markarian 501[8] and 1ES2344+514[9], revealed that, at least in flaring state, the peak of the synchrotron emission can reach very high energies, around 100 keV, with a consequently flat synchrotron x-ray spectral index. The term "extreme BL Lac" was introduced to describe such low luminosity BL Lacs whose synchrotron peak is located in the hard x-ray band[5].

3. Detected Sources of VHE γ-rays

All of the AGN detected sofar at TeV energies are BL Lacs. To date, from seven ground-based γ-ray observatories operating around the world, there have been eight claimed detections of sources of extragalactic γ-rays. The extragalactic TeV source catalog c.2002 is shown in Table 1. Currently, out of the eight claimed detections, four have been independently confirmed. The most recent results on the five Whipple-detected BL Lacs are summarized in the following sections.

Table 1.   The Catalog of Extragalactic TeV γ-ray sources c.2002.

Object	Discovery	Whipple	HEGRA	CAT	CANGAROO	Crimea	Durham	7 Telescope-Array	Independently Confirmed
HBLs									
Mrk421	Whipple[10]	√	√	√	√			√	YES
Mrk501	Whipple[11]	√	√	√		√	√	√	YES
1ES2344	Whipple[12]	√							yes^a
PKS2155	Durham[13]						√		no
1ES1959	Telescope Array[14]	√	√					√	YES
H1426	Whipple[15]	√	√	√					YES
LBLs									
3C66A	Crimea[16]					√			no
BL Lac	Crimea[17]					√			no

^{a} Confirmed by the group that originally reported the detection but, to date, not independently confirmed by another group.

3.1. Markarian 421 (Mrk421)

Markarian 421 (Mrk421), the first extragalactic source of VHE γ-rays ever discovered[10], at a redshift of 0.031, is perhaps the best studied of the TeV blazars. Since its detection, Mrk421 has exhibited extreme variability on timescales of minutes to years. In May 1996, a flare with doubling and decay time of less than 15 minutes was observed by Whipple[18].

Many multiwavelength campaigns have been carried out on Mrk421[19,20,21]. Evidence for correlated variability at x-ray and γ-ray has been observed during a number of these campaigns.

During March of 2001, Mrk421 went into an exceptionally long-lasting active state, with an average γ-ray rate of 3.7 times that of the Crab, enabling over 23,000 photons above 260 GeV to be detected at the Whipple Observatory. From this huge photon database, a very detailed spectrum was derived ($dN/dE \sim E^{-2.14\pm0.03}e^{-(E/E_0)}$ $m^{-2}s^{-1}TeV^{-1}$) and the first evidence for curvature in the spectrum of Mrk421 was seen[22]. The break energy of $E_0=4.3\pm0.3$ TeV is similar to that found for Mrk501 (Section 3.2).

Perhaps the most remarkable result to arise from this database however, was the indication of a correlation between the spectral index and flux[23]. The data were binned a priori into eight independent subsets with comparable numbers of excess events and average γ-ray rates, and a spectrum was derived for each of these datasets. A tight correlation between spectral index and flux was found. There is however, no evidence for variation of the cutoff energy with flux, all spectra are consistent with an average value for the cutoff energy of 4.3 TeV. The results are shown in Figure 1. Spectral measurements of Mrk421 from previous years by the Whipple collaboration are consistent with this flux-spectral index correlation, which suggests that this may be a constant or a long-term property of the source. If a similar flux-spectral index correlation were found for other γ-ray blazars, this could help disentangle the intrinsic emission mechanism from external absorption effects.

3.2. Markarian 501 (Mrk501)

Markarian 501 (Mrk501), at a redshift of 0.034, was the first ever source of extragalactic γ-rays to be discovered from the ground[11]. It can be classified as an extreme BL Lac because, during flaring episodes, the first peak in its SED shifts to above 10 keV. It was only during such flaring episodes that Mrk501 was detected by the EGRET instrument on board CGRO.

Mrk501 was the first TeV blazar discovered to have a curved energy

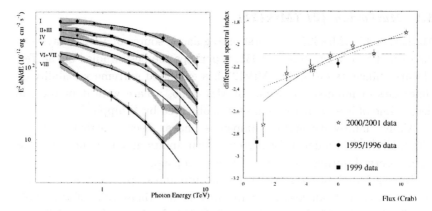

Figure 1. *Left*: Mrk421 spectra at different flux levels averaged for data over the 2000/2001 season. The shaded areas indicated the systematic errors on the flux measurements. *Right*: The flux-spectral index relation found in the Whipple Mrk421 data from 1995/1996, 1999 and 2000/2001. Both Figures are from Krennrich et al. (2002).

spectrum[24]. The initial discovery of curvature by Whipple was subsequently confirmed by the HEGRA group[25]; there is extremely good agreement between the derived spectra from both groups. The spectral break was found to occur at $\sim E_0 = 4.6 \pm 0.8$ TeV. This is very close to the energy at which the spectral break occurs for Mrk421 and therefore, since the two BL Lacs lie at very similar redshifts, it is difficult to say whether the break is intrinsic to this class of AGN or whether is due to these photons being attenuated by the infrared background photon field (discussed elsewhere in these proceedings[26]).

3.3. *1ES2344+514 (1ES2344)*

The detection of 1ES2344+514 ($z = 0.044$), the third AGN to be seen at TeV energies, was reported by the Whipple collaboration in 1998 [12]. Most of the emission came from a single night, on which a flux of approximately half that of the Crab Nebula was detected with a significance of 6σ. Recently, the Whipple collaboration have once again reported evidence for emission from 1ES2344+514, this time at the 3σ level [27]. This object has not been detected by any of the other VHE observatories to date.

3.4. *H1426+428 (H1426)*

H1426+428 (H1426), at a redshift of 0.129, is the most distant confirmed source of TeV γ-rays ever detected. In 1999, the *Beppo*SAX Collaboration carried out observations aimed at finding other objects as extreme as Mrk501 is in its flaring state[28,29]. Of the four candidate objects selected for observation, H1426 was singled out as the most likely among them to be a TeV emitter. This was because the peak of its x-ray emission was found to occur ~ 100 keV, even when it was not flaring. This indicates the presence of highly relativistic electrons which made H1426 a prime candidate for TeV γ-ray emission.

H1426 was observed at the Whipple Observatory, as part of a BL Lac monitoring campaign between 1995 and 1999; this was when the first evidence for TeV emission, in the form of a flare in March 1999, was detected[15]. This flare, and the indication from x-ray observations that H1426 would be a TeV emitter, prompted more extensive observations to be carried out at the Whipple Observatory in the 2000 and 2001 observing seasons. These observations led to the discovery of a weak, but statistically significant, TeV flux from H1426 during both observing seasons[30]. Since then, the TeV emission has been confirmed by both the HEGRA[31] and CAT[32] collaborations thus firmly establishing H1426 as a TeV source.

Using the 2001 data, the spectrum of H1426 was examined above 250 GeV[33]. The time-averaged spectrum was found to agree with a power law of the shape

$$(\frac{dF}{dE})(E) = 10^{-7.31\pm0.15_{stat}\pm0.16_{syst}} \cdot E^{-3.50\pm0.35_{stat}\pm0.05_{syst}} \, m^{-2}s^{-1}TeV^{-1}$$

The spectrum is consistent with the (non-contemporaneous) measurement by the HEGRA group[31] both in shape and normalization. Below 800 GeV, the data clearly favour a spectrum steeper than that of any other TeV blazars so far indicating a difference in the processes involved either at the source or in the intervening space.

3.5. 1ES1959+650 (1ES1959)

1ES1959+650 (1ES1959) is the most recently confirmed extragalactic source of VHE γ-rays. The Telescope Array collaboration initially reported the detection of TeV γ-rays from 1ES1959 in 1999[14]. This object, at a distance of $z = 0.048$, was confirmed as a TeV emitter by both the HEGRA[34] and Whipple[35] groups this year during a period of prolonged activity.

Acknowledgments

This work is supported in part by grants from the US Department of Energy, the National Science Foundation and the Smithsonian Institution.

294

References

1. T. C. Weekes, *Phys. Rep.* **160**, 1 (1988).
2. T. C. Weekes and K. E. Turver, *Proc. 12th ESLAB, Frascati*, **1**, 279 (1977).
3. P. J. Boyle, *these proceedings*.
4. P. Padovani, *ASP Conf. Ser. 159: BL Lac Phenomenon*, **159**, 339 (1999).
5. G. Ghisellini, *Astropart. Phys.*, **11**, (11).
6. P. Padovani and P. Giommi, *ApJ*, **444**, 567 (1995).
7. G. Fossati et al., *MNRAS*, **289**, 136 (1997).
8. E. Pian et al., *ApJL*, **492**, L17 (1998).
9. P. Giommi, P. Padovani and E. Perlman, *The Active X-ray Sky: Results from BeppoSAX and RXTE*, 407 (1998).
10. M. Punch et al., *Nature*, **358**, 477 (1992).
11. J. Quinn et al., *ApJ*, **456**, L83 (1996).
12. M. Catanese et al., *ApJ*, **501**, 616 (1998).
13. P. M. Chadwick et al., *ApJ*, **513**, 161 (1999).
14. T. Nishiyama et al., *AIP Conf. Proc. 516*, **26**, 369 (2000).
15. D. Horan et al., *AAS/High Energy Astrophysics Division*, **32**, 0503 (2000).
16. Y. I. Neshpor et al., *Astronomy Letters*, **24**, 134 (1998).
17. Y. I. Neshpor et al., *Astronomy Letters*, **27**, 228 (2001).
18. J. Gaidos et al., *Nature*, **383**, 319 (1996).
19. J. H. Buckley et al., *ApJL*, **472**, L9 (1996).
20. L. Maraschi et al., *ApJL*, **526**, L81 (1999).
21. G. Fossati, M. Jordan and J. Buckley, *AIP. Conf. Proc.* , **587**, 266 (2001).
22. F. Krennrich et al., *ApJL*, **560**, L45 (2001).
23. F. Krennrich et al., *ApJL*, **575**, L9 (2002).
24. F. Krennrich et al., *ApJ*, **511**, 149 (1999).
25. A. Konopelko et al., *Astropart. Phys.*, **11**, 135 (1999).
26. P. J. Boyle and D. Horan, *these proceedings*.
27. H. M. Badran et al., *Proc. 27th ICRC*, **7**, 2653 (2001).
28. L. Costamante et al., *Proc. of X-ray Astronomy '99 - Stellar Endpoints, AGN and the Diffuse Background*, (2000).
29. L. Costamante et al., *Astronomy and Astrophysics*, **371**, 512 (2001).
30. D. Horan et al., *ApJ*, **571**, 753 (2002).
31. F. A. Aharonian et al., *Astronomy and Astrophysics Letters*, **384**, L23 (2002).
32. A. Djannati-Atai et al., *Astronomy and Astrophysics Letters*, in press (astro-ph/0207618).
33. D. Petry et al., *ApJ*, in press (astro-ph/0207506).
34. A. Konopelko et al., *Bulletin of the APS*, **47**, No. 2, 4A3:002 (2002).
35. T. C. Weekes for the VERITAS Collaboration, IAUC 7903.

THE SCIENCE OF VERITAS

PATRICK J. BOYLE[1] AND DEIRDRE HORAN[2]

1. *Enrico Fermi Institute, University of Chicago*
933 E 56th Street, Chicago, IL 60637
E-mail: jojo@donegal.uchicago.edu
2. *Harvard-Smithsonian Center for Astrophysics*
P.O. Box 97, Amado, Arizona, 85645-0097
E-mail: dhoran@cfa.harvard.edu

The Very Energetic Radiation Imaging Telescope Array System (VERITAS) represents an important step forward in the study of extreme astrophysical processes in the universe. By employing an array of large atmospheric Čerenkov telescopes, VERITAS will combine the power of the imaging atmospheric Čerenkov technique with that of stereoscopic observations. The seven identical VERITAS telescopes, each of aperture 10m, which will be deployed in a hexagonal pattern of side 80m; each telescope will consist of 499 pixels with a field of view of 3.5°. VERITAS will substantially increase the catalog of very high energy (E > 100 GeV) γ-ray sources and will greatly improve measurements of established sources. This paper describes the main science goals of the VERITAS project.

1. Introduction

The field of ground-based γ-ray astronomy was revolutionized with the development of the imaging Čerenkov technique. This technique, which was largely developed by the Whipple γ-ray Collaboration[1], enabled many discrete sources of Very High Energy (VHE) radiation (E $\gtrsim$ 250 GeV) to be discovered. Although less than 1% of the sky has been surveyed using this technique, fifteen sources have now been reported by ground-based groups: three pulsar-powered nebulae, eight active galactic nuclei, three shell-type supernova remnants, and one X-ray binary system. These measurements have advanced our understanding of the origin of cosmic rays, the nature of AGN jets, the density of the extragalactic infra-red background radiation, and the magnetic fields within supernova remnants. The VERITAS array[2], with its substantially improved resolution and sensitivity, will significantly advance our understanding of existing TeV γ-ray emitters by allowing us to carry out deeper, more detailed studies as well as enabling the discovery of new sources and new fundamental physical phenomena.

2. Science Goals

2.1. *Active Galactic Nuclei (AGN)*

Active Galactic Nuclei (AGN) are galaxies in which a compact nucleus outshines the rest of the galaxy. Their broadband emission is dominated by non-thermal processes and is widely believed to be powered by accretion onto a supermassive black hole. To date, all AGN detected at TeV energies are members of the blazar subclass[3].

VHE observations of blazars, have proved critical in advancing our understanding of the physics of AGN. Whipple observations of the blazars Markarian 421 (Mrk 421) and Markarian 501 (Mrk 501) have revealed their emission to be extremely variable on timescales of years to minutes. Opacity arguments reveal that the variability observed on 15 minute timescales from Mrk 421 implies a compact γ-ray emitting region of 10^{-4} parsec, only an order of magnitude larger than the event horizon of a 10^8 solar mass black hole.

The only way to truly distinguish between different blazar emission models is through simultaneous observations of their flux at different energies. To date, the coverage provided by multiwavelength campaigns, has not been detailed enough to rule out any of the proposed emission models. With its increased sensitivity, VERITAS will both allow the γ-ray signals from AGN to be detected with much higher temporal resolution and will also enable us to detect new objects and the already detected objects when they are in lower flux states. With these capabilities both the sensitivity and the possible baselines of multiwavelength campaigns will be significantly increased.

To date, confirmed TeV emission has only been detected from four AGN: Mrk 421[4], Mrk 501[5], H1426+428[6] and 1ES1959+650[7]. In order to further understand and characterize the VHE emission from AGN, a much larger database of TeV-emitting objects is required. VERITAS, with its increased sensitivity, will enable us to detect nearby AGN that are intrinsically weaker at TeV energies and those whose emission is weaker because they lie at larger redshifts.

2.1.1. *Extragalactic Infra-red Background (IRB) Radiation*

The cross-section for high energy photons to pair-produce with the infrared background (IRB) photon field that permeates our universe, peaks at ~ 1 TeV. Hence the flux of photons of energy above about 1 TeV decreases rapidly as a function of redshift. In order to study this effect in detail and thus to indirectly gain valuable information about the density of the extragalactic IRB field, a search can be made for absorption signatures in the spectra of distant AGN. In order to decouple intrinsic cutoffs in AGN spectra from those due to attenuation by background IR photons however, the spectra of TeV sources at several different redshifts must be compared so that a relation between the energy at which a spectral break is found, and the distance to the source can be established.

To date, detailed spectral information has only been obtained for the two nearest TeV blazars: Mrk 421 (z=0.031) and Mrk 501 (z=0.034). Although there is evidence for a break in the VHE spectrum at around 4 TeV for both of these sources[8, 9], since they lie at similar redshifts, it is difficult to say whether the break is intrinsic to this class of AGN or whether is due to these photons being attenuated by the IRB photon field. The VHE spectra from Mrk 421 and Mrk 501 have been used however, to set upper limits on the IRB from 0.025 eV to 0.3eV[10,11,12]. At some wavelengths, these limits are as much as an order of magnitude below the upper limits set by the DIRE/COBE satellite. The current limits on the IRB density are ~ 10 times higher than those predicted from galaxy formation and evolution[13,14]. VERITAS will enable more detailed spectral information to be gathered on the known TeV sources and will also increase the TeV source catalog at both low and high redshifts. Thus, the energetics of many more AGN will be studied which will allow many more points to be added to the redshift versus spectral-cutoff curve. In this way, the density of the IRB photon field can be studied in detail.

2.2. Shell-type Supernova Remnants (SNRs)

Supernova Remnants (SNRs) are widely believed to be the sources of hadronic cosmic rays up to energies of approximately $Z \times 10^{14}$ eV, where Z is the nuclear charge of the particle. However, a clear indication for the acceleration of hadronic particles in SNRs is still missing. The existence of energetic *electrons* is well known from observations of synchrotron emission at radio and X-ray energies. Recently, the detection of TeV γ-rays from the shell-type remnants, SN 1006, RXJ 1713 and Cassiopeia A has been reported[15]. When information from radio and X-ray observations is also taken into account, it can be determined whether the γ-ray emission is of hadronic or leptonic origin.

To date however, γ-ray observations of the VHE spectra of SNRs are not sensitive enough to tell whether they are of hadronic or leptonic origin. The combined, multiwavelength observations can also give information about SNR shell environments such as the maximum particle energy and strength of the magnetic field. Both quantities are important but unknown parameters in shock acceleration. Previous Whipple upper limits for the SNR IC 443 already eliminate much of the allowed parameter space for γ-ray emission from these objects (from hadrons and electrons), and raise some questions about the validity of current models for the objects studied and even for the SNR origin of cosmic rays. Taking a typical SNR luminosity and angular extent and applying the hadronic model of Drury, Aharonian and Volk[16], one finds that VERITAS should be able to detect SNRs that lie within 4 kpc of Earth. Approximately twenty shell-type SNRs, with known distances, lie within this distance range thus providing a good selection of objects to be investigated with VERITAS.

2.3. Diffuse galactic γ-ray emission

High energy γ-rays traverse the Galaxy without significant attenuation implying that the diffuse emission traces high energy processes in the Galaxy as a whole. EGRET studies show generally good agreement with detailed theoretical models both in spatial and in spectral features. A striking exception is that there is a 40% excess in measured flux at the highest EGRET energies, with the measured spectrum systematically rising above predictions. In some models[17,18] this is attributed to the inverse Compton scattering of energetic SNR electrons although this appears to be in contradiction with upper limits at TeV energies[19]. Detailed spectral studies in the GeV-TeV energy bands are necessary to understand these discrepancies.

2.4. γ-ray pulsars

The attenuation of γ-rays by pair production interactions in the intense magnetic field near pulsars leads to a super-exponential cut-off in the spectra predicted by polar cap models. Because the outer gap models do not predict such sharp cut-offs, the detection of pulsed GeV-TeV γ-rays may be decisive in favouring the outer gap model over the polar cap model. The excellent sensitivity, energy resolution and the broad energy range of VERITAS will allow spectral measurements to be conducted even though these objects are expected to have rapidly falling spectra in the GeV-TeV range.

2.5. Particle Physics and Fundamental Physics

2.5.1. Cosmic Ray Composition

VERITAS, with its fine pixelation, large mirror area and stereoscopic capabilities will be able to measure the Čerenkov light emitted by primary cosmic ray nuclei before they interact with the atmosphere, thereby providing a high resolution ($\Delta Z/Z < 5\%$ for $Z > 10$) charge measurement of cosmic rays around the knee of the all-particle spectrum. This measurement will be essentially independent of any assumed nuclear interaction model; therefore VERITAS can provide a tagged nuclear beam that will determine the true air-nucleus interaction characteristics as a function of primary charge and energy. This should provide a method to eliminate interaction model-dependent biases in the interpretation of results, and will also provide other experiments with an experimentally determined interaction model with which to re analyze their data. In addition, VERITAS will be sensitive to nuclei heavier than iron in the PeV energy regime, and to exotic particle states such as magnetic monopoles or strange quark matter[20].

2.5.2. Neutralino annihilation in the galactic center

Current astrophysical data indicates the need for a cold dark matter component with $\Omega \approx 0.3$. A good candidate for this component is the neutralino, the lightest super-symmetric particle. If neutralinos do comprise dark matter and are concentrated near very massive astrophysical objects, like the centre of our Galaxy, their direct annihilation into γ-rays should produce a unique signal not easily mimicked by other astrophysical processes: a mono-energetic annihilation line with mean energy equal to the neutralino mass. Cosmological constraints and limits from accelerator experiments restrict the neutralino mass to within the range 30 GeV - 3 TeV. Thus, VERITAS and GLAST together will allow a sensitive search over the entire allowed neutralino mass range. Indeed, recent estimates of the annihilation line flux for neutralinos at the galactic centre using a galactic model, with central cusps in the density distribution of the dark matter halos, predict a γ-ray signal which may be of sufficient intensity to be detected with VERITAS and GLAST. The better sensitivity and lower energy threshold of VERITAS mean that a broad part of the allowed range of neutralino and dark matter parameters can be probed.

2.5.3. γ-ray bursts (GRBs)

Today, over thirty years after their discovery[21], the origin of γ-ray bursts (GRBs) still remains unknown. The detection of an 18 GeV photon from GRB940217 by EGRET more than 90 minutes after the GRB was detected by BATSE demonstrates that high energy γ-rays play an important role in the energetics of GRBs[22]. Indeed, many after-glow models[23, 24] predict a delayed TeV component in GRBs. In 1999, the Milagro Collaboration reported a weak ($\sim 3\sigma$) detection of TeV emission from a GRB raising the possibility that TeV photons may carry a significant fraction of the energy emitted in a GRB[25]. With its low energy threshold and fast slew capabilities ($\sim 1°s^{-1}$), VERITAS will be able to search for VHE emission from GRBs out to redshifts of about 1.

2.5.4. Quantum Gravity

Quantum gravity can manifest itself as an effective energy-dependence to the velocity of light in a vacuum caused by propagation through a gravitational medium containing quantum fluctuations on distance scales near the Planck length ($\simeq 10^{-33}$)[26]. If the quantum gravity correction to vacuum refractive index exists, it should appear at the energy scale comparable to Planck mass ($\approx 10^{19}$ GeV). Recent work within the context of string theory indicates, however, that the quantum gravity scale may occur at a much lower energy, perhaps as low as 10^{16} GeV [27]. TeV observations of variable emission from astrophysical objects provide a means of searching for the effects of quantum gravity[28]. VERITAS will significantly improve short timescale variability measurements and also will detect more distant objects. Variability on the short timescales from sources at z > 0.1 would be sensitive to quantum gravity effects and would provide a test of the validity of Lorentz symmetry at energies within a factor of five of the Planck mass (in some models).

2.5.5. Primordial black holes

Primordial black holes, if they exist, should emit a burst of radiation in the final stages of their evaporation[29]. In the standard model of particle physics, this last burst of radiation should release about 10^{30} erg in 1 s with the energy distribution peaked near 1 TeV [30]. In two years of operation VERITAS would be able to reach a sensitivity level for this type of evaporation of 700 pc^{-3} yr^{-1}. More extreme models[31] would produce lower energy events on much shorter time-scales which may be detectable with VERITAS using a special trigger and a flash ADC system[32].

3. Summary

The VERITAS system represents an important step forward in the study of extreme astrophysical processes in the universe. VERITAS will complement GLAST, the next generation space telescope, and will help close one of the last remaining windows of the electromagnetic spectrum.

Acknowledgments

This work is supported in part by grants from the US Department of Energy, the National Science Foundation, the Smithsonian Institution and the University of Chicago.

References

1. P. J. Boyle, *these proceedings*.
2. T. C. Weekes et al., *ApJ*, **17** 243 (2002).
3. D. Horan, *these proceedings*.
4. M. Punch et al., *Nature*, **358**, 477 (1992).
5. J. Quinn et al., *ApJ*, **456**, L83 (1996).
6. D. Horan et al., *ApJ*, **571**, 753 (2002).
7. T. Nishiyama et al., *AIP Conf. Proc. 516*, **26**, 369 (2000).
8. F. Krennrich et al., *ApJ*, **511**, 149 (1999).
9. F. Krennrich et al., *ApJ*, **560**, L45 (2001).
10. F. W. Stecker, O.C. deJager, *AsA* **334** L85 (1998).
11. S. D. Biller et al., *Phys. Rev. Lett.*, **80** 2992 (1998).
12. V. V. Vassiliev et al., *APh*, **12** 217 (2001).
13. J. R. Primack et al., *APh*, **11** 93 (1999).
14. M. A. Malkan and F. W. Stecker, *ApJ*, **496** 13 (1998).
15. T. C. Weekes, *TMAC*, **1** 3 (1999).
16. L. Drury, F. A. Aharonian and H. J. Volk, *AsA*, **287** 959 (1994).
17. M. Pohl and J. A. Esposito, *ApJ*, **507** 327 (1998).
18. I. V. Moskalenko and A. W. Strong, *ApJ*, **528** 357 (2000).
19. S. LeBohec et al., *ApJ*, **539** 209 (2000).
20. S. Banerjee et al., *APh*, **85** 1384 (2000).
21. R. W. Klebesadel, I. B. Strong and R. A. Olsen, *ApJ*, **182**, L85 (1973).
22. K. Hurley, *Nature*, **372** 652 (1994).
23. R. Sari and A. Esin, *ApJ*, **548**, 787 (2001).
24. B. Zhang and P. Mészáros, *ApJ*, **559**, 110 (2001).
25. J. McEnery et al., *TAMC*, **1** 243 (1999).
26. G. Amelino-Camelia et al., *Nature*, **383** 319 (1998).
27. E. Witten, *Nucl. Phys. B*, **471** 135 (1996).
28. S. D. Biller et al., *Phys. Rev. Lett.*, **83** 2108 (1999).
29. D. N. Page and S. W. Hawking, *ApJ*, **206** 1 (1976).
30. F. Halzen et al., *Nature*, **353** 807 (1991).
31. R. Hagedorn , *Nuovo Cimento*, **4** 1027 (1968).
32. F. Krennrich, S. Le Bohec and T. C. Weekes, *ApJ*, **529** 506 (2001).

EXPLORING THE GAMMA RAY HORIZON WITH THE NEXT GENERATION OF GAMMA RAY TELESCOPES

O. BLANCH BIGAS AND M. MARTÍNEZ

IFAE, Universitat Autonoma de Barcelona,
Building Cn,
08193, Bellaterra, SPAIN
E-mail: blanch@ifae.es, martinez@ifae.es

The physics potential of the next generation of Gamma Ray Telescopes in exploring the Gamma Ray Horizon is discussed. It is shown that the reduction in the Gamma Ray detection threshold might open the window to use precise determinations of the Gamma Ray Horizon as a function of the redshift to either put strong constraints on the Extragalactic Background Light modeling or to obtain relevant independent constraints in some fundamental cosmological parameters.

1. Introduction

Imaging Čerenkov Telescopes (ČT) have proven to be the most successful tool developed so far to explore the cosmic gamma rays of energies above few hundred GeV. A pioneering generation of installations has been able to detect a handful of sources and to start a whole program of very exciting physics studies. Nowadays a second generation of more sophisticated Telescopes is under construction and will provide soon with new observations. One of the main characteristics of some of the new Telescopes [1] is the potential ability to reduce the gamma ray energy threshold below $\sim 10-20$ GeV, helping to fill the existing observational energy gap between the detector on satellites and the ground-based installations.

In the framework of the Standard Model of particle interactions, high energy gamma rays traversing cosmological distances are expected to be absorbed by their interaction with the diffuse background radiation fields, or "Extragalactic Background Light" (EBL), producing e^+e^- pairs. The $\gamma_{HE}\gamma_{EBL} \to e^+e^-$ cross section is strongly picked to $E_{CM} \sim 1.8 \times (2m_e c^2)$. Therefore, there is a specific range in the EBL energy which is "probed" by each gamma ray energy.

1.1. Optical Depth

Gamma rays of energy E can interact with low-energy photons of energy ϵ from the EBL over cosmological distance scales. Then the flux is attenuated as a function of the gamma energy E and the redshift z_q of the gamma ray source. It can be parameterised by the optical depth $\tau(E, z_q)$, which is defined as the number of e-fold reductions of the observed flux as compared with the initial flux at z_q. This means that the optical depth introduces an attenuation factor $\exp[-\tau(E, z_q)]$ modifying the gamma ray source energy spectrum.

The optical depth can be written with its explicit redshift and energy dependence[2] as

$$\tau(E, z) = \int_0^z dz' \frac{dl}{dz'} \int_0^2 dx \frac{x}{2} \int_{\frac{2m^2c^4}{Ex(1+z')^2}}^\infty d\epsilon \cdot n(\epsilon, z') \cdot \sigma[2xE\epsilon(1 + z')^2] \quad (1)$$

where $x \equiv 1 - \cos\theta$ and $n(\epsilon, z')$ is the spectral density at the given z'.

1.2. Gamma Ray Horizon

For any given gamma ray energy, the Gamma Ray Horizon (GRH) is defined as the source redshift for which the optical depth is $\tau(E, z) = 1$.

In practice, the cut-off due to the Optical Depth is completely folded with the spectral emission of the gamma source. Nevertheless, the suppression factor in the gamma flux due to the Optical Depth depends only (assuming a specific cosmology and spectral EBL density) on the gamma energy and the redshift of the source. Therefore, a common gamma energy spectrum behaviour of a set of different gamma sources at the same redshift is most likely due to the Optical Depth.

1.3. Extragalactic Background Light

The actual value of the Optical Depth and the GRH horizon distance for gamma rays of a given energy depends on the number density of the diffuse background radiation of the relevant energy range that is traversed by the gamma rays. In the range of gamma ray energies which can be effectively studied by the next generation of Gamma Ray telescopes (from, say, 10 GeV to 50 TeV), the most relevant EBL component is the infrared contribution.

There exists observational data with determinations and bounds of the background energy density at $z = 0$ for several energies [3]. Based on this data, several models have been developed to predict that EBL density at redshift z[4,5].

2. Measurement of the GRH

Quantitative predictions of the Gamma Ray Horizon have already been made, but so far no clear confirmation can be drawn from the observations of the present generation of Gamma Ray Telescopes. The fact that the next generation of ČT will have a considerably lower energy threshold than the present one should be of paramount importance in improving the present experimental situation for, at least, two reasons:

- Lower energy points will allow to disentangle much better the overall flux and spectral index from the cutoff position.
- Sources at higher redshift should be observable, giving the possibility of observing a plethora of new sources that will allow unfolding the emission spectrum and the gamma absorption.

To understand the capability of the next generation of ČT to measure the GRH, several assumptions at the level of the detector and sources are needed.

We assumed a source with a similar spectrum to the current TeV extragalactic sources (Mkn501 and Mkn421): $dF/dN f_o \cdot E^{-\alpha}$, where $f_o \simeq 9 \cdot 10^{-11}$ and $\alpha \simeq 3.0$. We did not consider any energy cutoff since it comes from the GRH predictions and any intrinsic cutoff is neglected. Then, it is extrapolated at lower energy and higher redshift. The characteristics of the MAGIC Telescope [1] have been used for the energy extrapolation using its energy threshold (30 GeV) and its energy resolution.

Under these conditions the next generation of ČT should be able to measure the GRH up to high redshifts with a reasonable significance (Figure 1),That will allow us to exclude some of the EBL models. Actually, the gamma rays that are going to be observed by these detectors interact mainly with the unexplored infrared EBL providing the possibility of performing some measurements of the EBL at these wavelength.

3. Cosmological Parameters

Some fundamental cosmological parameters such as the Hubble constant and the cosmological densities play also an important role in the calculation of the GRH since they provide the bulk of the z dependence of the predictions:

$$\frac{dl}{dz} = c \cdot \frac{1/(1+z)}{H_0[\Omega_M(1+z)^3 + \Omega_K(1+z)^2 + \Omega_\Lambda]^{1/2}} \qquad (2)$$

306

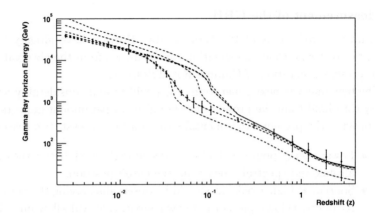

Figure 1. Points and error bars show the foreseen GRH values and its precision of a source like Mkn at some redshifts for a Cherenkov Telescope of 30 GeV threshold. Solid lines are the GRH predictions based on several of the current EBL models.

Over the last few years, the confidence in the experimental determinations of these cosmological parameters has increased dramatically. Table 1 shows best fit current values for these cosmological parameters[6].

Table 1.   Best current fit values for cosmological parameters with 2 σ confidence level.

Cosmological Parameter	Allowed range
H_0	68±6
Ω_Λ	0.65±0.15
Ω_M	0.35±0.1

Before discussing the impact of each one of these parameters in our predictions, we would like to see how the observables that will be measured (Optical Depths and GRH) depend on the redshift z. For that we have plotted the prediction for their z evolution in Figure 2. For comparison, the z variation of the Luminosity-Distance, used for the determination of the cosmological parameters using Supernova 1A and of the Geodesical-Distance are shown. One can see that each observable behaves differently with z. Hence, any measurement done with the GRH or the Optical Depth is complementary with the current ones.

The sensitivity of the measurement of the GRH energy as a function of the redshift z on each one of the parameters has been computed and is plotted in Figure 3. In that figure the sensitivity for each parameter p is

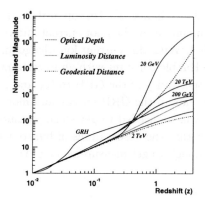

Figure 2. Redshift dependence of different observables. The predictions are normalized to their value at $z = 0.01$. The solid lines correspond to the Optical Depth prediction for gamma rays of different energies (20 GeV to 20 TeV) while the dashed line is the prediction for a flat νI_ν EBL spectrum. The GRH curve gives the z dependence of the inverse of the GRH energy.

actually defined as

$$S_p(z) \equiv p\, \frac{dE_{GRH}(z)/E_{GRH}(z)}{dp} \tag{3}$$

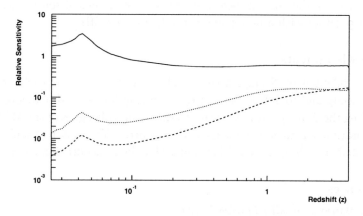

Figure 3. Sensitivity of the GRH energy to relative variations in H_0 (solid line), Ω_M (dashed line) and Ω_Λ (dotted line).

Then one can foresee the relative precision in the single-parameter determination of p from the sensitivity and the uncertainty in the estimation of the GRH energy (shown in Sec. 2). On top of that, the systematic uncertainty, which comes mainly from the poor current knowledge of the infrared EBL, should be added. With this current knowledge and one source Mkn-like, one will get the Hubble constant with an uncertainty around 30%. On the other hand, the sensitivity of the GRH to Ω_Λ and Ω_M is much smaller and both better precision in the GRH energy and more precise knowledge of the infrared EBL will be needed to get some measurements. Although, one should remember that the observations of tens of sources is very likely and then it will be enough to get measurements of Ω_Λ and Ω_M with errors around 50%.

4. Conclusions

Assuming sources Mkn-like will be found, the next generations of ČT should be able to measure the Gamma Ray Horizon up to high redshifts, with a reasonable significance, although this depends on the number of sources.

The determination of the GRH will provide the possibility of excluding some EBL models, since the actual EBL changes the GRH prediction. The dependence on the cosmological parameters gives a method to calculate them that is independent on the current ones. This method will allow to get the Hubble constant with precisions below 30% with very few sources. The cosmological densities Ω_Λ and Ω_M may be also measured with precision at the order of the current measurements but some tens of sources will be needed together with a better understanding of the EBL.

Acknowledgments

We are indebted to N.Magnussen for his cooperation in the early stages of this study and to T.Kneiske and K.Mannheim for many discussions and providing the EBL spectra. We want to thank our colleagues of the MAGIC collaboration for their comments and support. We want to thank also G.Goldhaber, P.Nugent and S.Perlmutter for their encouraging comments.

References

1. J.A. Barrio et al, MPI-PhE/98-5 (1998).
2. F.W. Stecker and O.C. De Jager, Space Sci.Rev. 75, 401-412 (1996).
3. R. Gispert, G. Lagache and J.L. Puget, Astron. Astrophys. 360, 1-9 (2000).
4. E. Dwek et al, The Astrophysical Journal 508, 106-122 (1998).
5. T. Kneiske, K. Mannheim and D. Hartmann, Astron. Astrophys. 386, 1 (2002).
6. L.M. Krauss, World Scientific, 1 (2001).

THE PRESENT STATUS OF THE MAGIC TELESCOPE

J. LOPEZ FOR THE MAGIC COLLABORATION

Institut de Física d'Altes Energies,
Universitat Autònoma de Barcelona,
08193, Bellaterra, Barcelona, SPAIN,
E-mail: jlopez@ifae.es

MAGIC is a new generation Imagining Air Cherenkov Telescope which is placed in the the observatory of *El Roque de los Muchachos* at the Canary island of *La Palma*. MAGIC is in the commissioning phase, becoming operative before the end of year 2002. An overview of the status of the Telescope will be reported, giving special importance to the status of every part of it.

1. Introduction

Imaging Air Cherenkov Telescopes (IACTs) have turned out to be the most efficient technique to detect Gamma Rays in the so-called Very High Energy range, i.e. above 300 *GeV*. In the last decades there has been a first generation of IACTs which have established without any doubt the detection of few VHE γ-rays sources. The MAGIC Telescope [1] is a new generation IACT whose main goal is to achieve a energy threshold of 30 *GeV* to be able to explore the virtually unobserved energy range from 10 *GeV* to 300 *GeV* and match the IACT and satellite observations.

2. The MAGIC Project Status

In order to reach lower energy threshold and also higher sensitivity than the previous generation IACTs, many new elements are incorporated in the MAGIC Telescope design. They are explained in the following sections.

2.1. *The Structure*

Two main characteristics define the 10 *tons* carbon fiber structure of MAGIC: lightness and strength. The first one allows a fast repositioning of the Telescope (less than 30" for 180°) which in the case of short-time events such as Gamma Ray Bursts detection [2] is very important. The second one guarantees a small deformation of the reflector disk (less than 3 mm of sagging) during the Telescope movement. This is necessary for a permanent good alignment between the mirrors and the camera.

The MAGIC structure is already constructed and assembled since the end of year 2001 (see the figure 1).

Figure 1.   The MAGIC Telescope structure at *El Roque de los Muchachos* site.

The space frame is steered using a drive system composed by two independent motors, one for azimuthal motion and one for zenithal movement. The installation of the Drive system has started during summer of this year 2002 and is planned to be finished by next autumn.

2.2. *The Reflector*

Due to a 17 m diameter reflector, MAGIC will collect 3 times more light than a typical first generation 10 m diameter IACT. The surface of this reflector has a parabolic shape, i.e. isochronous for light collection, which will improve the background rejection power of MAGIC. An Active Mirror Control system [3] is used to guarantee the maximum stability of the reflector shape. This means that every four mirrors are controlled using stepper motors in order to correct possible deformations in the space frame.

All the components of the reflector surface, 980 50x50 m^2 aluminum, diamond milled, quartz coated mirrors elements [4], are in mass production. Moreover, there are already installed 45 m^2, i.e. around 20% of the final reflector surface. It is planed to install around half of the reflector before the end of the year 2002.

2.3. The Camera

The camera of 1.5 m diameter is composed by 577 pixels with a total Field of View of 4° (see the figure 2). Compact photo-multipliers [5] with quantum efficiency of 20% in the 300-500 nm range and ultra-fast and very low noise pixel pre-amplifier [6] are used.

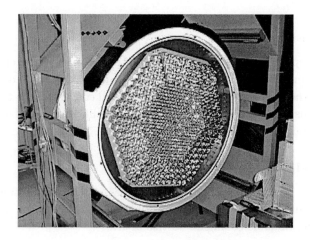

Figure 2.   Front view of the MAGIC camera.

From the point of view of the structure, the camera is required to be watertight and robust, and include a water-cooling system. On the other hand, due to its very high position, about 17 m above the reflector, it has to be as light as possible. Our final version takes into account all these requirements being only 500 Kg weight.

The mechanics of the camera were finished by the beginning of this year 2002. All the electronic components have been already produced and installed in the camera. The camera ready to be assembled in the Telescope frame will be sent to the site in October of year 2002.

In the future it is planned to construct a second MAGIC Telescope in order to do stereoscopic observations. The camera of this new MAGIC will be equipped with Hybrid Photo Detectors with a QE around 40% in a wider wavelength range, i.e. 330-660 nm.

2.4. *The Readout*

Cherenkov light is produced in very short flashes, i.e. few *ns*, in Gamma ray air showers. MAGIC introduces some improvements in order to record this time information to use it in the reduction of background. The signal produced in the camera is transmitted using analog optical fiber [7] which reduce the cable weight and also the noise related with the transmission. This signal is digitized using 300 *MHz* Flash-ADCs [8]. Finally this data is recorded by a DAQ system at 1 *Kevent/s*.

All the components of the readout chain are nowadays in mass production.

3. Conclusions

We have reported the present status of the MAGIC Telescope project. MAGIC is in the commissioning phase and it is planned to be operative by the end of the year 2002. Next year 2003 we will concentrate in the MAGIC physics goals which are wide and covers subjects like Active Galactic Nuclei, SNRs, Gamma Ray Burst, Pulsars, etc.

Acknowledgments

I would like to thank my colleagues of the MAGIC collaboration for their comments and support. And also I would like to acknowledge Manel Martinez for his valuable comments to this article.

References

1. J.A. Barrio, "The MAGIC Telescope", *Max-Planck-Institut Report MPI-PHE/98-5*.
2. D. Petry, "The MAGIC Telescope - Pospects for GRB research", *Astron.Astrophys.Suppl.Ser* 138-601 (1999).
3. A. Wacker et al, *Procs "Towards a Major Atmospheric Cerenkov V" - Kruger Park, south Africa* 358 (1997).
4. *Procs "Towards a Major Atmospheric Cerenkov V" - Kruger Park, south Africa* 358 (1997).
5. A. Ostankov et al, *Nucl. Instrum. Methods* A **442**, 117 (2000).
6. G. Blanchot et al, accepted by *1998 IEEE Nuclear Science Symposium and Medical Imagining Conference, Toronto* (1998).
7. R. Mirzoyan et al, *Procs "Towards a Major Atmospheric Cerenkov VI" - Snowbird, Utah*, 358 (1999).
8. J. Cortina et al, *Procs "Towards a Major Atmospheric Cerenkov VI" - Snowbird, Utah*, 363 (1999).

THE MAGIC TELESCOPE AS A DETECTOR OF GAMMA-RAY PULSARS ABOVE 10-30 GEV

M. LÓPEZ MOYA[1], V. FONSECA[1], O.C. DE JAGER[2]

[1] Fac. de C.C. Físicas, Univ. Complutense de Madrid, 28040 Madrid, Spain
[2] Unit for Space Physics, Potchefstroom Univ., Potchefstroom 2520, South Africa
E-mail: marcos@gae.ucm.es

To date, no confirmed pulsed γ-ray emission from EGRET pulsars has been seen from any ground-based γ-ray telescope, which suggests that their pulsed spectra terminate at energies below a few hundred GeV. Only with next generation of ground-based γ-ray telescopes we can expect to detect such pulsed emission. The 17 meter MAGIC Telescope is one of such instruments, with the capability to trigger on 10 GeV cosmic γ-rays and to overlap with EGRET in the 10-30 GeV range. This paper investigates the potential of MAGIC to detect GeV pulsed emission from EGRET pulsars, given the assumption of super exponential cutoffs as expected from polar cap emission and using the spectral information obtained by EGRET. We find that MAGIC should be able to detect pulsed emission from Crab and PSR B1951+32 within a few hours, if background rejection based on shower *size* and *distance* cuts is applied. In addition, some hard-spectrum unidentified EGRET sources may also be pulsars, and if their spectra extend to the 10 - 30 GeV range, searches for pulsations would also be possible.

1. Introduction

Observations with the CGRO/EGRET instrument[a] during its mission between 1991 and 2000 have led to the detection of seven γ-ray pulsars and a few more likely candidates[1]. Whereas some of these pulsars are amongst the brightest sources in the 1-30 GeV range, only their plerions appear to be visible at TeV energies. The non-detection by current Čerenkov telescopes of pulsed sub-TeV γ-rays from EGRET pulsars, proves that their pulsed spectra should terminate at energies below a few hundred GeV. This is not unexpected, since both *polar cap*[2] and *outer gap*[3] models predict that the spectra of γ-ray pulsars should cut off at energies between a few GeV and a few tens of GeV. In the polar cap model, electrons are accelerated above

---

[a]EGRET is the high-energy γ-ray telescope on board *Compton Gamma-Ray Observatory*.

the polar cap radiating γ-rays via synchro-curvature radiation and inverse Compton scattering on thermal photons emitted by the hot surface of the neutron star. And since these γ-rays are created in superstrong magnetic fields, magnetic pair production is unavoidable. This produces electron-positron pairs which in turn radiate more γ-rays, and γ-ray/$e^{+/-}$ cascade develops. Only those secondary photons which survive pair creation (a few GeV for typical pulsars) escape to infinity as an observed pulsed emission. A natural consequence of the polar cap process is a superexponential cutoff of the spectrum above a characteristic energy E_o, as discussed by Nel & de Jager[4]. In the outer gap model γ-ray production is expected to occur near the pulsar light cylinder. In this case the cutoff is determined by photon-photon pair production, which has a weaker energy dependence compared to magnetic pair production, and therefore a larger E_o may be observable. Based on these considerations, a generic model (polar cap and/or outer gap) for the tails of pulsed differential spectra is given by de Jager[6] as:

$$dN_\gamma/dE = K_1 E^{-\Gamma_1} exp(-(E/E_2)^b) + K_2 E^{-\Gamma_2} exp(-(E/E_2)^c) \quad (1)$$

where the second component would be absent in the case of pure polar cap γ-ray emitters.

Next generation of ground-based γ-ray telescopes, and specially MAGIC[5] with its low energy threshold of 10-30 GeV, should be able to overcome the superexponential cutoffs expected near 10 GeV and detect pulsed γ-rays. This would allow to measure the spectral shape of the pulsed emission in the relevant energy range (above 10 GeV), and therefore to discriminate between polar cap and outer gap models.

In this paper we investigate the capability of MAGIC for detecting γ-ray pulsars above 10 GeV, by calculating the pulsed rates and minimal detection times required for MAGIC to detect such sources.

2. Detection Capability of MAGIC for EGRET Pulsars

2.1. MAGIC Detection Rates for Pulsed Emission

To obtain conservative estimates for the detection rates, we have to employ the most conservative model for the pulsar spectra above 1 GeV. Following the procedure described by de Jager et al.[6], we will assume that the polar cap mechanism is the only responsible for the pulsed γ-ray emission. As discussed above, outer gap model would lead to larger values of the energy cutoff, and thus to most optimistic detection rates. We therefore model the pulsar spectra above 1 GeV as a power law times an exponential cutoff with cutoff energy E_0:

$$\frac{dN_\gamma}{dE} = K\,(E/E_n)^{-\Gamma}\exp\left(-(E/E_o)^b\right). \qquad (2)$$

The constant K represents the monochromatic flux at the normalising energy $E_n \ll E_o$. We will normalise spectra at E_n near 1 GeV. The strength of the cutoff is determined by the index b.

Nel & de Jager[4] were able to constrain some of the parameters of equation 2 by fitting the total pulsed spectra of the six brightest EGRET γ-ray pulsars. But only for the case of Vela and Geminga the cutoffs are well defined by the EGRET data. For Crab and PSR B1055-52 an evidence of a turnover is seen above 10 GeV, but it is difficult to obtain a reliable measure of E_0. In the case of PSR B1951+32 and PSR B1706-44 no evidence of a turnover was seen up to 30 GeV in the EGRET data, and a minimum value of $E_0 = 40$ GeV (consistent with EGRET) was selected. For those unconstrained pulsars where E_0 is not well defined we have selected $b = 2$ (a typical value in a polar cap scenario) to obtain conservative detection rates. Table 1 shows the spectral parameters for the EGRET pulsars for $E > 1$ GeV, which will be used below for the calculation of detection rates.

The expected rate of triggers R_p for pulsed γ-rays is calculated by integrating the product of the energy dependent collection area $A(E)$, with the differential pulsed spectrum, which includes the cutoff:

$$R_p = \int A(E)\frac{dN_\gamma}{dE}dE \qquad (3)$$

The background rate was calculated assuming incident cosmic ray showers with the known cosmic ray spectrum, obtaining a value of $R_B \sim 200$ Hz.

2.2. Detection Times for Pulsars

Next we calculate the minimal observation times required for MAGIC to detect the EGRET pulsars at a given significance level. As the expected values for the cutoff energies E_0 are near the detection threshold of MAGIC (10-30 GeV), we assume that we will have no imaging capability. In this situation we have to rely on timing analysis, using a periodicity test.

To calculate detection sensitivities for periodicity searches we calculate the basic scaling parameter $x = p\sqrt{N}$ which holds for any test for uniformity on a circle[7], where $p = R_p/(R_B + R_p)$ is the pulsed fraction and $N = (R_p + R_B)T$ is the total number of events, with T the observation time (R_p and R_B are the pulsed and background rates respectively). As it was shown by Thompson[1] that the pulse profiles above 5 GeV consist mostly of one strong narrow peak, as opposed to the two peaks at lower energies, we will assume that only a single sharp peak with a duty cycle $\delta = 5\%$ survives above 10 GeV. In this case the Z_m^2-test with number of harmonics $m = 1/(2\delta) = 10$ should be optimal[7], with expected value $< Z_m^2 >= x^2\Phi + 20$, where $\Phi = 5.8$ for a single peak (Gaussian) with a 5% FWHM. A DC excess of $x = 3\sigma$ above the sky background in an spatial analysis, should give in timing analysis a detection with $< Z_{10}^2 >= 72$, which gives a chance probability of 8×10^{-8} ($\sim 9\sigma$) if the period is known.

If the period is unknown, we have to multiply the probabilities by the number of trials. For instance, for a 6 hour observation and searching for periods as short as 33 ms, for the same DC excess of 3σ we would have a chance probability of 0.5 after multiplying with all the trials $M = \eta T \Delta f = 6.5 \times 10^6$ (with $\eta \sim 10$ the factor of oversampling[7]). This should bury the true frequency amongst one of many candidate frequencies. A detection within a single night restricts the number of independent frequencies to be searched, which enables the identification of a single frequency, or at least, a number of candidate frequencies which can be confirmed within a few days of follow up observations.

Table 1 shows the expected pulsed rates and required observation times for the EGRET pulsars, calculated assuming a DC excess of $x = 3\sigma$ and solving for T. The time T_{200} correspond to the background rate of 200 Hz, whereas the time T_{25} is calculated assuming some degree of background rejection, based on *size* and *distance* cut on the events. Specifically, we assume a final background rate of $R_B \simeq 25$ Hz (or equivalently, an increase in the signal-to-noise ratio by a factor $Q = 3$), but still detailed simulations are required to determine the best rejection factor against background.

From Table 1 it is clear that we can detect Crab and PSR B1951+32 within a single night, if we can reach a background rate of less than 25 Hz after making suitable *size* and *distance* cuts. These two pulsars also transit

Table 1. Assumed γ-ray spectral parameters above 1 GeV and corresponding MAGIC rates and observation times for detection.

Object	k ($\times 10^{-8}$) ($/cm^2/s/GeV$)	Γ	E_o (GeV)	b	R_p (hr^{-1})	T_{200} (hour)	T_{25} (hour)
Crab	24.0	2.08	30	2	455	31	4
Vela	138	1.62	8.0	1.7	14	326	36
Geminga	73.0	1.42	5.0	2.2	1	10^7	10^5
PSR B1951+32	3.80	1.74	40	2	440	33	4
PSR B1055-52	4.00	1.80	20	2	62	10^3	180
PSR B1706-44	20.5	2.10	40	2	670	14	2

close to La Palma, so that the minimum threshold energy can be realised at transit.

3. MAGIC Sensitivity for Unidentified EGRET Sources

In addition to the seven well established γ-ray pulsars, EGRET discovered several hard-spectrum unidentified sources which are thought to be γ-ray pulsars, for which the EGRET statistics are too small to resolve the periodicity. The recent discovery of three young energetic radio pulsars associated with unidentified EGRET sources[8,9] seems to confirm this general belief. If the spectra of the unidentified sources extend to the 10 - 30 GeV range, searches for pulsations with MAGIC would be possible. But, since in general their periods are unknown in advance, the constraint for this capability is the detection within a single night, with confirmation runs the following few nights (as discussed above).

From the GeV source catalogue[10], we find that the fluxes of the galactic unidentified EGRET sources range from $F(> 1 \text{ GeV}) = 1$ to $25 \cdot 10^{-8} \text{cm}^{-2} \text{s}^{-1}$. Assuming that the photons seen by EGRET from these sources are pulsed, we calculate the MAGIC sensitivity for pulsed detection. For that, as we don't know what the cutoff energy is, we calculate the required observation times as a function of the cutoff energy and their fluxes. Figure 1 gives the MAGIC sensitivity for a wide range of possible pulsar photon spectral indices between 1 and 2, requiring a detection within T=3 to 6 hours for a minimum "DC significance" of $x = 3\sigma$ (assuming a final background of 25 Hz). We see that even weak EGRET sources may be detectable within a single night, provided that E_o exceeds the levels prescribe by Fig. 1 (above ~ 20 GeV for moderate spectral indices of $\Gamma = 1.5$).

4. Conclusions

The construction of a 17 meter class telescope such as MAGIC allows the detection of showers induced by γ-rays as low as 10 GeV, covering the so far unexplored gap between 30 and 250 GeV, and overlapping with EGRET in the 10-30 GeV range.

Assuming a polar cap scenario to obtain conservative spectral expectations above 10 GeV, we show that MAGIC should be able to detect Crab and PSR 1951+32 from its location at La Palma, within a few hours. The condition for such a detection within a single night is the reduction of the background rate to less than 25 Hz, which could be achieved after making suitable shower *size* and *distance* cuts.

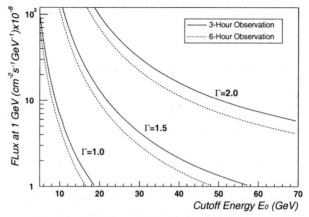

Figure 1. MAGIC sensitivity for the detection of unknown pulsars within one night in the parameter space K vs. E_o, using a timing analysis approach and assuming a DC excess of $x = 3\sigma$. Three different photon spectral indices of 1, 1.5 and 2 are used.

Finally, some unidentified EGRET sources may also be detectable by MAGIC. The detection of pulsations from these sources would allow to determine whether some of the unidentified EGRET sources are indeed radio quiet pulsars, either due to unfavorable radio beaming or very high radio dispersion in a molecular cloud.

Acknowledgments

I would like to thank to Prof. Maurice Shapiro for the invitation to attend to the *International School of Cosmic Ray Astrophysics*.

References

1. Thompson, D.J. 2001, in Proc. International Symposium, Heidelberg 26-30 June 2000, F.A Aharonian, H.J. Völk American Institute of Physics 558, 103.
2. Daugherty, J.K. & Harding, A.K. (1996), *ApJ*, **458**, 278.
3. Hirotani, K. (2001), *ApJ*, **549**, 495.
4. Nel, H.I. & de Jager, O.C. (1995), *Astr. Space Science*, **230**, 299.
5. Barrio, J.A. et al. (1998), MPI-PhE/98-5.
6. de Jager, O.C., Konopelko, A., Raubenheimer B.C., & Visser, B. 2001, in Proc. International Symposium, Heidelberg 26-30 June 2000, Germany, Eds. F.A. Aharonian, H.J. Völk American Institute of Physics 558, 613.
7. de Jager, O.C., Swanepoel, J.W.H, Raubenheimer B., (1989), *A&A*, **170**, 187.
8. D'Amico, N. et al. (2001), *ApJ*, **552**, L45.
9. Halpern, J.P. et al. (2001), *ApJ*, **552**, L125.
10. Lamb, R.C. & Macomb, D.J., (1997), *ApJ*, **488**, 872

DEVELOPMENT OF THE IMAGING ATMOSPHERIC CHERENKOV TECHNIQUE AT THE WHIPPLE OBSERVATORY

PATRICK J. BOYLE*

Enrico Fermi Institute, University of Chicago
933 E 56th Street, Chicago, IL 60637
E-mail: jojo@donegal.uchicago.edu
* *for the VERITAS Collaboration*

The energy range between 100 GeV and 100 TeV is the domain of very high energy (VHE) γ-ray astronomy. This band of the electromagnetic spectrum is essential for the multifrequency study of extreme astrophysical sources. Determination of the spectra of detected gamma-rays is necessary for developing models for acceleration, emission, absorption and propagation of VHE particles at their sources and in space. The Whipple Collaboration has pioneered the imaging Atmospheric Cherenkov technique over the last 30 years and in the process established the field of VHE γ-ray astronomy. This paper offers a brief chronology of the major milestones in this development and offers a glimpse at some important results obtained in the process. Next generation telescopes based on the Whipple 10m are under construction and will increase dramatically the knowledge available at this extreme end of the cosmic electromagnetic spectrum.

1. Introduction

The detection of VHE γ-rays uses ground based sampling of the Cherenkov light generated by extensive air-showers (EAS). EAS comprise the secondary particles generated as primary γ-ray and cosmic rays interact with the Earth's atmosphere. The Cherenkov signature of EAS carries information about the primary particle's direction, energy and nature. First proposed in 1977[1], the imaging atmospheric Cherenkov technique distinguishes γ-ray induced showers from the more numerous background induced by cosmic rays by utilizing differences in the distribution of the Cherenkov light. Due to its excellent imaging properties, the Whipple Observatory 10 meter reflector has been instrumental in pioneering the IACT. The Whipple Observatory, located on Mount Hopkins in southern Arizona (elevation 2.3 km, latitude 31°.5), is ideally suited for VHE γ-ray astronomy as it is at a dark location with a high percentage of clear skies. The reflector was originally intended to search for the origin of cosmic radiation by observing γ-rays produced through hadron acceleration, however, every γ-ray source that has been discovered can be explained as a source of cosmic electron acceleration and interaction[2]. The VHE γ-rays have, instead, been indicative of a whole new set of astrophysical processes whose study has been interesting and rewarding. Thus, although the origin of the highest energy cosmic rays is still unknown, VHE γ-ray astronomy has been established as a field in its own right.

2. Whipple Observatory γ-ray Telescope

2.1. 10m Optical Reflector

The Whipple Observatory 10m optical reflector has a unique optical structure based on a design by Davies-Cotton[3,4]. Its 248 hexagonal mirror facets each have a spherical figure with a 14.6 meter radius of curvature and are individually mounted on a 7.3 m radius spherical support structure. Each facet then functions as an off-axis spherical mirror focusing light parallel to the optic axis of the reflector to the center of the 7.3 m sphere. The spherical design gives superior off-axis properties, however, it introduces a spread of about 6ns for the arrival times for photons from different mirrors. The full width half maximum of the on-axis point spread function of the reflector is ~ 0.15°.

2.2. First Generation : 1968 - 1976

The first camera mounted on the Whipple 10m reflector consisted of a single 12.5 cm phototube which gave a field of view (fov) of 1.0° and a collection area of 75m². Although the camera had no imaging properties, 90% of the reflected light from a distant point source on the axis of the reflector fell within a circle of 5 cm diameter in the focal plane.

In general, the stability of atmospheric Cherenkov detectors is increased if two or more light receivers are operated in coincidence. Since only one reflector was available at the time, a two channel system was achieved by refocusing the elements of the reflector so that there were two focal points, 30 cm (2.4° apart), in a plane perpendicular to the reflector axis. Using this experimental setup, upper limits were presented for twenty seven celestial sources ranging from Supernova remnants and Magnetic Variables to Galaxies and Quasars[5]. After three years of observations Fazio[6] reported a γ-ray detection of the Crab Nebula above 250 GeV at 3 standard deviations above background.

2.3. First Imaging Camera : 1982 - 1987

In 1982, using a grant of $10,000 from the Irish National Board of Science and Technology, the first imaging camera was built and implemented on the 10m reflector. The camera consisted of 37 PMTs arranged in a hexagonal pattern. The diameter of each PMT was 5 cm and the spacing between pixel centers was 6.25 cm. The full aperture of the camera was 3.5°.

In 1985, Micheal Hillas presented a paper at the ICRC in La Jolla[8] detailing the ability to distinguish between background hadronic showers and VHE γ-ray showers on the basis of the shape of their image. This allowed for the rejection of hadronic showers with an efficiency of over 99.7%. To date this is one of the most important milestones in the development of γ-ray astronomy. Although the imaging technique was originally proposed in 1977, the technique was not demonstrated until 10 years later when, using the method presented by Hillas, the Crab Nebula was detected. The Crab detection was at 9 standard deviations above background after 90 hours of data taking at an energy threshold of 700 GeV[7].

2.4. High-Resolution Camera : 1988 - 1996

The next milestone was the development of the High-Resolution Camera (HRC) which was installed on the 10m reflector in April 1988. The HRC consisted of 91 pixels of 0.25° (2.5 cm diameter phototubes) surrounded by an outer ring of 18 pixels of 0.5° (5cm diameter tubes) giving a full field of view of 3.75° diameter. The smaller pixel size resulted in a lower energy threshold for γ-ray detection than that reported for the original imaging camera. The overall improvement in signal-to-noise compared with the original camera was of the order of 3-4[9]. With this improvement the HRC detected the Crab at 20 standard deviations in 65 hours with an energy threshold of 400 GeV[10].

The BL Lac object Markarian 421 ($z = 0.031$) was detected as the first extragalactic source in 1992[11]. This discovery was followed by a number of papers proposing the use of VHE γ-ray spectra from these extragalactic sources to constrain the extragalactic infrared radiation field[12]. VHE γ-ray emission was discovered from the second closest BL Lac Markarian 501 ($z = 0.034$) in 1995[13]. Mrk 501 was not detected as a significant source of γ-rays by EGRET, so this was the first object to be discovered as a γ-ray source from the ground. A second BL Lac not detected by EGRET, 1ES2344+514 ($z = 0.044$), was detected by the Whipple Observatory in 1995[14]. Most of the emission came from a single night in which a flux of approximately half the Crab was detected. The hypothesis that the γ-ray emission of BL Lacs could flare on sub-day timescales was borne out in spectacular fashion with observations of two flares from Mrk 421 in 1996[15]. Variability on a 15 minute time-scale observed in the second flare implies a compact emission region of dimension $R \leq 10^{-4}$ parsec which is only an order of magnitude larger than the event horizon for a 10^8 solar mass black hole. This flare also allowed the most stringent limit to date to be inferred for the energy scale of Quantum Gravity ($> 4 \times 10^{16}$ GeV)[25]. Hence, ground based γ-ray astronomy was established as a legitimate channel of astronomical investigations in its own right, and not just an adjunct of high-energy observations from space.

2.5. High-Resolution Camera II : 1996 - 1999

In the summer of 1996 the first stage of the HRC upgrade was completed. The camera was expanded from 109 to 151 2.5 cm phototubes giving a fov of 3.3°. As in the original HRC, only the inner 91 PMTs were used to trigger the camera. In the summer of 1997 the camera was expanded further to 331 PMTs with a fov of 4.8°. The introduction of a pattern selection trigger[21] allowed the entire fov to act as an active trigger, this allowed for the routine detection of the Crab at 5σ above background in one hour. With the expanded fov, the emphasis for this period shifted from point sources to extended objects such as supernova remnants[17] and the Galactic plane[18] and to objects where the position is uncertain such as EGRET unidentified objects[19] and gamma-ray bursts[20].

Upper limits were obtained from observations on six shell-type SNRs (IC 443, γ Cygni, W44, W51, W63, and Tycho) selected as strong candidate γ-ray emitters based on their radio properties, distance, small angular size and possible association with a molecular cloud[23]. The upper limits in some (e.g IC 443) are significantly below the predicted fluxes from the model of Drury et al[24].

For an understanding of the mechanisms at work in AGN jets, multi-wavelength campaigns are needed. Using the HRC II a number of simultaneous X-ray and VHE γ-ray observational campaigns were carried out; these are reported elsewhere[16]. Multiwavelength observations of Mrk 501 during its high emission in 1997 revealed for the first time, clear correlations between its VHE γ-ray and X-ray emission. The results from this campaign show that for Mrk 501, like Mrk 421, the VHE γ-rays and the soft X-rays vary together and the variability in the synchrotron emission increases with increasing energy. These observations also showed that hard X-ray observations hold the promise of identifying a new class of extreme blazars which would have detectable and variable VHE γ-ray emission.

2.6. Granite III : 1999 - present

During the summer and fall of 1999, a 490 HRC was installed on the 10m reflector. The camera consisted of 379 1.3 cm PMTs constituting a 2.6° inner camera. This is surrounded by 111 2.5 cm PMTs which fills out the fov to 4.0° diameter. The system is triggered on the innermost 331 pixels. The peak energy response to a Crab-like spectrum is 390±80 GeV, however, it still has a good response down to ≈ 200 GeV[22].

In 2000, the BL Lac H1426+428 ($z = 0.129$) was singled out by the *BeppoSAX* collaboration as a probable VHE γ-ray emitter because of the high frequency of its synchrotron peak. Whipple observations of H1426 in 1999 revealed marginal evidence for a flaring signal[27]. Subsequent observations at Whipple during 2000 and 2001 led to the discovery of a weak, but statistically significant TeV flux during both seasons[28]. These observations are an important addition to the catalog of VHE emitting objects as H1246 is the most distant BL Lac detected thus far. Stronger detections of such sources, which allow an accurate measure of the VHE energy spectrum, may place significant limits on the density of the extragalactic infrared background. Also that the source was predicted to be a VHE emitter based on its X-ray spectrum signifies the maturity of the observational techniques and the theoretical understanding of BL Lac objects. It reinforces the symbiosis between observations at X-ray wavelengths and those at VHE energies, particularly those with sensitivity below 1 TeV. Also the existence of a population of sources whose most prominent emission is at energies of 10-100 keV and 300-1000 GeV points to a fruitful overlap between the next generation of ground based atmospheric telescopes and the future hard X-ray experiment, EXIST.

3. Future of VHE γ-ray Astronomy

The Whipple Observatory γ-ray telescope has firmly established ground based VHE γ-ray astronomy as a highly regarded discipline. The imaging technique has now been utilized by seven other groups throughout the world. Although less than 1% of the sky has been surveyed at 300 GeV, fifteen sources, have now been reported by ground-based groups using the imaging techniques : three pulsar-powered nebulae, eight BL Lacs, three shell-type supernova remnants and one X-ray binary system. These measurements, while they have not provided conclusive evidence for the origin of cosmic rays, have advanced our understanding of the nature of AGN jets, the density of the extragalactic infrared background, quantum gravity and magnetic fields within supernova remnants.

324

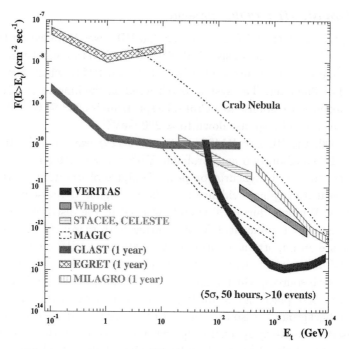

Figure 1. Comparison of point source sensitivity of next generation gamma-ray tele-scopes : VERITAS, MAGIC, CELESTE/STACEE, GLAST, EGRET and MILAGRO (see Weekes, 2002 and references therein)

However, to exploit fully the potential of ground-based γ-ray astron-omy detection techniques must be improved. The next few years will see the completion of several new imaging Atmospheric Cherenkov telescopes, which will significantly improve the scientific potential of the discipline. Three new facilities, VERITAS[29], in the northern hemisphere, and HESS[30] and SuperCANGAROO[31], both in the southern hemisphere, will consist of arrays of telescopes based on the Whipple γ-ray telescope. All three projects are similar to one another in concept and will cover the energy range 50 GeV - 100 TeV. Another northern hemisphere project is MAGIC[32] which consists of a single 17m imaging telescope and will utilize new de-tector technologies to cover a lower energy range of 10 - 1000 GeV. The next generation γ-ray space telescope, GLAST is scheduled for launch in 2006[33]. GLAST will cover the energy range 20 MeV - 300 GeV comple-menting ground based γ-ray telescopes and closing one of the last windows of the electromagnetic spectrum.

Acknowledgments

This work is supported in part by NSF grant PHY-0079793 and funds from the University of Chicago.

References

1. T.C. Weekes and K.E. Turver, *Proc 12th ESLAB, Frascati*, **1** 279 (1977).
2. T.C. Weekes,*Hegr. Proc.*, **1** 15W (2001).
3. J.M. Davies and E.S. Cotton, *J. Solar Energy*, **1** 16 (1957).
4. G.H. Reike, *SAO Report*, 301 (1969).
5. T.C. Weekes et al,*Ap. J.*, **174** 165 (1972).
6. G.C. Fazio et al, *Ap. J.*, **175** L117 (1972).
7. T.C. Weekes et al, *Ap. J.*, **342** 379 (1989).
8. A.M. Hillas, *Proc 19th ICRC, La Jolla*, **3** 445 (1985).
9. M. F. Cawley et al, *Exper, Astr.*,, **1** 173 (1990).
10. G. Vacanti et al, *Ap. J.*, **377** 467 (1989).
11. M. Punch et al, *Nature*, **358** 477 (1992).
12. F.W. Stecker and O.C. DeJager, *Ap. J.*, **390** L49 (1992).
13. J. Quinn et al, *Ap. J.*, **456** L83 (1996).
14. M. Catanese et al, *Ap. J.*, **505** 616 (1996).
15. J. Gaidos et al, *Nature*, **383** 319 (1996).
16. M. Catanese and T.C. Weekes, *PASP*, **111** 1193 (1999).
17. R.W. Lessard et al, *Proc 25th ICRC, Durban*, **3** 233 (1997).
18. S. LeBohec et al, *Ap. J.*, **539** 209L (2001).
19. S.J. Fegan et al, *AIP Proc.*, **587** 296 (2001).
20. P.J. Boyle et al, *Proc 25th ICRC, Durban*, **3** 61 (1997).
21. S.M. Bradbury et al, *Proc 26th ICRC, Salt Lake*, **5** 263 (1999).
22. J.P. Finley et al, *Proc 27th ICRC, Hamburg*, **OG2.5** 2827 (2001).
23. J.H. Buckley et al, *A & A*, **329** 639 (1998).
24. L. O'C. Drury et al, *A & A*, **287** 959 (1994).
25. S.D. Biller et al, *Phys. Rev. Lett*, **83** 2108 (1999).
26. L. Costamente et al, *A & A*, **371** 512 (2001).
27. D. Horan et al, *AAS/HEAD*, **32** 503 (1999).
28. D. Horan et al, *Ap. J.*, **571** 753 (2002).
29. T.C. Weekes et al, *A. Ph.*, **17** 221 (2002).
30. W. Hoffman et al, *Towards a Major Atmos. Cherenkov Det.*, **1** 500 (2000).
31. M. Mori et al, *Towards a Major Atmos. Cherenkov Det.*, **1** 485 (2000).
32. E. Lorenz et al, *Towards a Major Atmos. Cherenkov Det.*, **1** 510 (2000).
33. N. Gehrels and P. Michelson, *A. Ph.*, **11** 227 (1999).

GAMMA RAYS AND NEUTRINOS FROM BLAZARS

C. D. DERMER*

Code 7653, Naval Research Laboratory
4555 Overlook Ave. SW
Washington, DC 20375-5352
E-mail: dermer@gamma.nrl.navy.mil

Blazar active galactic nuclei are thought to be a subset of radio galaxies where the jets are pointed towards us. A brief review of the properties of blazars is presented, and the central features of leptonic and hadronic models are summarized. Differences between BL Lac objects and flat spectrum radio quasars (FSRQs) are considered with regard to dominant radiation processes and an evolutionary scenario for blazar evolution. Features of a proton blazar model are described, including a calculation of hadronic cascade radiation from photomeson processes. The AGN broad line region (BLR) provides an intense external radiation field in FSRQs that is important for producing detectable fluxes of neutrinos. Neutral beams formed through hadronic processes in the inner jets of blazars could power the extended jets and account for morphological differences between BL Lac objects and FSRQs.

1. Introduction

Large-scale jets of radio galaxies are thought to be powered by mass accretion onto supermassive black holes which drive collimated relativistic outflows in the poleward directions.[1,2] In advection-dominated accretion disk models,[3] transitions between spectral states follow a progression due to changes in the Eddington ratio $\ell_{\mathrm{Edd}} = \eta \dot{m} c^2 / L_{\mathrm{Eddd}}$, where η is the efficiency to transform the energy of matter accreting at the rate $\dot{m}$ into radiant energy, and L_{Edd} is the Eddington luminosity. States with $\ell_{\mathrm{Edd}} \gtrsim 1$ represent the soft high state, with the optically-thick disk pushing close to the innermost stable orbit. The convectively unstable, low Eddington luminosity ($\ell_{\mathrm{Edd}} \lesssim 0.1$) regime is thought to give rise to black-hole sources with radio-emitting jets. Here we give a brief review of blazar jet physics, with emphasis on the role that black hole growth and evolution play, the prospects for high-energy neutrino detection of blazars, and the conjecture

---

*Work supported by the Office of Naval Research and NASA DPR # S-15634-Y.

that neutral beams of high-energy neutrons and gamma-rays transport energy to form the extended jets of radio galaxies.

2. Active Galactic Nuclei: A Brief Review

The observational characteristics of an AGN are primary, and the pictures or models that we carry around in our heads to understand these sources are simply mental aids that, we hope, are in at least some crude correspondence to reality.

2.1. Radio-Quiet AGNs

If the flux density F (e.g., in units of ergs cm^{-2} s^{-1} Hz^{-1}) of a sample of AGNs is measured at radio and optical frequencies, then there is a tendency for the ratio $R = F(5 \text{ GHz})/F(\text{B band})$ to cluster at values $R \lesssim 10$ and $R \gtrsim 100$ in optically selected samples of quasars. For example, analysis of the 114 quasars in the Palomar Bright Quasar Survey[4] shows that 85-90% of these objects are radio-quiet with $R < 10$. By contrast, a radio-selected sample at some limiting flux magnitude, for example, the 1 Jansky NRAO-MPI catalog of extragalactic radio sources,[6] will preferentially identify radio galaxies with $R \gg 100$, and typically in the range 10^3-10^5. Although it is conventionally stated that $\sim 10\%$ of active galaxies are radio-loud with $R > 10$, it is important to note that such statements depend on the selection and search criteria and, indeed, the criteria that define AGNs.

Radio-quiet AGNs are typically classified according to their optical/UV lines. Seyfert galaxies, which show a bluish compact nucleus, are divided into Seyfert 1 and Seyfert 2 galaxies.[5] In the former case, broad ($\sim 2{,}000$-20,000 km s^{-1}) permitted lines and narrow (~ 500 km s^{-1}) permitted and forbidden lines are seen, whereas only narrow permitted and forbidden lines are measured from Seyfert 2 galaxies. The conventional explanation is that the BLRs of Sy 2s are obscured from our line-of-sight by a gaseous torus.[7] Quasi-stellar objects (QSOs) are scaled-up Seyfert galaxies, with the dividing line conventionally set at 10^{45} ergs s^{-1}, though Seyfert AGNs and QSOs appear to represent a continuous sequence.[8] Radio-quiet Seyfert galaxies and QSOs show intense hard X-ray emission that is truncated above several hundred keV, and there is no evidence for $\gg$ MeV emission from these sources. The hard X-rays are generally thought to originate from a hot, optically thin accretion disk.

2.2. Radio Galaxies

Radio emission from a radio galaxy originates primarily from directed plasma outflows produced by a central active black-hole core. Fanaroff and Riley[9] identified a striking relationship between the morphology and radio power of radio galaxies. They correlated the separation between the points of peak intensity of the extended radio galaxy jet emission with radio power. Radio galaxies with a separation smaller than half the largest size of the source, as represented by the edge-darkened twin jet sources, are primarily low luminosity sources, with 178 MHz radio power $\ll 2 \times 10^{25}$ Watts/Hz ($\cong 4 \times 10^{40}$ ergs s^{-1}). These twin jet radio sources have low radio powers. Radio galaxies with radio jets larger than half the largest size of the source, which includes the classical radio doubles such as Cygnus A and galaxies with edge-brightened hot spots and radio lobes, have large radio powers. This morphology/radio-power correlation is very striking, and represents an important clue to the nature of radio galaxies. The twin jet sources are referred to as FR1 galaxies, and the high radio-power, lobe-dominated sources are referred to as FR2s. Optical emission lines in FR2s are brighter by an order of magnitude than in FR1s for the same galaxy-host brightness, suggesting greater dust and gas near the central black hole. The host galaxies of radio galaxies are generally found to be elliptical or disturbed (tidally interacting or merging) systems.

2.3. Blazars

Blazars comprise sources that exhibit some or all or the following properties: extreme and rapid optical variability; flat ($F_\nu \propto \nu^{-\alpha}$ with $\alpha \leq 0.5$) radio spectra; strong ($> 3\%$) linear polarization; superluminal motion; and broadband nonthermal continuum radiation extending from the radio though 100 MeV-GeV gamma ray energies.[10] Blazar emission may extend to higher photon energies, but pair production attenuation of high-energy gamma rays by the diffuse intergalactic infrared radiation field hides TeV blazar emission at redshifts $z \gtrsim 0.2$.[11,12] TeV-PeV neutrinos suffer no such attenuation, and will give us an important new channel of information about black-hole jet sources.

The blazar class divides into two subclasses. The first comprises the nearly lineless **BL Lac** objects which display a featureless continua, and are technically defined as sources with equivalent optical-emission line widths < 5 Å. The second subclass comprises the **FSRQs**, which have strong emission lines as defined by equivalent widths > 5 Å. Dilute gas surrounding the central active nuclei in BL Lac objects, and dense BLR clouds near the central powerhouse in FSRQs, could qualitatively account for the differences in the strengths of the atomic lines of these two subclasses.

A number of lines of evidence indicate that blazar emission is produced by collimated relativistic outflows of radiating plasma. The first evidence is provided by the so-called Compton catastrophe. By measuring the radio flux and the synchrotron self-absorption frequency, the magnetic field can be inferred from Compton-synchrotron theory. The size scale of the region can be determined by direct observation or through inferences from the variability time scale. These results yield the ratio of the synchrotron to magnetic field energy density, which implies large fluxes of unobserved synchrotron self-Compton X-rays. Bulk relativistic motion provides an explanation of these results.

Observations of apparent superluminal motion provides additional support for this inference. Superluminal motion has now been detected from scores of extragalactic radio sources, with apparent superluminal speeds typically between 1 and 10, though with a few reaching ~ 20.[13] These observations are explained in terms of bulk relativistic radiating regions moving at Lorentz factors $\Gamma \sim 10$-30, and directed at an angle $\theta \sim 1/\Gamma$ to our line of sight. Further arguments for relativistic motion in blazars include the Elliot-Shapiro relation,[14] where the variability time scale is used to derive the maximum size of the emitting region, which should not to be smaller than the Schwarzschild radius of a central black hole. The black hole mass implies the Eddington luminosity. Super-Eddington luminosities can be avoided by arguing that the radiating region is moving at relativistic speeds. An additional arguments[16,15] for relativistic bulk motion in blazars include the requirement that the radiating region is optically thin to $\gamma\gamma$ pair production attenuation.

Various lines of evidence based upon the statistics of radio galaxies and blazars indicate that FR2 radio galaxies are the parent population of FSRQs,[2] so that the major difference between these two classes of sources is the orientation of the jets to the observer. Orientation effects are also thought to unify FR I radio galaxies with BL Lac objects. The BL-Lac/FR-I subclasses have lower average luminosities than the radio-quasar/FR-II subclasses. There appears to be an almost continuous sequence of properties from FSRQs through LBLs to HBLs. This trend is characterized by decreasing bolometric luminosities, a shift of the peak frequencies of their broadband spectral components towards higher values, and a decreasing fraction of power in γ rays compared with lower-frequency radiation.[17,18]

3. Gamma Rays from Blazars

A major discovery of the *Compton Gamma Ray Observatory* is that blazars are prominent sources in the gamma-ray sky. The Third EGRET (Energetic Gamma Ray Experiment Telescope) catalog[19] lists 66 high ($> 5\sigma$) confidence and 27 lower ($4\text{-}5\sigma$) confidence detections of blazars. Nonsimultaneous and simultaneous multiwavelength observations reveal two broadband features in the spectral energy distributions (SEDs) of blazars. The lower energy feature is interpreted as synchrotron radiation from nonthermal electrons. The νF_ν peak photon energies E_{pk} of the synchrotron emission radiated by FSRQs are in the far IR/optical range. BL Lacs have synchrotron E_{pk} values in the infrared/X-ray range. The E_{pk} values for the gamma-ray component are at $\gtrsim$ 10-100 MeV for the FSRQs and at 100 GeV- TeV energies for BL Lac objects. The intrinsic values of E_{pk} for the gamma-ray component are not well known, however, because of $\gamma\gamma$ attenuation within the sources and due to the diffuse extragalactic infrared and optical radiation fields. The intrinsic γ-ray spectra of FSRQs is likely to extend to photon energies $\gg$ 100 GeV.

The γ-ray components are generally thought to originate from Compton scattering by the same population of nonthermal electrons that radiate the nonthermal synchrotron continua. A wide variety of photon sources is possible. In the high-energy peaked BL Lac objects (HBLs), where E_{pk} is in the X-ray range, a simple one-zone synchrotron self-Compton model is usually adequate to fit the SED data, and correlated X-ray and TeV γ-ray observations suggest[20] a purely leptonic model for BL Lacs such as Mrk 421 and Mrk 501, at least during some periods of activity.

In contrast to the BL objects, a Compton-scattering component that arises from photons which originate from outside the jet is required to model the broadband SEDs of FSRQs in leptonic models. This is because the gamma-ray component can be a factor of $\sim$ 10 more luminous than the synchrotron feature, and can display rather sharp spectral breaks that are difficult to model with a synchrotron self-Compton model.[21] External photon fields include the cosmic microwave background radiation field,[22,23], the accretion-disk radiation field,[24,22] a scattered radiation field due to surrounding gas and dust,[25,26] infrared emissions from hot dust or a molecular torus,[27,28,29] reflected synchrotron radiation,[30,31] and BLR atomic-line radiation.[32]

Blazar models where nonthermal hadrons are the primary radiating components have been devised, as discussed in Section 4. Although emissions from nonthermal hadrons can, and probably do, contribute to blazar emission at gamma-ray energies, a directly accelerated nonthermal lepton population seems required to form the synchrotron component at low energies.[33] This is because cascade secondaries form single component gamma-ray spectra with photon spectral indices between -1.5 and -2, without forming two distinct components.

Fig. 1 shows the separate spectral components that comprise a standard-model FSRQ blazar,[34] as explained in the figure caption. For this calculation, it is assumed that relativistic plasma with Lorentz factor $\Gamma = 20$ is expelled from a central supermassive black hole with mass $M_{BH} = 10^9$ $M_\odot$ at redshift $z = 1$ that accretes near the Eddington limit. The radiation is viewed at an angle $\theta = 1/\Gamma$, and the effective Thomson scattering optical depth of the BLR clouds is $\tau_{sc} = 0.01$ in a region of radius 0.1 pc. Power-law nonthermal electrons are injected with comoving power of 10^{44} ergs s^{-1} into a spherical ball of ball of plasma with a size corresponding to a measured variability time scale of 1 day (a jet power corresponding to $\ell_{\rm Edd} \approx 4\%$). The minimum and maximum Lorentz factors of the injected electron distributions are 10^3 and 10^5, respectively, and the electrons are injected with $dN/d\gamma \propto \gamma^{-2.3}$. The injection is uniform between 1000 and 10000 gravitational radii r_g ($= GM_{BH}/c^2$). The disk radiation spectrum is described by a thermal Shakura-Sunyaev spectrum as shown in the figure.

Such blazar models describe the basic qualitative features of the overall SEDs of FSRQs, including the double-peaked SEDs, and the luminosity and peak frequencies of the separate components. When viewing at large angles, the beamed components become much weaker, especially the Compton component from external photons due to its narrower beaming factor,[35] and the disk emission becomes pronounced. A third component thought to be accretion-disk radiation is observed in the SED of some blazar sources such as 3C 273, which is thought to be a slightly misaligned blazar.

Fig. 2 shows how the SED changes with distance from the black-hole engine for this blazar model. The duration as measured by an observer at $\theta \approx 1/\Gamma$ of an episode where the relativistic plasma is uniformly energized between 10^3 and $10^4 r_g$ is 24 ksec. A pivoting behavior about 1 GeV is observed due to the decline in importance of the accretion disk radiation, whereas the quasi-isotropic scattered radiation field remains at roughly the same intensity within this region of the jet. Observations with the Gamma Ray Large Area Space Telescope (GLAST), now scheduled for launch in 2006, are well-suited to explore this timescale and energy range. Different injection profiles can change the detailed behavior, but the discovery of

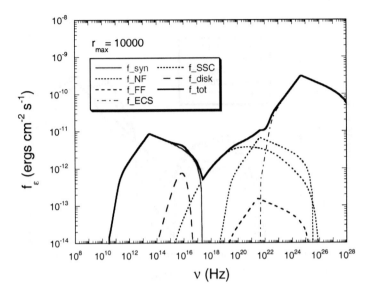

Figure 1. Spectral components in a FSRQ blazar model. Plotted are the nonthermal synchrotron radiation (syn), the Compton-scattered accretion disk radiation, divided into near-field (NF) and far-field components, a Compton-scattered quasi-isotropic radiation component (ECS) associated with the accretion-disk radiation scattered by BLR clouds, the synchrotron self-Compton radiation (SSC), the disk radiation field (disk), and the total radiation field (tot). Parameters of the calculation are given in the text.

γ-ray spectral components that pivot in this manner would provide strong evidence for the effects of the accretion-disk radiation on the jet.

As summarized above, the FSRQs are distinguished by strong atomic lines in their optical/UV spectra, whereas BL Lac objects are weak or nearly lineless sources. This would mean that the spectral component associated with the quasi-isotropic radiation field would be absent or very weak in lineless or weakly-lined BL Lac objects. It is indeed the case that this component is not needed to fit BL Lac objects where E_{pk} is in the X-ray range, although an external Compton scattering component is needed in the case of BL Lac, which has E_{pk} in the optical range.[36]

In a recently-proposed scenario,[37,38] BL Lac objects are AGN jet sources at a stage in their lives where the fueling is in decline and the black-hole engine is most massive. This evolutionary scenario links FSRQs, low-frequency peak BL Lac objects (LBLs), and high-frequency peak BL Lac objects (HBLs) through gradual depletion of the circumnuclear environment of a supermassive black hole. The formation of radio jets in blazars

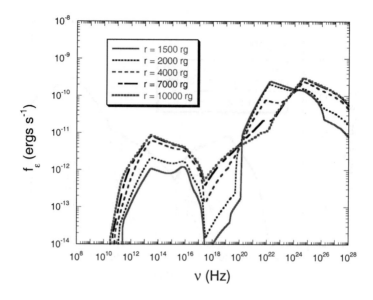

Figure 2. Multiwavelength variability behavior for a uniform injection event for the model in Fig. 2, beginning when the jet is at 1000 gravitational radii. Note the decline of the direct disk-radiation field component with distance from the black hole.

and radio galaxies, if related to an advection-dominated accretion mode in the inner portions of the accretion flow, is triggered by a decreasing Eddington ratio.[37] In the scenario of [38], the Blandford-Znajek mechanism plays a major role in jet activity. The decline of ℓ_{Edd} might be due to a combination of a decreasing accretion rate and an increasing black-hole mass. A sequence of blazar spectra, starting with the parameters derived for 3C 279, is shown in Fig. 3. The accretion-disk luminosity and the efficiency for this radiation to be reprocessed into an external radiation field, are parameterized by the Thomson depth τ_{repr} of the circumnuclear material. The magnetic field is chosen to be a constant fraction of the equipartition magnetic field. The resulting sequence of broadband spectra due to a reduction in τ_{repr} is shown in Fig. 3. This behavior provides a quantitative explanation for the observed[17,18] trend of luminosities and peak photon energies in the FSRQ→ LBL → HBL sequence.

An implication of this scenario is that the masses of the central black holes in galaxies which host BL Lacs should, on average, be greater than the masses of black holes in galaxies which host FSRQs or QSOs. Moreover, subclasses at earlier stages in the blazar sequence should exhibit increas-

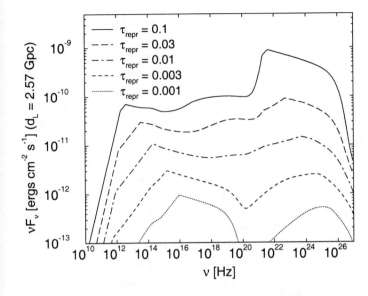

Figure 3. One-parameter model sequence of broadband blazar spectra that reproduces the trend in the spectral energy distributions of FSRQs, LBLs, and HBLs. Starting with typical FSRQ parameters, the sequence is generated by reducing the optical depth τ_{repr} of the circumnuclear scattering material, and assuming that the accretion disk luminosity is proportional to τ_{repr}. The comoving magnetic field $B = 0.3 B_{eq}$, where B_{eq} is the equipartition magnetic field with electrons.

ingly stronger cosmological evolution. Evidence for negative cosmological evolution of X-ray selected BL Lac objects,[39] and for positive cosmological evolution in a radio-selected BL Lac sample,[40] in accord with this picture.

4. Neutrinos and Neutral Beams from Blazars

The multiwavelength observations of blazars, and the models designed to explain their SEDs, convincingly demonstrate that intense radiation fields and efficient particle acceleration to high energies are found in these sources. Based on Fermi acceleration theory, acceleration of hadrons is expected with at least the same power as that of the leptons. Acceleration of hadrons in blazar jets would be directly confirmed with the detection of neutrinos, which are produced provided that there are significant interactions of accelerated hadrons with ambient material or photon fields. Proton synchrotron emission would also be radiated at TeV energies by ultrarelativistic protons and ions (with no associated neutrino emission), but this requires extremely strong magnetic fields ($\sim$ 20-100 G) in BL Lac objects.[41,33]

One group of hadronic blazar models invokes interactions with ambient matter[42,43] through the process $p + p \to \pi^{\pm} \to \nu, e^{\pm}$. Nuclear interaction models require, however, large masses and kinetic energies.[44] A second group of hadronic models is based upon photomeson interactions of relativistic hadrons with ambient photon fields in the jet. The relevant proton-photon processes are $p + \gamma \to p + \pi^0$ followed by $\pi^0 \to 2\gamma$, and $p + \gamma \to n + \pi^+$, followed by $\pi^{\pm} \to \mu^{\pm} + \nu_{\mu} \to e^{\pm} + 2\nu_{\mu} + \nu_e$. In about half of these inelastic collisions, the primary relativistic proton will be converted to a relativistic neutron.

Most of the models of this type take into account collisions of high-energy protons with the internal synchrotron photons,[45,46,47] while others also take into account external radiation that originates either directly from the accretion disk[48] or from disk radiation that is scattered by surrounding clouds to form a quasi-isotropic radiation field.[49] BL Lac objects have weak emission lines, so in these sources the dominant soft photon field is thought to be the internal synchrotron emission. The strong optical emission lines from the illumination of BLR clouds in FSRQs reveal luminous accretion-disk and scattered disk radiation.

In the case of internal synchrotron radiation, the energy output of secondary particles formed in photohadronic processes is generally peaked in the energy range from $\approx 10^{16}$-10^{18} eV in either low- or high-frequency peaked BL Lac objects,[47] which implies that such models can only be efficient if protons are accelerated to even higher energies. This demand upon proton acceleration for efficient photomeson production on the *internal* synchrotron photons also holds for FSRQs, which have similar nonthermal soft radiation spectra as low-frequency peaked BL Lac objects. The presence of the isotropic external radiation field in the vicinity of the jets of FSRQs increases the photomeson production efficiency and relaxes the very high minimum proton energies needed for efficient production of secondaries.[49]

In the model of Atoyan and Dermer,[49,44] protons are assumed to be accelerated in an outflowing plasma blob moving with bulk Lorentz factor Γ along the symmetry axis of the accretion-disk/jet system. The relativistic protons are assumed to have an isotropic pitch-angle distribution in the comoving frame of a plasma blob, within which is entrained a tangled magnetic field. The high-energy neutrino flux is calculated under the assumption that the power to accelerate relativistic protons is equal to the power injected into nonthermal electrons which explains the observed gamma-ray emission.

Fig. 4 shows the integrated neutrino fluences over the time it takes for the blob to pass through the BLR. Model parameters relevant to the 1996 February 4-6 flare from 3C 279 detected by EGRET[50] are used. The solid and dashed curves show the fluences calculated for $\delta = 6$ and 10, respectively. The thick and thin curves represent the fluences of neutrinos produced by photopion interactions inside and outside the blob, respectively. For the spectral fluences shown in Fig. 4, the total number of neutrinos that could be detected by a 1 km^3 detector such as IceCube, using calculated neutrino detection efficiencies,[52] are 0.29 and 0.078 for $\delta = 6$, and 0.13 and 0.076 for $\delta = 10$, where the pair of numbers refer to neutrinos formed inside and outside the blob, respectively.

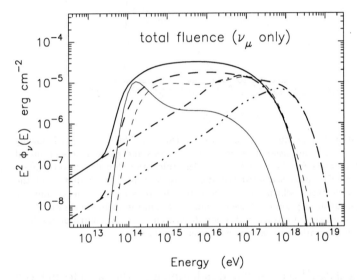

Figure 4. Fluences of neutrinos integrated over several days in the observer frame determined by the time for the blob to pass through the BLR. The solid and dashed curves show the fluences calculated for $\delta = 6$ and 10, and the thick and thin curves represent the fluences of neutrinos produced by photopion interactions inside and outside the blob, respectively. The dot-dashed and 3-dot – dashed curves show the fluences due to $p\gamma$ collisions if external radiation field is not taken into account.

Over the course of one year, several neutrinos should be detected from FSRQs such as 3C 279 with km-scale neutrino detectors, and possibly many more neutrinos if the efficiency to accelerate hadrons is much greater than for electrons. The presence of a quasi-isotropic external radiation field enhances the neutrino detection rate by an order-of-magnitude or more over the case where the field is absent, so we predict that FSRQs will be

338

detected with km-scale neutrino detectors, whereas BL Lac objects are not promising for neutrino detection.

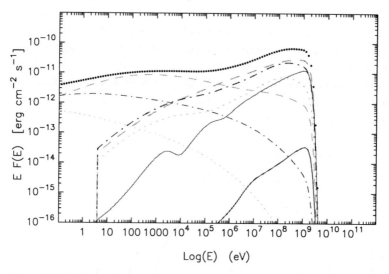

Figure 5. Radiation flux produced in and escaping from the blob (full dots) following the electromagnetic cascade initiated by energetic electrons and gamma rays produced in photopion interactions for the case $\delta = 10$. The thick and thin curves correspond to synchrotron and Compton scattered radiation, respectively. The radiation of the first generation of electrons, which includes both the electrons from $\pi^\pm$ decay and the electrons produced by absorption of π^0-decay gamma rays in the blob, are shown by the solid curves. The dashed, dot-dashed and 3-dot–dashed curves show contributions from the 2d, 3d and 4th generations of cascade electrons, respectively.

We also calculated[44] the associated radiation from the cascades induced by photohadronic processes. Fig. 5 shows the multiple generations of cascade radiation and the total emergent flux associated with the neutrino production given in Fig. 4. This illustrates the difficulty of explaining the SEDs of blazars using a purely hadronic model. As can be seen, the efficiency of accelerating protons is limited by the intensity of the cascade radiation. In particular, X-ray observations will limit the proton-to-electron ratio in the accelerated particle distribution in order that the cascade radiation does not exceed the observed flux. By comparing with the SED of 3C 279 during its flaring state, Fig. 5 indicates that this ratio cannot exceed ≈ 10.

The model also takes into account the effects of relativistic neutrons, which can escape from the blob unless they are converted back to protons due to decay or further photohadronic collisions inside the blob. Ultra-high energy gamma-rays can escape from the BLR and avoid $\gamma\gamma$ attenuation when they are produced by $n + \gamma \to \pi^0 \to 2\gamma$ reactions near the edge of the BLR. The escaping neutron and ultra-high energy gamma ray beams transport energy to large distances from the compact nucleus following neutron decay and $\gamma\gamma \to e^{\pm}$ attenuation[51,44] on the cosmic microwave background radiation field. This method of energy transport avoids quenching of the jet by BLR material and explains the appearance of extremely straight jets as seen in Cygnus A and Pictor A. The more luminous radio lobes in FR2 galaxies than in FR1 galaxies is, in this model,[44] a consequence of the more intense neutral beams formed in FSRQs than BL Lac objects. High-energy neutrino and *GLAST* gamma-ray observations will test this scenario.

Acknowledgments

Collaboration with Armen Atoyan and Markus Böttcher is gratefully acknowledged.

References

1. Begelman, M. C., Blandford, R. D., & Rees, M. J. 1984, Rev. Modern Phys., 56, 255
2. Urry, C. M. & Padovani, P. 1995, Proc. Astron. Soc. of the Pacific, 107, 803
3. Esin, A. A., McClintock, J. E., & Narayan, R. 1997, Astrophys. J., 489, 865
4. Kellermann, K. I., Sramek, R., Schmidt, M., Shaffer, D. B., & Green, R. 1989, Astronomical Journal, 98, 1195
5. Netzer, H. 1990, Saas-Fee Advanced Course 20, (Berlin: Springer-Verlag), 57
6. Kühr, H., Witzel, A., Pauliny-Toth, I. I. K., & Nauber, U. 1981, Astronomy and Astrophysics Supplement Series, 45, 367
7. Antonucci R., 1993, Annual Reviews of Astronomy and Astrophysics, 31, 473
8. Reynolds, C. S. 1996, Ph.D. Thesis, University of Cambridge
9. Fanaroff, B. L. & Riley, J. M. 1974, Monthly Notices of the Royal Astronomical Society, 167, 31P
10. Impey, C. 1996, Astronomical Journal, 112, 2667
11. Salamon, M. H. & Stecker, F. W. 1998, Astrophys. J., 493, 547
12. Primack, J. R., Bullock, J. S., Somerville, R. S., & MacMinn, D. 1999, Astroparticle Phys., 11, 93
13. Vermeulen, R. C. & Cohen, M. H. 1994, Astrophys. J., 430, 467
14. Elliot, J. L. & Shapiro, S. L. 1974, Astrophys. J., 192, L3
15. Dermer, C. D. & Gehrels, N. 1995, Astrophys. J., 447, 103
16. Maraschi, L., Ghisellini, G., & Celotti, A. 1992, Astrophys. J., 397, L5
17. Sambruna, R., et al., 1997, Astrophys. J., 474, 639
18. Fossati, G., Maraschi, L., Celotti, A., Comastri, A., & Ghisellini, G., 1998, Monthly Not. Roy. Astron. Soc., 299, 433

19. Hartman, R. C. et al. 1999, Astrophysical Journal Supplements, 123, 79
20. Krawczynski, H., Coppi, P. S., & Aharonian, F. 2002, Monthly Not. Roy. Astron. Soc., 336, 721
21. Hartman, R. C. et al. 2001, Astrophys. J., 553, 683
22. Dermer, C. D., and Schlickeiser, R. 1993, Astrophys. J., 416, 458
23. Tavecchio, F., Maraschi, L., Sambruna, R. M., & Urry, C. M. 2000, Astrophys. J., 544, L23
24. Dermer, C. D., Schlickeiser, R., and Mastichiadis, A. 1992, Astron. & Astrophys., 256, L27
25. Sikora, M., Begelman, M. C., and Rees, M. J. 1994, Astrophys. J., 421, 153
26. Dermer, C. D., Sturner, S. J., and Schlickeiser, R. 1997, Astrophys. J. Supp., 109, 103
27. Protheroe, R. J., and Biermann, P. 1997, Astroparticle Phys., 6, 293
28. Blazejowski, M., Sikora, M., Moderski, R., and Madejski, G. M. 2000, Astrophys. J., 545, 107
29. Arbeiter, K., Pohl, M., & Schlickeiser, R. 2002, Astron. Astrophys., 386, 415
30. Ghisellini, G., and Madau, P. 1996, Monthly Not. Roy. Astron. Soc., 280, 67
31. Böttcher, M., and Dermer, C. D. 1998, Astrophys. J., 501, L51
32. Koratkar, A., Pian, E., Urry, C., & Pesce, J. E. 1998, Astrophys. J., 492, 173
33. Mücke, A. & Protheroe, R. J. 2001, Astroparticle Physics, 15, 121
34. Dermer, C. D., and Schlickeiser, R. 2002, Astrophys. J., 575, 667
35. Dermer, C. D. 1995, Astrophys. J., 446, L63
36. Böttcher, M. & Bloom, S. D. 2000, Astrophys. J., 119, 469
37. Böttcher, M. & Dermer, C. D. 2002, Astrophys. J., 564, 86
38. Cavaliere, A. & D'Elia, V. 2002, Astrophys. J., 571, 226
39. Bade, N., et al. 1998, Astron. & Astrophys., 334, 459
40. Stickel, M., Fried, J. W., Kuehr, H., Padovani, P., & Urry, C. M. 1991, Astrophys. J., 374, 431
41. Aharonian, F. A. 2000, New Astronomy, 5, 377
42. Beall, J. H. & Bednarek, W. 1999, Astrophys. J., 510, 188
43. Schuster, C., Pohl, M., & Schlickeiser, R. 2002, Astron. Astrophys., 382, 829
44. Atoyan, A., & Dermer, C. D. 2002, Astrophys. J., submitted (astro-ph/0209231)
45. Mannheim, K., and Biermann, P.L. 1992, Astron. Astrophys. 253, L21
46. Mannheim, K. 1993, Astron. Astrophys. 269, 67
47. Mücke, A., Protheroe, R.J., Engel, R., Rachen, J.P., Stanev, T. 2002, Astropart. Phys., in press (astro-ph/0206164)
48. Bednarek, W., and Protheroe, R. J. 1999, Monthly Not. Roy. Astron. Soc., 302, 373
49. Atoyan, A., & Dermer, C. D. 2001, Phys. Rev. Letters, 87, 221102
50. Wehrle, A. E. et al. 1998, Astrophys. J., 497, 178
51. Neronov, A., Semikoz, D., Aharonian, F., & Kalashev, O. 2002, Phys. Rev. Letters, 89, 51101
52. Gaisser, T. K., Halzen, F., & Stanev, T. 1995, Phys. Reports, 258, 173

HIGH ENERGY NEUTRINO ASTRONOMY

TODOR STANEV

Bartol Research Institute, University of Delaware,
Newark, DE 19711, USA

Neutrino astronomy is a very young and fast developing field of cosmic ray astrophysics. We give a very basic introduction of what it is and links to the web pages of the new experiments that also contain some of the theoretical ideas.

1. Why would one want one more astronomy ?

The usual argument for extending the frequency range of telescopes and creating a new type of astronomy is that multiwavelength observations reveal much better the total energy output, as well as the dynamics of an astrophysical object. The argument for neutrino astronomy is based on two very important arguments. The emission of many luminous astrophysical systems is absorbed from intervening bodies or clouds, which are often related to the emission of the system itself. A typical example would be a star with very heavy stellar winds that shield the star emission in many frequencies. Multiwavelength observations help, but generally the optical depth of an astrophysical object does not change drastically with frequency.

Let me explain what a drastic change means here: if the typical electromagnetic cross section is Thomson's cross section $\sigma_T = 6.65 \times 10^{-25}$ cm^2, neutrinos of energy 1 GeV have interaction cross section of 5×10^{-39} cm^2. GeV neutrinos can propagate without interactions not only through the densest molecular clouds, but also through whole stars. The best example here are the MeV solar neutrinos (that have even smaller interaction cross sections). Solar neutrinos are generated in nuclear processes in the core of the Sun and are not absorbed by interactions in the matter of the star. I will not discuss here solar neutrinos and the great successes in understanding the dynamics and structure of the Sun by studies of solar neutrinos.[a] I

---

[a]Ray Davis Jr received the Nobel price for physics in 2002 for his pioneering observation of solar neutrinos.

will concentrate on neutrinos of much higher energy.

The neutrino cross section grows with energy, but never becomes even close to σ_T. The neutrino nucleon cross section is proportional to the neutrino energy. The relation is expressed as

$$\sigma_\nu \propto E_\nu \frac{M_W^4}{M_W^2 + Q^2}, \tag{1}$$

where Q^2 is the momentum transfer in the interaction and M_W is the mass of the intermediate vector boson, which carries the interaction. When the neutrino energy is much smaller than M_W^2 (81 GeV/c^2) the cross section is really proportional to E_ν. At energies above M_W^2 the momentum transfer can reach very high values and the cross section increase slows down.

So the very large penetration ability of neutrinos make them important for the observation of hidden sources and processes. There is a catch, however. Neutrinos are produced only in hadronic processes, that are practically not studied in astrophysics. If we turn the argument around, we can state that the possible future detection of astrophysical neutrinos will reveal the importance of hadronic processes in the dynamics of the source system.

1.1. How are these neutrinos generated

Neutrinos are generated in the decay chain of charged pions, that are the most common product of hadronic interactions. Other meson decays also contribute at high energy, but the typical process is

$$p + p \rightarrow p(n) + m\pi^0 + 2m\pi^\pm$$

the production of neutral and charged secondary pions. The number of charged pions is roughly twice as large. At moderate energies the interaction proton loses on the average one half of its energy and the other half is distributed between the secondary particles. Charged pions decay into a muons and muon neutrinos, and in the astrophysical environment muons also always decay into a muon neutrino, electron neutrino and an electron. So imagine an interaction where one secondary pion of every charge is produced.

π^0	π^+	π^-	p
$\frac{1}{6}$	$\frac{1}{6}$	$\frac{1}{6}$	$\frac{1}{2}$
$\gamma\ \gamma$	$e^+\ \ \nu_e\ \ \nu_\mu\ \ \bar{\nu}_\mu$	$e^-\ \ \bar{\nu}_e\ \ \bar{\nu}_\mu\ \ \nu_\mu$	

The two γ-rays share equally the π^0 energy and the four particles from the charge pion - muon decay chain get approximately $1/4$ of the the $\pi^\pm$ energy. So the total amount of energy that goes into electromagnetic particles (γ-rays and $e^\pm$) is about equal to energy in neutrinos, while the energy of the individual neutrinos is smaller that that of γ-rays. Folded with the steep cosmic ray spectrum $E^{-\alpha}$ the flux of neutrinos is smaller than that of γ-rays -

$$\phi_\nu = \phi_\gamma (1 - r_\pi)^\alpha , \qquad (2)$$

where $r_\pi = (m_\mu / m_\pi)^2$.

The π^0 gamma ray spectrum peaks at one half m_{π^0} - about 70 MeV. The neutrino spectra peak at about $m_{\pi^\pm}/4$.

1.2. How are the pions produced

Pions and other mesons that contribute to the neutrino flux at higher energies are generated in two types of interactions - nucleus-nucleus interactions, which in astrophysical environment are mostly pp with about 10% contribution from He, and photoproduction interactions $p\gamma$.

The cross section for pp interactions σ_{pp} is about 3×10^{-26} cm^2 at 100 GeV and increases as $s^{0.03-0.04}$ at higher energy. s is the total interaction energy in the center of mass system. The mean free path for interaction λ_{pp} is then 60 g/cm^2. The neutrino production then requires targets of thickness of order λ_{pp}, which are infrequent in astrophysical systems. Such targets can be found in accretion disks and in the companion star in binary star systems. Magnetic fields, that are needed to accelerate protons anyway, help since they scatter the charged protons and effectively increase their pathlength inside and around the system.

The photoproduction cross section $\sigma_{p\gamma}$ is much smaller - it reaches maximum of 5×10^{-28} cm^2 at the Δ^+ resonance peak, falls to 10^{-28} cm^2 after that and increases approximately as σ_{pp} at higher energy. The mean free path $\lambda_{p\gamma} = (\sigma_{p\gamma} n_\gamma)^{-1}$ depends on the ambient photon density and energy spectrum. For the best known photon background - the 2.7° microwave radiation with density 400 cm^{-3} - $\lambda_{p\gamma}$ is of order 3 - 10 Mpc. The threshold energy for photoproduction interactions is high

$$E_p^{thr} \simeq \frac{(m_p + m_\pi)^2}{2\varepsilon} , \qquad (3)$$

where ε is the energy of the ambient photon. The thresholds of interactions of different photon fields are very different - from 10^{20} eV for interactions on 10^{-3} eV photons to 10^{17} eV for interactions on optical (1 eV) photons to 10^{14} eV for interactions on KeV X-rays. The situation is very different from pp interactions with a threshold (1.232 GeV) slightly above the proton mass.

Photoproduction interactions need both high energy protons and high energy photon seed radiation. We know, however, that high energy nonthermal radiation is observed from many powerful astrophysical systems.

2. Sources of astrophysical neutrinos

Since no extraterrestrial neutrinos of energy above 50 MeV have been detected yet we can only discuss the sources from theoretical point of view. We can roughly divide them in Galactic and extragalactic sources. The current prejudice is that the Galactic sources are based on pp interactions. Typical examples of such sources are powerful binary systems, where one of the objects (neutron star or black hole) is powered by accretion of its companion matter and supernova remnants, where we assume all Galactic cosmic rays are accelerated.

In the binary systems the assumption is that protons are accelerated in the magnetic field of the neutron star and interact either in the accretion disk or the companion star. Supernova remnants may present enough matter density for interactions of the accelerated cosmic rays, especially if they are associated with dense molecular clouds [1]. The EGRET instrument of the Compton gamma ray observatory detected emission of several supernova remnants [2] that may be of π^0 origin, since the best fits of its spectral shape identify a π^0 peak [3]. Such objects have been suggested as possible neutrino sources long ago [4,5]. Another obvious source, that should exist, is the Galaxy itself. GeV cosmic ray diffuse for 10^6 - 10^7 years, during which they cross more than 10 g/cm^2 of Galactic matter. Diffuse Galactic γ-rays has also been observed by EGRET [6] and there are no doubts about a corresponding diffuse neutrino flux. Similar neutrino fluxes certainly also come from some of the nearby galaxies. These 'certain' neutrinos are, however, of minor astrophysical interest.

More recent astronomical discoveries of microquasars, galactic objects with extended jets, may suggest neutrino production via photoproduction interactions.

Extragalactic neutrino sources are expected to utilize the photoproduction interaction. The expectations are based on the fact that many extragalactic objects are much more powerful than our own Galaxy. The sources that have been theoretically analysed are active galactic nuclei (AGN) and gamma ray bursts (GRB). There are at least two ways to produce neutrinos at an AGN: near to the central black hole engine [7] or in the AGN jet [8].

Almost nothing is know about the region close to the AGN nucleus except for its very high power. The models thus can not be compared to other observations and constrained in this way. AGN jets are better observed and the models based on their features suggest that certain number of neutrinos will be produced. Only the most important question - how large the flux is - does not yet have an answer.

GRBs are in many respects similar to AGN jets. The basic difference is that while AGN jets can be active for long periods (10^8 years), GRBs last on the hour to day scales. The GRB signals are observed at Earth on the 1 to 10 second time scale, although some high energy (20 GeV) γ-rays have been seen with hour delays. From observational viewpoint this is good, as the timing could reveal the neutrino–GRB connection.

2.1. Diffuse extragalactic neutrino fluxes

Even if we can not observe source neutrinos there is a chance to see the diffuse neutrino flux from unidentified sources. each of which is not powerful enough to be observed individually. A simple example is to imagine 50 sources in the Universe, each of which generates only one event in a detector. We can never be certain that this neutrino event came from an astrophysical source, but could define the diffuse flux of 50 events if they had a distinguished signature. Such signature is most likely their energy spectrum.

The neutrinos generated by cosmic rays in the atmosphere - atmospheric neutrinos - have a steep energy spectrum. It is partially due to the steep galactic cosmic ray spectrum, but also to the small dimensions of the atmosphere. High energy pions start interacting before they decay and high energy muons hit the ground and lose their energy. In astrophysical environment secondary interactions are not very likely and all unstable particles decay. Thus the diffuse neutrino background is expected to have indeed much flatter energy spectrum.

A reliable estimate of the diffuse neutrino flux is not easy because one has to account not only for the processes that happen in individual sources, but also for the luminosity spectrum of these sources and their cosmological evolution. The diffuse flux is calculated as

$$\frac{dF}{dE} = \frac{c}{H_0} \frac{1}{ER_0^3} \int dL_x \int dz \rho(L_x, z)(1+z)^{-\alpha} \frac{dL}{dE}(E_z, L_x) \qquad (4)$$

Here H_0 is the Hubble constant and ρ is the luminosity evolution of the sources with redshift z. $(1 + z)^{-\alpha}$ is the matrix element with $\alpha = 5/2$ for the Einstein-deSitter Universe.

2.2. Neutrinos from propagation of ultra high energy cosmic rays

The existence of such neutrinos was proposed independently by V.S. Berezinsky and by F.W. Stecker a few years after the microwave background was discovered. In Fig. 1 we show the results of a recent calculation [9].

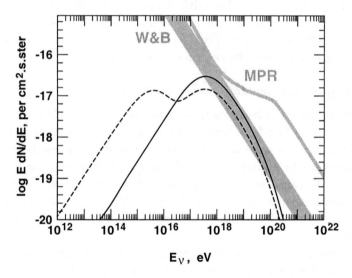

Figure 1. Fluxes of neutrinos from propagation of UHECR. Solid line shows the sum of $\nu_\mu + \bar{\nu}_\mu$ and the dashed shows $\nu_e + \bar{\nu}_e$. The shades show the limits of Waxman&Bahcall and of Mannheim, Protheroe & Rachen.

Muon neutrino fluxes peak at about 10^{17} eV. At lower energy the flux is cut-off by the decreasing effective photoproduction cross section and on the higher end by the assumption for a cutoff of the proton spectrum. Electron neutrinos exhibit a double peak. The higher energy peak coincides with the ν_μ spectral shape and shows ν_e from muon decay. The lower peak contains $\bar{\nu}_e$ from neutron decay. The calculation uses cosmological neutrino of the cosmic ray sources proportional to $(1 + z)^3$.

The two shaded bands show the upper limits of extragalactic source neutrinos calculated from estimates of the luminosity of cosmic ray sources. The wide band is from Waxman&Bahcall [10] and the narrow one - from Mannheim, Protheroe&Rachen [11]. The idea is that if extragalactic neutrinos are produced in the sources of the ultra high energy cosmic rays (UHECR), one can set a limit of the magnitude of their fluxes using the flux of UHECR itself. W&B limit is obtained using a flat $\alpha = 2$ proton spectrum and a luminosity of 4.5×10^{44} ergs/Mpc^3/year. The lower end of the band shows the case of no cosmological evolution and the upper one - with $(1 + z)^3$ evolution. The limit of MPR is not that restrictive - it accounts for many uncertainties in the modeling of the neutrino flux.

3. Detection of extraterrestrial neutrinos

The classical method for neutrino astronomy is the detection of upward going neutrino induced muons. A fraction of the muon neutrinos interact in the Earth and produce muons. Downgoing neutrino induced muons are swamped by the flux of atmospheric muons that is higher by at least 5-6 orders of magnitude. Upward going neutrino induced muons have to background - only neutrinos can penetrate the Earth. The disadvantage is that only muon neutrinos are detected this way - electrons from ν_e interactions are absorbed very fast, and that only 2π solid angle is available for observation. The advantage is that the effective volume of the detector is determined by the range of the muon R_μ, i.e. the amount of matter that a muon can penetrate without stopping. GeV muons lose roughly 2 MeV/(g/cm^2) on ionization, i.e. a 100 GeV muon can penetrate through 50,000 g/cm^2 or 0.5 km water equivalent. In standard rock with density 2.65 g/cm^3 this would correspond to a length of almost 200 meters. A detectors of instrumented volume $50 \times 50 \times 50$ m^3 turns into a $50 \times 50 \times 200$ m^3, i.e. the effective volume increases by a factor of 4. For higher energy muons the increase is bigger. At energy above 500 GeV, however, muon radiation losses start dominating the energy loss and R_μ grow slower with energy.

The flux of upward going muons can be calculated easily using the probability $P_{\nu\mu}$ that a neutrino of energy E_ν can generate a muon of energy bigger than E_μ at the detector. The flux of upward going muons can then be obtained by an integration over the neutrino energy spectrum:

$$F(> E_\mu) = \int_{E_\mu} P_{\nu\mu} \frac{dF_\nu}{dE_\nu} dE_\nu \tag{5}$$

$P_{\nu\mu}$ is calculated by folding the fraction of neutrino energy that a muon receives in the neutrino interaction with the muon range. Fig. 2 shows its values for muons of energy above 1 GeV and 1 TeV. The flux of upward

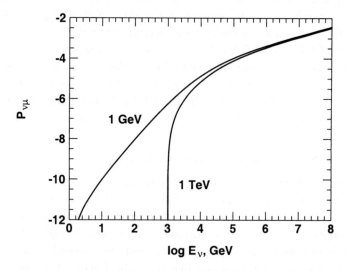

Figure 2. Probability for generation of muons of energy above 1 GeV and 1 TeV at the detector in standard rock as a function of the neutrino energy.

going muons is not very sensitive to the muon energy in the GeV range. The reason is that both the neutrino cross section and the muon range increase linearly with energy. For these reasons a power law neutrino flux $E_\nu^{-\alpha}$ will generate a power law muon flux $E_\mu^{-(\alpha-2)}$ at muon energies up to 100 GeV or so. At higher energies σ_ν and R_μ increase slows down and the muon spectrum is steeper. Nevertheless existing and planned neutrino telescopes aim at the detection of higher energy muons to be well above the atmospheric neutrino background.

The growth of the neutrino cross section makes the Earth opaque to neutrinos at high energy. Fig. 3 shows the mean free path for neutrinos as a function of energy and compares it to the column density through the Earth as a function of the zenith angle θ. Neutrinos of energy 10^5 GeV are absorbed if they penetrate the Earth vertically upwards, where the column density is about 10^{10} g/cm^2. At energies above 4×10^9 GeV the Earth is opaque at all zenith angles. There are some secondary effects that decrease the effect. One is the existence of the neutral current (NC) neutrino interactions that generate hadrons and a secondary neutrino. In such interactions the neutrino energy is only decreased. Part of the neutrino

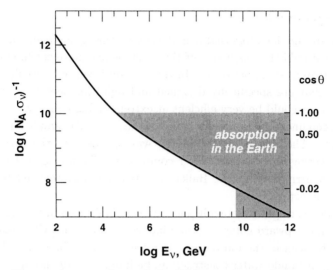

Figure 3. The neutrino mean free path for CC interactions is shown as a function of the neutrino energy and is compared to the column density of the Earth.

flux is recovered and the loss cone in the Earth is decreased. Fluxes of ν_τ that are expected to exist because of ν_μ oscillations are also absorbed differently because of the τ energy loss and decay time. But generally with increased neutrino energy neutrino astronomy becomes more difficult with upward going muons.

At such energies, as is the case with the diffuse neutrinos in Fig. 1 one has to use other methods. One is the detection of very high energy cascades generated by ν_e's. The total energy of an electron neutrino goes into a cascade in a CC neutrino interaction. Atmospheric neutrinos have a very steep energy spectrum and vanish before ν_μ do. Extraterrestrial electron neutrinos have the the same spectra as muon neutrinos and there is no background if the detectors can measure the cascade energy.

Many current ideas about the detection of ultra high energy neutrinos are linked to the oscillations of muon into tau neutrinos. Although the ratio of different neutrino flavors at production is $\nu_e : \nu_\mu : \nu_\tau = 1 : 2 : 0$ the fluxes arriving at Earth will have approximately $\nu_e : \nu_\mu : \nu_\tau = 1 : 1 : 1$. Tau neutrinos and mesons create unique spectacular signatures that could identify them immediately. The number of events with such signatures is not, however, expected to be significant.

3.1. *Neutrino telescopes*

All underground detectors that can identify upward going neutrino induced muons have published skymaps of their muons and have set limits on the flux of extraterrestrial neutrinos. In this section I will only briefly mention detectors that are specifically designed and build for neutrino astronomy or others that could be very efficient at extremely high energy.

The deep underwater/ice detectors are the classical example for neutrino telescopes. The first one, DUMAND, was proposed in the early 1970s. This pioneering detector was never completed. The first one that indeed detected neutrinos is in lake Baikal and is commonly referred to as the BAIKAL detector.

The idea is to put photomultiplier tubes (PMT) deep in the ocean (lake in this case). Upward going neutrino induced muons emit Cherenkov light that can be seen in the darkness of the deep ocean. There are, of course, problems with underwater construction, bioluminescence and other sources of light. Currently there are three such projects in the Mediterranean - NESTOR [12], ANTARES [13] and Nemo. All three of them aim at a sensitive area of about 0.1 km² by instrumenting strings of PMTs that are anchored at the bottom of the sea. NESTOR is the oldest one off the coast of Greece. It uses 'towers' containing sets of PMTs on the same level instead of strings. ANTARES is the best developed as a concept and will start deployment in 2003. Nemo is a still young italian effort off the coast of Sicily that is expected to join the efforts of the other groups in building even bigger 1 km³ detector.

The example for such a detector is ICECUBE [14] at the South Pole that is already proposed and partially funded. ICECUBE is a km³ detector that follows on the success of the AMANDA detector of effective area 0.1 km³. This is the same idea as in underwater experiments. The PMTs this time are deployed in the holes drilled in clean ice at depths 1.5 - 2.5 km. Ice has the advantage of relatively easy construction - there are no waves on the surface. It has no background light and long absorption length for the Cherenkov light. The ice is, however, less uniform than water and the light scatters relatively more than in sea water. Detectors cannot be put deeper than 2.5 km and thus have relatively high background.

The AMANDA collaboration was able to overcome these and other problems related to the South Pole location and prove the concept of neutrino detection in the ice. If there are no surprises ICECUBE will be completed and will start full operation in 2008/9. It will collect data during deployment together with the AMANDA strings. On top of ICECUBE is the km^2 air shower detector ICETOP that will be used for calibration and veto shielding for cascade events in addition to other cosmic ray science. Ultrahigh energy neutrino interactions will be difficult to contain and measure even in a km^3 detector. Besides such detector may be too small for 10^9 GeV and higher energy neutrinos, that could be related to the sources of UHECR and many exotic, not very likely but very important, phenomena. Such neutrinos can be detected either by giant air shower arrays or by radio detectors. The best example of a giant air shower array is the Auger Observatory [15] that is being built in Argentina. It has an area of 3,000 km^2. As a neutrino detector Auger will have target mass of about 30 km^3 of water and a threshold of about 10^{19} eV. The EUSO [16] project, optical air shower detector, is being considered for a flight on the Space station. EUSO will have much higher sensitivity as well as energy threshold as a neutrino detector.

The detection of radio signals (a fraction of the Cherenkov radiation is in that frequency) is another alternative method for detection of neutrino induced cascades. The RICE detector [17] was deployed with AMANDA in the South Pole ice, and was successful as a prototype experiment that set important limits on the fluxes of very high energy neutrinos. Another experiment, ANITA, will circle the Antarctic in a balloon and look for radio signals from neutrino interactions in the ice.

With so many projects being developed and constructed now the neutrino astronomy is the critical stage of its growth. It will open a new window to the Universe, through which we will see one of its unexplored sides.

Acknowledgments This research is funded in part by US DOE contract DE-FG02 91ER 40426 and NASA grant NAG5-10919.

References

1. E.G.Berezhko & H.J. Völk, Astropart. Phys, **14**, 201 (2000).
2. J.A. Esposito et al., Ap. J., **461**, 820 (1996).
3. T.K. Gaisser, R.J. Protheroe & T. Stanev, Ap. J. **492**, 219 (1998)
4. V.S. Berezinsky, C. Castagnoli & P. Galeotti, Nuov. Cim. C, **8**, 185 (1985).
5. F.W. Stecker, M.M. Shapiro & R. Silberberg, Proc. 18th ICRC (Kyoto), **10**, 346 (1981).
6. S.D. Hunter et al., Ap. J., **481**, 205 (1997).
7. F.W. Stecker et al, Phys. Rev. Lett., **66**, 2697 (1991); see also A.P. Szabo & R.J. Protheroe, Astropart. Phys., **2**, 375 (1994).
8. For a discussion of models and ideas , see J.G. Learned & K. Mannheim, Ann. Rev. Nucl. Part. Sci, **50**, 679 (2000) and T.K. Gaisser, F. Halzen & T. Stanev,

Phys. Reports., **258**, 173 (1995).

9. R. Engel, D. Seckel & T. Stanev, Phys. Rev. D**64**:093010 (2001).
10. E. Waxman & J.N. Bahcall, Phys. Rev. D**59**: 023002 (1999).
11. K. Mannheim, R.J. Protheroe & J. Rachen, Phys. Rev. D**63**: 023003 (2001).
12. see *www.nestore.org.gr*
13. see *antares.in2p3.fr*
14. see *icecube.wiscd.edu*
15. see *www.auger.org*
16. see *ifcai.pa.cnr.it/ EUSO/*
17. one can find information on *kuhep4.phsx.ukans.edu/ iceman/*

THE RADIO ICE CERENKOV EXPERIMENT (RICE)

S. SEUNARINE*

Department of Physics and Astronomy,
The University of Canterbury,
Private Bag 4800,
Christchurch, New Zealand
E-mail: physsse@phys.canterbury.ac.nz

The RICE detector is an array of radio antennas deployed in the ice at the geographic South Pole. The array is tuned to detect Ultra High Energy neutrinos($E_\nu > 10^{15} eV$) using the radio Cerenkov technique. An Ultra High Energy neutrino interacting in Antarctic ice produces a electron which carries away about eighty percent of the neutrino's energy. The electron will initiate an electromagnetic cascade which develops a net negative charge as atomic electrons are knocked out. Each charged track emits Cerenkov radiation which is coherent at radio wavelengths. We describe the general concept of radio detection and details of the RICE experiment

1. Introduction

Ultra High Energy neutrinos which travel through large distances without scattering may prove to be useful probes of astrophysics and cosmology. They can also be used to study the Standard Model of particle physics at energies which are orders of magnitude beyond those reached by terrestrial particle accelerators. The flux of neutrinos falls steeply with energy so detectors with large target volumes are required to capture enough events for any physics study. The RICE experiment is designed to use the concept of coherent Cerenkov radio emission to extend the energy range of detectable neutrinos to PeV energies. The large target mass is an approximately kilometer ice cube of sensitive volume surrounding an array of radio receivers. When a high energy neutrino interacts with a nucleon in the ice, the secondary lepton carries away, on average, 80% of the neutrino's energy. In the case of high energy electron neutrinos, an energetic electron is created

---

*For the RICE collaboration

which then generates an electromagnetic cascade. A negative excess charge develops as atomic electrons are knocked out and transported forward with the cascade. Each charged track emits broadband Cerenkov radiation and, because of the compact size of the cascade, the radiation is coherent at radio wavelengths corresponding to frequencies of $200MHz - 1GHz$. This was first predicted by Askaryan[1] and the effect was recently experimentally observed at a test beam[2]. Figure 1 depicts this sequence of events which lead to the detection of a neutrino by RICE. Detailed Monte Carlo[3] studies have confirmed the feasibility of using the Askaryan effect to efficiently study ultra high energy neutrinos.

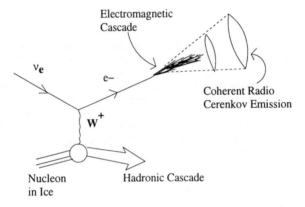

Figure 1.   Events leading to the detection of an ultra high energy neutrino signal(not to scale).

2. Radio Signal From Electromagnetic Cascades

As described above, when a high energy neutrino interacts in the target volume of ice it produces an electron which initiates a cascade(shower) that develops primarily in the direction of the incident neutrino. At the beginning of the shower, bremsstrahlung and pair production are the dominant processes and the number of particles increases exponentially. At high energies the population of photons quickly dominates that of electrons or positrons. The exponential production of particles is halted when the charged particles reach a critical energy where ionization loss overtakes radiation loss as the most important electron energy loss mechanism. The particle population also reaches its maximum at this point. Below the critical energy, charged particles lose their energy mostly via ionization of the medium and the number of particles decline. Multiple Coulomb scattering is responsible for the transverse spread of the developing shower. The shower core is populated by the highest energy particles. Other processes, Compton, Moller and Bhabha scattering, and positron annihilation, build up a net excess of electrons in the shower. Each charged particle which travels with velocity, v, greater than the speed of light in the medium, of refractive index n, emits broadband Cerenkov radiation. The electric field at a distance R given by[3],

$$R\vec{E}(\omega) = \frac{\mu_r}{\sqrt{2\pi}}\left(\frac{e}{c^2}\right)e^{i\omega\frac{R}{c}}e^{i\omega(t_1 - n\vec{\beta}\cdot\vec{r}_1)}\vec{v}_T\frac{(e^{i\omega\delta t(1 - \hat{n}\cdot\vec{\beta}n)} - 1)}{1 - \hat{n}\cdot\vec{\beta}n}, \qquad (1)$$

where μ_r is the relative permeability. The condition $1 - \hat{n}\cdot\vec{\beta}n = 0$ defines the Cerenkov angle θ_c as $\cos\theta_c = 1/n\beta$. Equation (1) is valid when the distance from the track to the observer is large compared to the length of the track and the wavelength is of order the track length or greater. The net electric field from a shower is the contribution from the excess charged tracks in the shower. An appreciable signal can be obtained at the Cerenkov angle, θ_c. Figure 2 shows the angular distribution of the average radio signal from

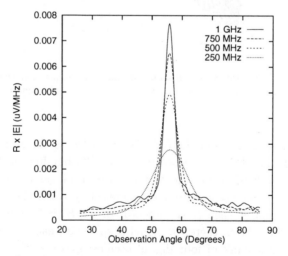

Figure 2. Angular pulse distribution, at four frequencies, of a 100 GeV shower (averaged over 50 showers) with 0.611 MeV total energy threshold from GEANT.

100GeV showers. The signal clearly peaks at the Cerenkov angle in ice of 55.8°.

3. The RICE Detector: Hardware and Triggers

In its current configuration, Figure 2, RICE consists of a 16-channel array of dipole receivers deployed within a $200m \times 200m \times 200m$ cube between $100 - 300m$ depths. Most channels are co-located in holes drilled for the AMANDA experiment[a]. Each antenna is a fat dipole with a peak response at $250MHz$ consistent with the expected signal and compatible with the rest of the data acquisition system(DAQ). The signal from each channel

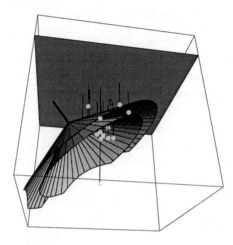

Figure 3. The RICE array, viewed from below and drawn to scale, showing cables and antenna locations. Two cones, drawn at $\theta_c \pm 3°$, depict an approximate 3 dB range of signal strength.

is boosted by an in-ice amplifier($36dB$) and is fed to the surface through coaxial cable. The DAQ is housed in the MAPO building on the surface. The signal below $200MHz$ is filtered to suppress both galactic noise and the impulsive low frequency background generated by AMANDA phototubes. The filtered signal is then re-amplified(52 or $60dB$) before being fed into a CAMAC crate. From there the signal is split with each of two identical copies going to a bank of four digital oscilloscopes and the trigger logic hardware respectively. There are also four transmitters in the ice which are used for calibration and three surface horn antennas used to reject surface noise. For details of the numerous calibration procedures done for RICE,

---

[a]http://amanda.physics.wisc.edu/

see I. Kravchenko, et. al.[4].

There are three sets of criteria which can initiate the recording of an event. The main trigger requires at least four channels hit above a predetermined threshold and all hits occurring within a $1.2\mu s$ time window. This time is set by the geometry of the array and the travel time of a signal across it, and four hits is the minimum number required for vertex reconstruction. Data is also taken if at least one RICE module is hit in coincidence (within $1.25\mu s$) with a big SPASE[b] or AMANDA trigger. The third trigger is a forced trigger designed to take random noise 'events' used for calibration.

4. Analysis

The vertex reconstruction routine searches a grid around the RICE array to find a location where the times of hits are consistent with the a radio signal originating from that point. The changing refractive index, due to the temperature gradient in the ice, is taken into account in this procedure. If there are more than four hits for an event, and the χ^2 formed by the timing residuals between the found vertex and the measured times is large, one of the channels is removed and the vertex location is re-calculated. This accounts for the possibility that a random noise excursion in one of the channels may have occurred in coincidence with a valid trigger. Once a suitable vertex is found based on timing, the pattern of hits is checked for consistency with a Cerenkov cone. The configuration of the RICE array has been dynamic since initial deployment. This reflects the continuing process of understanding noise backgrounds and expected event signal characteristics and improvements in the DAQ hardware and software. For consistent data analysis we look at the data in subsets corresponding to frozen settings of the whole detector. I. Kravchenko et. al.[5] contains a detailed description of the analysis of the data taken in August 2000. This analysis found no events consistent with an ultra high energy neutrino. However, even a single event found in this data would have severely challenge most source models of Ultra High Energy neutrinos. One further remarkable feature of radio technology will be mentioned. The attenuation length of radio in ice is long[5] and a single radio receiver can probe a large volume of ice. The effective volume averaged over the August 2000 exposure is shown as the bold curve in Figure 4. For $E_s = 300 PeV$, $V_{eff} \sim 1km^3$. For energies

---

[b]South Pole Air Shower Experiment

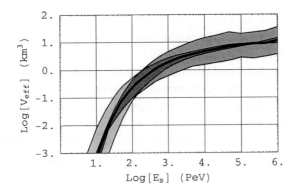

Figure 4. RICE effective volume, as a function of shower energy, for the August 2000 configuration. The nominal result corresponds to the bold curve. The region in medium gray spans variations in response due to variation in the attenuation length by factors of (0.5-2.0). The region in light gray the to changes in signal strength by (0.5-2.0). The region shaded in dark gray is both variations.

below about $50PeV$, $V_{eff} \sim E^3$ as a result of the nearly linear scaling of the signal strength with shower energy[3]. Above $50PeV$ other effects become important and the linear scaling breaks down[5]. Nonetheless, at $300PeV$ for the above configuration, the ratio of effective volume to instrumented volume is $1km^3/.008km^3 \sim 10^2$.

Acknowledgments

The author thanks M.Shapiro and the organizers of the International School of Cosmic Ray Astrophysics for local financial support which made participation in the school possible. This work is supported in part by a Marsden Grant.

References

1. G.A. Askaryan, *Zh. Eksp. Teor. Fiz.* **41**, 616 (1961); G.A. Askaryan, *Soviet Physics JETP* **14**, 441 (1962).
2. D. Saltzberg, et al *Phys.Rev.Lett.* **86**, 2802 (2001)
3. E. Zas, F. Halzen and T. Stanev *Phys.Rev.* **D45**, 362 (1992); G. M. Frichter, J. P. Ralston and D. W. McKay, *Phys. Rev.* **D53**, 1684 (1996); S. Razzaque, S. Seunarine, D.Z. Besson, D. W. McKay, J. P. Ralston and D. Seckel *Phys.Rev.* **D65**, 103002 (2002);
4. I. Kravcenko et al, astro-ph/0112372
5. I. Kravcenko et al, astro-ph/0206371

LIST OF PARTICIPANTS

Ames, Susan   sames@astro.uni-bonn.de
Andersen, Victor   andersen@shasta.phys.uh.edu
Ave Pernas, Maximo   ave@ast.leeds.ac.uk
Baret, Bruny   baret@isn.in2p3.fr
Bartosik, Marek   mbart@kfd2.fic.uni.lodz.pl
Becuci, Alexandru   bercuci@ik3.fzk.de
Bednarek, Wlodek   bednar@fizwe4.fic.uni.lodz.pl
Blanch Bigas, Oscar   blanch@ifae.es
Blome, Hans   blome@fh-aachen.de
Buesching, Ingo   ib@tp4.ruhr-uni-bochum.de
Bohacova, Martina   bohacova@fzu.cz
Boyle, Patrick   jojo@donegal.uchicago.edu
Canoa Roman, Veronica   canoa@fpaxp1.usc.es
Case, Gary   case@phunds.phys.lsu.edu
Cazon Boado, Lorenzo   lorenzo@fpaxp1.usc.es
Chiosso, Michela   chiosso@to.infn.it
Cohen, Fabrice   cohen@cdf.in2p3.fr
Danziger, John   DANZIGER@TS.ASTRO.IT
Dermer, Charles   dermer@gamma.nrl.navy.mil
Domingo Santamaria, Eva   domingo@ifae.es
Duperray, Remy   duperray@isn.in2p3.fr
Gaug, Markus   markus@ifae.es
Giller, Maria   mgiller@kfd2.fic.uni.lodz.pl
Gonzalez Sanchez, Maria Magdalena   magda@titus.physics.wisc.edu
Goodman, Jordan   goodman@umdgrb.umd.edu
Hansen, Patricia   hansen@particle.kth.se
Horan, Deirdre   horan@egret.sao.arizona.edu
Ionica, Maria   maria.ionica@pg.infn.it
Jedrzejczak, Karol   kj@zpk.u.lodz.pl
Kacperczyk, Andrzej   andy@kfd2.fic.uni.lodz.pl
Karczmarz, Piotr   pkarczmarz@kfd2.fic.uni.lodz.pl
Kildea, John   kildea@ferdia.ucd.ie
Kiraly, Peter   pkiraly@sunserv.kfki.hu

Lachaud, Cyril	lachaud@cdf.in2p3.fr
Lee, Hyesook	hslee@mpifr-bonn.mpg.de
Lee, Kerry	ktlee@ems.jsc.nasa.gov
Lopez, Javier	jlopez@ifae.es
Lopez Moya, Marcos	marcos@gae.ucm.es
Maier, Gernot	Maier@ik.fzk.de
Marchesini, Maria	mm@ast.leeds.ac.uk
Marques, Enrique	marques@fpaxp2.usc.es
Natale, Sonia	sonia.natale@cern.ch
Nerling, Frank	nerling@ik.fzk.de
Ortiz, Jeferson Antenhofen	jortiz@ifi.unicamp.br
Pesce, Joseph	pesce@physics.gmu.edu
Popov, Sergey	polar@sai.msu.ru
Priester, Wolfgang	priester@astro.uni-bonn.de
Radu, Aurelian Andrei	andrei@oddjob.uchicago.edu
Reitz, Jessica Kristin	jessica@physics.gmu.edu
Richardson, Jeremy	lee.richardson@colorado.edu
Sambruna, Rita	rms@physics.gmu.edu
Saucedo, Julio	jsaucedo@cajeme.cifus.uson.mx
Scherini, Viviana	viviana.scherini@mi.infn.it
Scholz, Joachim	joachim.scholz@ik.fzk.de
Sefako, Ramotholo R	rocky@fskrrs.puk.ac.za
Seunarine, Surujhdeo	physsse@cantua.canterbury.ac.nz
Sevilla, Ignacio	ignacio.sevilla@ciemat.es
Shapiro, Maurice	mmshapiro@mailaps.org
Shaviv, Nir	shaviv@phys.huji.ac.il
Shaviv, Giora	gioras@physics.technion.ac.il
Sierpowska, Agnieszka	asierp@kfd2.fic.uni.lodz.pl
Smith, Arthur	a.smith1@physics.ox.ac.uk
Stamerra, Antonio	stamerra@pi.infn.it
Stanev, Todor	stanev@bartol.udel.edu
Teufel, Andreas	ate@tp4.ruhr-uni-bochum.de
Van Speybroeck, Leon	lvs@cfa.harvard.edu
Vitale, Vincenzo	VITALE@MPPMV.MPG.DE
Watson, Alan A.	a.a.watson@leeds.ac.uk
Wefel, John	wefel@phunds.phys.lsu.edu
Wilson, Thomas	twilson@ems.jsc.nasa.gov